AF323278

Photosynthesis and Bioenergetics

Photosynthesis and Bioenergetics

Edited by

James Barber
Imperial College London, UK

Alexander V Ruban
Queen Mary University of London, UK

World Scientific

NEW JERSEY · LONDON · SINGAPORE · BEIJING · SHANGHAI · HONG KONG · TAIPEI · CHENNAI · TOKYO

Published by

World Scientific Publishing Co. Pte. Ltd.

5 Toh Tuck Link, Singapore 596224

USA office: 27 Warren Street, Suite 401-402, Hackensack, NJ 07601

UK office: 57 Shelton Street, Covent Garden, London WC2H 9HE

British Library Cataloguing-in-Publication Data
A catalogue record for this book is available from the British Library.

PHOTOSYNTHESIS AND BIOENERGETICS

Copyright © 2018 by World Scientific Publishing Co. Pte. Ltd.

All rights reserved. This book, or parts thereof, may not be reproduced in any form or by any means, electronic or mechanical, including photocopying, recording or any information storage and retrieval system now known or to be invented, without written permission from the publisher.

For photocopying of material in this volume, please pay a copying fee through the Copyright Clearance Center, Inc., 222 Rosewood Drive, Danvers, MA 01923, USA. In this case permission to photocopy is not required from the publisher.

ISBN 978-981-3230-29-3

Typeset by Stallion Press
Email: enquire@stallionpress.com

Printed in Singapore

Foreword by James Barber FRS

This book is a tribute to three outstanding scientists, Professors Jan Anderson FRS, Leslie Dutton FRS and John Walker FRS, Nobel Laureate. I have had the privilege of being a close friend to all of them and it was a joy to both edit this book together with Professor Alexander Ruban, and write this Foreword.

Jan Anderson passed away in August 2015. Just one month before her untimely death, she was in the UK and attended a meeting I had organised at the Royal Society on Artificial Photosynthesis. Despite being 83 she attended every lecture. This was typical of Jan who was devoted to science and photosynthesis in particular. Indeed she was the first to discuss how the photosynthetic electron chain was inserted *via* protein complexes into the thylakoid membrane in her influential review in BBA (Anderson, 1975). Her ideas were built on her pioneering biochemical work with Keith Boardman (Boardman and Anderson, 1964) and formed the basis of the concept of lateral heterogeneity of photosystem I (PSI) and photosystem II (PSII) in the thylakoid membranes of plants (Andersson and Anderson, 1980) which she developed with Bertil Andersson.

Like Jan Anderson, Leslie Dutton and John Walker continue to push forward their subjects despite being in their mid-70s. As explained in the Preface, the idea of this book was in part to celebrate their 75th birthdays in 2015. Les Dutton continues to work on maquettes to mimic electron transfer in bioenergetics systems including PSII as emphasised in his chapter in this book. Of course, he has also focused his talents on electron tunnelling in redox active proteins for many years and has done so with close reference to Marcus theory. This resulted in seminal papers in *Nature* (Moser and Dutton, 1992; Page *et al.*, 1999) and the establishment of the Dutton ruler. It is highly appropriate that Nobel Laureate Rudy Marcus (awarded in 1992) has produced a chapter in this book and despite being in his 90s, continues to contribute to theoretical aspects of bioenergetics.

John Walker is also Nobel Laureate, awarded for his outstanding contribution to understanding how cells make ATP from ADP using the energy of a transmembrane proton gradient (Abrahams *et al.*, 1994). Although his Nobel Prize was awarded in 1997, he has continued vigorously to work on ATP synthase and recently finally obtained a complete structure of this amazing rotatory enzyme by X-ray crystallography, thus revealing the positioning and interactions of all its subunits (Morales-Rios *et al.*, 2015).

Abrahams, J.P., Leslie, A.G., Lutter, R. and Walker, J.E. (1994). Structure at 2.8 A resolution of F1-ATPase from bovine heart mitochondria, *Nature*, 370, 621–628.

Anderson, J.M. (1975) The molecular organization of chloroplast thylakoids, *Biochim. Biophys, Acta -Reviews on Bioenergetics*, 416, 191–235.

Andersson, B. and Anderson, J.M. (1980). Lateral heterogeneity in the distribution of chlorophyll-protein complexes of the thylakoid membranes of spinach chloroplasts, *Biochim. Biophys. Acta — Bioenergetics*, 593, 427–440.

Boardman, N.K. and Anderson, J.M. (1964). Isolation from spinach chloroplasts of particles containing different proportions of chlorophyll *a* and chlorophyll *b* and their possible role in the light reactions of photosynthesis, *Nature*, 203, 166–167.

Morales-Rios, E., Montgomery, M.G., Leslie, A.G. and Walker, J.E. (2015) Structure of ATP synthase from *Paracoccus denitrificans* determined by X-ray crystallography at 4.0 Å resolution, *Proc. Nat. Acad. Sci.*, 112, 13231–13236.

Moser, C.C. and Dutton, P.L. (1992). Nature of biological electron transfer. *Nature*, 355, p.796.

Page, C.C., Moser, C.C., Chen, X. and Dutton, P.L., (1999). Natural engineering principles of electron tunnelling in biological oxidation–reduction, *Nature*, 402, 47–52.

Preface

This book covers some of the most recent advances in the fields of Bioenergetics and Photosynthesis presented at an international workshop held in the Institute of Advanced Studies (IAS), Nanyang Technological University (NTU), Singapore, from 21st to 23rd March, 2016. The meeting was held as a tribute to Jan Anderson FRS who passed away in August 2015 and to celebrate the 75th birthdays of Leslie Dutton FRS and John Walker FRS, Nobel Laureate. Contributions to the workshop and to this book, are from outstanding scientists which includes the Nobel Laureate Rudolph Marcus (Chapter 2 by Marcus and Volkán-Kacsó) who created a theory of electron transport reactions. Marcus first provided the kinetic and thermodynamic description of the movement of electrons between molecules — a chemical processes at the very heart of the biological electron transfer in the major classes of bioenergetic systems such as bacterial cells, mitochondria and chloroplasts. Not surprisingly, the theme of electron transport reactions in the key redox enzymatic complexes of mitochondria and chloroplasts is covered in the majority of chapters of this book. Indeed, Chapters 3 (Wikström) and 4 (Rich) are dedicated to cytochrome *c* oxidase — a complex that terminates the respiratory electron transport chain producing water as a by-product. The most recent advances in studies of cytochrome *c* oxidase, photosystem II (PSII), that splits water and produces oxygen, are thoroughly reviewed in Chapters 5 (Kaucikas *et al.*) and 6 (Barber). The recent structural advances in understanding photosystem I that functions in series with PSII, are covered in Chapter 7 (Nelson). In modern times the science of bioenergetics attained an extraordinary progress with development and applications of state-of-the-art techniques, such as femtosecond crystallography (Chapter 5, Kaucikas *et al.*); single molecule microscopy and spectroscopy,

(Chapter 2, Marcus and Volkán-Kacsó); combination of electron and confocal fluorescence microscopy (Chapter 9, Ruban; Chapter 10, Koochak *et al.*). A significant pool of chapters is dedicated to the regulatory processes that take place in the photosynthetic electron transport chain owing to the dynamic nature of the photosynthetic membrane organisation reviewed in Chapter 10 by Koochak *et al.* These mechanisms include regulation of the photosynthetic light harvesting (Chapter 9, Ruban), cyclic electron transport (Chapter 12, Kou *et al.*; Chapter 13, Hanke and Schreibe and Chapter 14, Larkum *et al.*) and the whole oxygenic electron transport chain (Chapter 11, Järvi *et al.*). Chapter 8 by Shivhare and Mueler-Cajar, summarises the recent progress in understanding the protein *rubisco activase* that regulates the most abundant enzyme, Rubisco (ribulose-1,5-bisphosphatecarboxylase/oxygenase) in the biosphere and is responsible for carbon fixation. And finally, the ideas and the most recent developments in the field of applications of the knowledge of Bioenergetics and Photosynthesis are discussed in Chapter 1 (Ennist *et al.*) and Chapter 6 (Barber).

We recommend this book to specialists in Bioenergetics and Photosynthesis, researchers and postgraduate students who are working and studying various aspects of respiratory and photosynthetic electron transport chains as well as dynamic regulation of the light harvesting and electron transport events in oxygenic photosynthesis.

We would like to acknowledge IAS and NTU for supporting and hosting the workshop in March 2015 and encouraging us to produce this book.

J. Barber and A. V. Ruban
London, 7 June 2017

Contents

Foreword by James Barber FRS   v

Preface   vii

Chapter 1   Maquette Strategy for Creation of Light- and Redox-active Proteins   **1**

Nathan M. Ennist, Joshua A. Mancini, Dirk B. Auman, Chris Bialas, Martin J. Iwanicki, Tatiana V. Esipova, Bohdana M. Discher, Christopher C. Moser and P. Leslie Dutton

1. Introduction   2
2. Maquette protein strategy   3
    2.1. Recognizing natural complexity   3
    2.2. α-Helical scaffolds common in natural protein structures reveal the basics of how cofactors are ligated and positioned   5
    2.3. First-principles of *de novo* designed protein structures outlined   6
    2.4. Electron transfer in protein understood for engineering   12
3. Practical Strategies for Development of Functional Light- and Redox-active Proteins   14
    3.1. Lessons learned from the first heme protein maquettes   14
    3.2. Characterization of one maquette promotes development of others   16
        3.2.1 Charge-activated conformational switch in a homodimeric four-α-helix heme B protein [Grosset *et al.*, 2001]   16
        3.2.2. Electron tunneling between monolayers of linked-homodimeric heme protein and gold electrodes [Chen *et al.*, 1999; 2002]   17

3.2.3. Linked-homodimeric heme protein oxygen transporter
[Koder *et al.*, 2009]   19

4. Single-chain Four-α-Helix Maquettes   20

4.1. From disulfide linked homodimeric bundles to single-chain
four-α-helix structures   20

4.2. Compatibility of single-chain four-α-helix maquettes
with natural proteins   22

4.2.1. Single-chain four-α-helix heme B maquette interactions
with natural cytochrome *c*   22

4.2.2. *In vivo* single-chain four-α-helix protein covalently
linking heme C: a novel cytochrome *c*   24

5. Prospects and Previews   25

Acknowledgements   27

References   27

**Chapter 2 Free, Stalled, and Controlled Rotation Single Molecule
Experiments on F_1-ATPase and their Relationships   35**

Sándor Volkán-Kacsó and Rudolph A. Marcus

1. Introduction   35

2. Description of Three Single-Molecule Experiments   37

2.1. Free rotation experiments: Stepping rotation
and concerted kinetics   37

2.2. Stalling experiments   38

2.3. Controlled rotation experiments   40

3. Group Transfer Theory in Stalling and Controlled
Rotation Experiments   40

3.1. Angle dependent rate constants and free energies   41

3.2. Application to ATP binding in the overlapping θ-range   43

3.3. Turnover, near symmetry and long binding events
in the controlled rotation experiments   44

4. Application of the Theory to Free Rotation   46

4.1. Rate constant of a free rotation experiment   46

4.2. Relation between controlled and free rotation experiments   47

4.3. Use of the rate constant *versus* rotor angle data
to predict the step size and dwell angles in free rotation   49

5. Concluding Remarks   50

Acknowledgements   50

Appendix   50

Evaluation of Eq. (8)   50

References   51

Chapter 3 The Role of the H-Channel in Cytochrome c Oxidase: A Commentary **55**

Mårten Wikström

1. Introduction 55
2. Conservation of the H-Channel Structure 56
3. The Linkage of Proton Pumping to Individual Steps of the Catalytic Cycle 57
4. The Proton Pump Cycle 59
5. Conclusions 60
Acknowledgements 61
References 61

Chapter 4 Cytochrome *c* Oxidase: Insight into Functions from Studies of the Yeast *S. cerevisiae* Homologue **65**

Peter R. Rich

1. Introduction: Catalysis, Coupling and Efficiency 65
2. Core Structural and Functional Variations in the HCO Superfamily 67
3. Yeast CcO — A Link Between Mammalian and A1 Bacterial CcOs 69
   3.1. The H channel and its possible roles 70
   3.2. Supernumerary and additional subunits 71
4. Concluding Remarks 74
Acknowledgements 75
References 75

Chapter 5 Femtosecond Infrared Crystallography of Photosystem II Core Complexes: Watching Exciton Dynamics and Charge Separation in Real Space and Time **81**

Marius Kaucikas, James Barber, Thomas Renger and Jasper J. van Thor

1. Introduction 82
2. Femtosecond Infrared Crystallography 83
3. Summary of Theory for Exciton Dynamics and Trapping by Electron Transfer for the Crystallographic Case 85
4. Femtosecond Infrared Crystallography of Exciton Dynamics 99
5. Structural Measurement of the Charge Separated State P_{680}^+/Pheo$^-$ 102
   5.1 Vibrational mode assignments of the P_{680}^+/P_{680} and Pheo$^-$/Pheo spectra 105
   5.2 Fitting the P_{680}^+/P_{680} and Pheo$^-$/Pheo FTIR spectra 109

5.3 Fitting the 832 ps TR-IR crystallography spectra ... 109

6. Implications for Light Harvesting Function of
PSII Core Complexes ... 110

References ... 114

**Chapter 6 Bioenergetics, Water Splitting and
Artificial Photosynthesis** ... **117**

James Barber

1. Introduction ... 118
2. Photosystem II (PSII) ... 120
    2.1. The catalytic centre ... 121
    2.2. Synthetic cubane mimics ... 125
3. Mechanism of Water Splitting ... 126
    3.1. Base-catalysed nucleophilic attack mechanism ... 126
    3.2. Indirect support for the nucleophilic mechanism
for O–O bond formation in PSII ... 128
    3.3. Comparison with Fe–Ni carbon monoxide dehydrogenase
(CODH) ... 129
    3.4. Ligands to the $Mn_4 Ca^{2+}O_5$ cluster ... 131
4. Synthesized Chemical Model Systems ... 132
    4.1. Mechanism of O_2 production from organo-Ru
complexes ... 132
    4.2. Mechanism of O_2 production from organo-Mn complexes ... 134
5. Alternative Mechanisms are Less Compelling ... 135
6. Artificial Photosynthesis ... 136
7. Conclusions ... 138
References ... 140

**Chapter 7 A Quest for the Atomic Resolution of
Plant Photosystem I** ... **149**

Nathan Nelson

1. Introduction ... 149
2. The Structure of Plant PSI ... 150
3. The Light Harvesting Complex of Plant PSI ... 154
Acknowledgements ... 155
References ... 155

Chapter 8 Rubisco Activase: The Molecular Chiropractor of the World's Most Abundant Protein **159**

Devendra Shivhare and Oliver Mueller-Cajar

1. Introduction 159
2. The Most Abundant Protein on Earth is a Poor Catalyst 160
   2.1. The structure and diversity of Rubisco 161
   2.2. Reaction mechanism: co-factor binding and catalysis 163
   2.3. Rubisco is prone to inhibition by sugar phosphates 164
3. Rubisco Activase, the Molecular Chiropractor 167
   3.1. Rubisco activases are members of the AAA+ family of proteins 168
   3.2. Oligomeric state and mechanism of the Rubisco activase 171
   3.3. Regulation of Rubisco activase 172
   3.4. The role of Rubisco activase in regulating photosynthesis at elevated temperatures 174
4. Outlook 175
References 177

Chapter 9   Adaptive Reorganisation of the Light Harvesting Antenna **189**

Alexander V. Ruban

1. Introduction 190
2. The Structure of LHCII Complex and the Landscapes of the Photosystems 191
3. Evidence for the Fast Dynamics of the Photosynthetic Membrane 195
4. LHCII Composition and Membrane Dynamics 197
5. State Transitions are Affected by PSII Antenna Composition 198
6. Zeaxanthin and Non-Photochemical Quenching Modulate LHCII Dynamics 200
   6.1. LHCII clustering *versus* mobility 201
   6.2. NPQ control by LHCII antenna hydrophobicity 203
7. Formation of the NPQ Quencher(s): Inner LHCII Complex Dynamics 207
8. Physiological Implications of the Dynamic Properties of LHCII 210
Acknowledgements 212
References 212

Chapter 10   Thylakoid Membrane Dynamics in Higher Plants   **221**

Haniyeh Koochak, Meng Li, Helmut Kirchhoff

1. Introduction   221
2. Supramolecular Dynamics   223
   2.1. Protein order and disorder in thylakoid membranes   223
      2.1.1. Significance of mesoscopic dynamics for energy transformation and its regulation   223
      2.1.2. Types of semi-crystalline protein arrays in grana thylakoids   224
2.2. Potential advantages and disadvantages of semi-crystalline protein arrays   225
   2.2.1. PSII repair   226
   2.2.2. Diffusion in the lipid matrix   227
   2.3. Factors controlling mesoscopic protein arrangements   228
      2.3.1. Lipids and fatty acids   228
      2.3.2. Protein phosphorylation   229
   2.4. The elusive mesoscopic level   230
3. Thylakoid Membrane Dynamics   230
   3.1. Types of thylakoid lumen swelling and shrinkage   230
   3.2. Factors controlling thylakoid membrane dynamics   232
      3.2.1. Ion transporters   232
      3.2.2. Protein kinases and phosphatases   233
      3.2.3. Membrane curvature proteins   233
   3.3. Controversies of observed thylakoid dynamics   234
Acknowledgements   235
References   235

Chapter 11   Oxygenic Photosynthesis — Light Reactions within the Frame of Thylakoid Architecture and Evolution   **243**

Sari Järvi, Marjaana Rantala and Eva-Mari Aro

1. Introduction   244
2. Thylakoid Membrane Heterogeneity — From Cyanobacteria to Higher Plants   244
3. Key Regulatory Mechanisms of Thylakoid Electron Transfer Reactions   246
   3.1. No clear model organism for regulation of light reactions   246
   3.2. Evolution of key regulatory mechanisms of thylakoid electron transfer reactions   247
4. Factors Regulating the Thylakoid Architecture in Oxygenic Photosynthetic Organisms   250

5. Revealing the Dynamics of Thylakoid Architecture
in Higher Plant Chloroplasts   251
   5.1. Methods to study thylakoid heterogeneity —
Dark acclimated plants   251
   5.2. Thylakoid fractionation from light-acclimated plants
demonstrates light-dependent, reduced stringency of
lateral heterogeneity   256
   5.3. Significance of thylakoid plasticity for photosynthetic
light reactions and beyond   258
Acknowledgments   259
References   259

Chapter 12   Estimation of the Cyclic Electron Flux around Photosystem I in Leaf Discs   265

Jiancun Kou, Duncan Fitzpatrick, Da-Yong Fan,
Shunichi Takahashi, Riichi Oguchi and Wah Soon Chow

1. Introduction   266
2. Partitioning of Absorbed Light Between the Two Photosystems   267
3. Cyclic Electron Flux in Spinach Leaves   268
4. Cyclic Electron Flux in *Arabidopsis* Leaves   269
   4.1. *CEF* in wild type *Arabidopsis*   269
   4.2. *CEF* in the *pgr5* mutant of *Arabidopsis*   270
   4.3. *CEF* in the *ndh* mutant of *Arabidopsis*   270
   4.4. The antimycin A-sensitive component of *ETR1* in *Arabidopsis*   270
5. Concluding Remarks   272
Acknowledgements   273
References   273

Chapter 13   The Contribution of Electron Transfer after Photosystem I to Balancing Photosynthesis   277

Guy Hanke and Renate Scheibe

1. Introduction   278
2. The Hierarchy of Electron Donation   281
3. Diversity Among Electron Carriers   285
4. Regulation of Sink Demand Beyond the First Acceptor   288
5. Flexible Adjustment of Target-Enzyme Activity According
to Metabolic Demand   290
6. Signaling upon Redox-Imbalances for Long-Term Adaptation at the
Transcriptional Level   291
References   292

Chapter 14   Cyclic Electron Flow in Cyanobacteria and Eukaryotic Algae   **305**

A. W. D. Larkum, M. Szabo, D. Fitzpatrick and J. A. Raven

1. Introduction — 306
2. *In Vivo* Evidence of Cyclic Photophosphorylation in Algae — 306
   2.1. Overview — 306
   2.2. Energetics of the assimilation of exogenous glucose and acetate powered by CEF and their mechanistic implications — 310
3. Detailed Mechanism of Cyclic Electron Transport and ATP Formation in Oxygenic Photosynthetic Organisms — 315
   3.1. Overview — 315
   3.2. Non-cyclic (= linear) electron flow, cyclic electron flow, the Q cycle and ATP generation — 316
   3.3. Early evolution of NAD dehydrogenases — Bacteria, cyanobacteria, plastids and mitochondria. — 317
   3.4. Mechanisms for CEF in Cyanobacteria. — 317
   3.5. Detailed mechanism and inhibitors of the NDH system — 319
   3.6. Towards a better understanding of cyanobacterial CEF — 320
4. Cyclic Electron Transport Complexes of Chlamydomonas and Other Green Algae — 322
   4.1. Introduction — 322
   4.2. PGR5/PGRL1 complexes in thylakoids — 323
   4.3. Reductases of thylakoid membranes and their role in CEF — 324
5. Evidence for Cyclic Electron Transport in Eukaryotic Algae — 326
   5.1. Pyrrophyta, Dinophyta (Dinoflagellata) — 326
       5.1.1. *Symbiodinium* sp., the coral endosymbiont algae — 326
   5.2. Haptophyta — 327
       5.2.1. Prymnesiophyceae — 327
   5.3. Ochrophyta — 327
       5.3.1. Bacillariophyceae — 327
       5.3.2. Eustigmatophyceae — 328
       5.3.3. Phaeophyceae — 328
   5.4. Rhodophyta — 328
       5.4.1. Bangiophyceae — 328
       5.4.2. Floridiophyceae — 328

6. The Role of CEF in Eukaryotic Algae in a Range of Habitats   329

    6.1. Desiccation   329

    6.2. Variations in salinity   329

    6.3. Low temperature   329

    6.4. High temperature   330

    6.5. High light   330

    6.6. Nitrogen deficiency   330

    6.7. Iron deficiency   330

    6.8. Summary and Conclusions   331

Acknowledgements   332

References   332

Index   345

Chapter 1

Maquette Strategy for Creation of Light- and Redox-Active Proteins

Nathan M. Ennist, Joshua A. Mancini, Dirk B. Auman,
Chris Bialas, Martin J. Iwanicki, Tatiana V. Esipova,
Bohdana M. Discher, Christopher C. Moser and P. Leslie Dutton*

*The Johnson Research Foundation, University of Pennsylvania,
Perelman School of Medicine, Philadelphia, PA 19104, USA*
**dutton@mail.med.upenn.edu*

Twenty five years ago we began utilizing the maquette strategy long used by sculptors and architects to develop first-principles *de novo* proteins that could perform functions inspired by natural proteins. This brief account traces our approach to engineering light- and redox-driven activities in structurally elementary maquette proteins that reflect little or no primary sequence relationship with those of the inspiring natural system. We follow first-principle maquette protein development from their original production by chemical synthesis of individual helix peptides with di-cystenyl-linked assembly into four-helix bundle proteins, to their bacterial expression as single chain maquette proteins and recently to cellular assembly with selected light- and redox-active cofactors. This last advance includes integration of expressed maquettes with the natural machinery of cofactor maturation and transmembrane transport and extends to maquette fusion with natural cofactor-equipped proteins in cells. Growing experience has led to the recognition that mechanistic components developed *en route* to one selected maquette function can be retained and applied to the engineering of maquettes designed for other activities. This flexibility plus extraordinary compatibility of maquette proteins with cellular operations explains the large number of maquettes currently in development directed toward performance of a broad range of light and redox driven functions *in vitro* and *in vivo*.

1. Introduction

The Oxford English Dictionary definition of maquette [OED Online, 2017], ranging from its Latin origins to the 1994 recognition of our own extended use of it, is shown in Figure 1. The original definition of the word and the accompanying extended definitions make clear that a maquette is a model that lends itself to

Oxford English Dictionary | The definitive record of the English language

---

maquette, *n.*

Pronunciation: Brit. /maˈkɛt/, U.S. /mæˈkɛt/

Frequency (in current use):

Origin: A borrowing from French. **Etymon:** French *maquette*.

Etymology: < French *maquette* artist's preliminary sketch or model (1752) < Italian *macchietta* sketch, outline (18th cent.; 1598 in sense speck, little spot) < *macchia* spot (12th cent.; < classical Latin *macula* MACULA *n.*; compare MACCHIA *n.*) + *-etta* (see -ET *suffix¹*).

1. *Art.* A small preliminary sketch, or wax or clay model, from which a
work (usually in sculpture) is elaborated.

> 1880   J. B. MATTHEWS *Theatres of Paris* xii. 196   A *maquette*..is the miniature model..which the scene painter prepares.
>
> 1903   *Athenæum* 24 Jan. 122/3   M. J. B. E. Detaille has, after a long delay, executed four *maquettes*, each comprehending three large panels.
>
> 1926   W. J. LOCKE *Stories Near & Far* 78   The maquette or model in clay.
>
> 1951   H. READ *Meaning of Art* (ed. 3) II. 240   The sculptor's maquette, or model, was reproduced, generally by other hands, either by being cast in bronze, or by being reproduced to scale by mechanical methods in marble.
>
> 1970   *Country Life* 31 Dec. 1280/3   This was the noble terra-cotta of a mourning woman..the maquette for the figure of the wife on the Westminster Abbey monument to the poet Nicholas Rowe.
>
> 1988   F. SPALDING *Brit. Art since 1900* i. 29   Much Edwardian sculpture is preferable in maquette form, for on this intimate scale it can display a fluency and subtlety of mode lling.

2. In extended use: a small-scale model or template for something.

> 1958   *Times* 8 Oct. 6/4   One might describe his art as a prolonged *maquette* for some ultimate synthesis or other.
>
> 1982   F. RAPHAEL *Byron* (1988) 49   The Spenserian stanza was borrowed from *The Faerie Queene* —thanks to his anthological volume *Elegant Extracts* he had the maquette with him.
>
> 1994   *New Scientist* 4 June 15/3   Dutton and colleagues see their maquette as a model for enzymes such as haemoglobin peroxidase and especially cyctochromes.

Figure 1.   The Oxford English Dictionary definition of maquette [OED Online, 2017] recognizes the use of the word applied to a model protein designed to facilitate the study of natural protein functions.

progressive stepwise alteration to achieve an acceptable level of completion of a product or a component of a larger project: whether it be a building, structural work of art, or an airplane. Maquettes for these examples are developed by drawing on general structural and engineering information known at a level of understanding that after a number of trials, and likely some reversible errors, eventually can be expected to reach the performance or aesthetics hoped-for. The maquette strategy applied to the development and creation of working proteins inspired by nature includes additional substantial challenges. While the approach applied to biology enjoys many role models drawn from energy conversions and chemical transformations that offer efficiencies and performance unmatched outside the cell, they are entangled with an unknown number of other functions and the historical vicissitudes of natural selection. These complexities render many of the individual elements of protein structural engineering and accompanying physical-chemical choreography of enzymatic mechanisms in natural proteins obscure and unresolved; by the standards of Richard Feynman's principle, they are insufficiently understood to inform creation.

What I cannot create I do not understand. [Paz, 1989]

2. Maquette Protein Strategy

2.1. *Recognizing natural complexity*

The experiences of many researchers have proven that good intentions to abstract and transfer established structures and physical chemistries efficiently supporting function in a complex natural protein into simpler designed frames for further research examination or application are often met with disappointment. Those familiar with using mutagenesis to learn about an enzyme mechanism are not surprised by results of even a single amino acid change that have no apparent connection with the intent of the selected mutational change.

Figures 2–4 identify sources recognized for over a century that restated at the molecular level stand as a caution against mimicry or naïve abstraction of functional parts of a natural protein. Figure 2 is a molecular level representation of Darwin's principle of multiple utility [Darwin, 1872] that presents some of the roles any single amino acid or peptide motif can play in the lifetime of a natural protein in a cell from translation to final breakdown and removal for recycling [Moser *et al.*, 2006a; Brisendine and Koder, 2016]. A particular amino acid may be critical not only in promoting the chemical catalysis for which the protein is named, but also and certainly not only, in folding and unfolding the protein and hence in establishing stability and dynamics. Figure 3A illustrates how in the

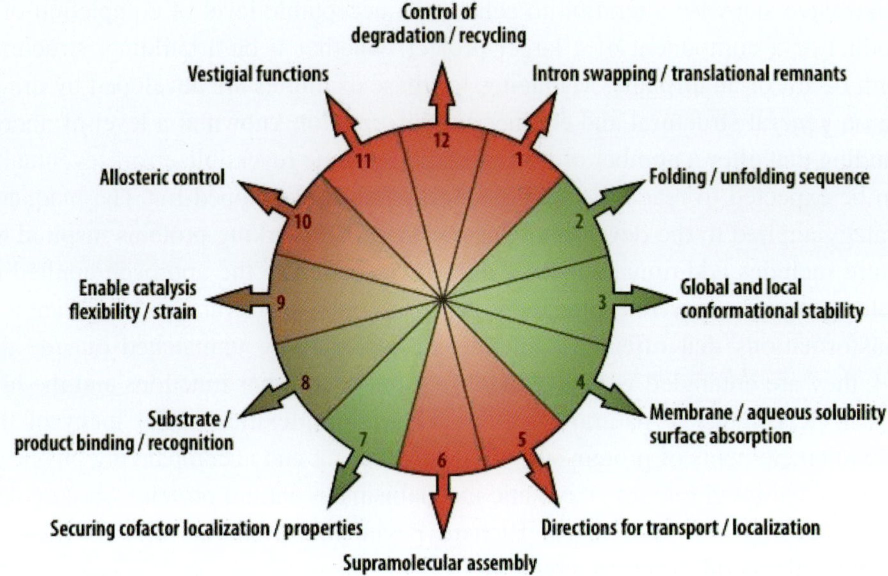

Figure 2. Darwin's principle of multiple utility at the protein level. In maquette design red segments 1, 11 and 12 would be excluded, as initially would 5 and 6. Green segments 2, 3, 4 and 7 represent minimal utilities for the simplest maquette incorporating a bioinorganic cofactor. Green/red segments 8, 9 and 10 represent catalytic utilities.

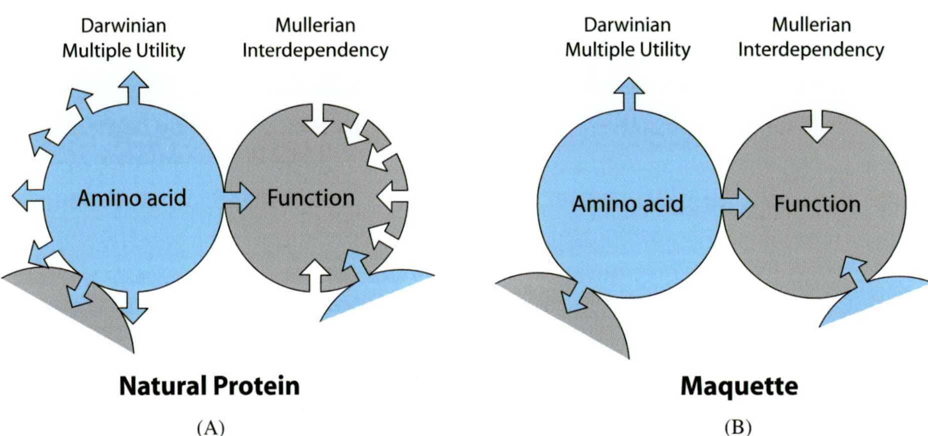

Figure 3. Complexity derived from Mullerian and Darwinian sources in (A) natural proteins and in (B) simple proteins engineered from scratch (maquettes). The figure explains that any amino acid serves a number of utilities in a protein. It also demonstrates that an amino acid or a function in a protein displays interdependencies, close and far, with a number of other amino acids. The maquette approach to protein construction diminishes both the number of utilities and interdependencies [Darwin, 1872; Muller, 1964].

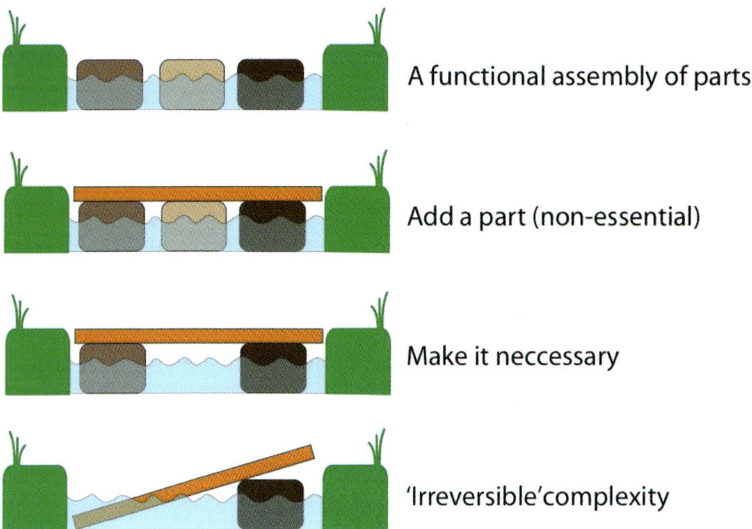

Figure 4. Muller's ratchet explains the source of fragility, amino acid interdependence and irreversible complexity in proteins.

course of evolution this multi-purpose utility of amino acids leads them to become increasingly interdependent on others [Bridgham *et al.*, 2009; Finnigan *et al.*, 2012], while Figure 4 illustrates how in a different and more dramatic way, with the protein as an equivalent to Muller's genetic ratchet [Muller, 1918, 1964], the interdependency can become irreversible.

An escape from the many complexities intrinsic to evolved natural proteins came with the emergence beginning in the 1980s and 1990s of three biochemical and biophysical developments. These developments at that time provided the necessary insights with which to dodge some of the complexity arising from multiplicity of utilities in which an amino acid may be engaged (Figure 2) as well as reducing the number of accompanying interdependencies (Figure 3B) and to avoid or deliberately make use of Muller's ratchet. They also provided the basis to identify functional/mechanistic gaps in understanding between natural and maquette proteins at any stage in development, required if a logical progression is to be made.

2.2. α-*Helical scaffolds common in natural protein structures reveal the basics of how cofactors are ligated and positioned*

The first of the developments that helped make the maquette strategy a viable proposition was the successful crystallization and determined high-resolution structures of a large number of membrane protein structures engaged in photosynthetic and respiratory energy conversion [Deisenhofer *et al.*, 1995; Iwata *et al.*, 1995; McDermott *et al.*, 1995; Tsukihara *et al.*, 1995; Ben-Shem *et al.*, 2003; Gao *et al.*, 2003; Kurisu *et al.*, 2003; Ferreira *et al.*, 2004; Sun *et al.*, 2005; Umena *et al.*, 2011; Zhu *et al.*, 2016; Baradaran *et al.*, 2017]. Figure 5A shows the backbone structure of the water oxidizing photosystem II (PSII) [Deisenhofer *et al.*, 1995; Ferreira *et al.*, 2004] while Figure 5C shows PSII as a membrane spanning protein companion with photosystem I (PSI) [Ben-Shem *et al.*, 2003] and plastoquinone-plastocyanin oxidoreductase (cytochrome b_6f) [Kurisu *et al.*, 2003] common in photosynthetic and respiratory (quinone-cytochrome c oxidoreductase; cytochrome bc_1) energy conversions. These and many related transmembrane energy conversion protein structures are, like PSII, dominated by α-helices. Helical structures have proven to offer relatively straightforward ways to view how pigment cofactors (chlorins, bilins and carotenoids) and the redox driven electron tunneling mediated charge-separating cofactor chains (chlorins, hemes, metal clusters and redox active amino acids) are ligated and located within the helix columns (Figure 5B). Also indicated in Figures 5C–5F are examples of the essential supporting, small water-soluble proteins that serve as electron transporters (cytochromes c, ferredoxins, and flavodoxins) that diffusively connect light-activated oxidants and reductants generated in photosystems to other non-photosynthetic energy converting membrane-protein complexes and oxidoreductases (Figure 5G).

This visualized organization of natural structures supporting light- and redox-driven catalyses has collectively stimulated a new wave of kinetic and physical-chemistry explorations and in depth analyses that have grown into an enormous resource of information readily translatable into maquette development. This resource has also provided unparalleled reference and support for logical testing in maquettes of structural motifs, cofactor assembly and functional motifs.

2.3. *First-principles of de novo designed protein structures outlined*

The second advance offered a way to diminish the massive structural complexity problem intrinsic to natural proteins without compromising the size or the stability of an initiating maquette design. This advance, made largely independently of the rise of molecular level structures including the helix dominated membrane proteins, derives from first-principle, largely empirical demonstrations that focused on the variable tendencies naturally displayed by individual amino acids to promote folding into a helical structure. As shown by Regan and DeGrado (1988), a sequence of polar (P) and non-polar, hydrophobic (H) amino acids in the heptad (for instance HPPHHPP realized as LEELLKK where L is leucine, E is glutamate

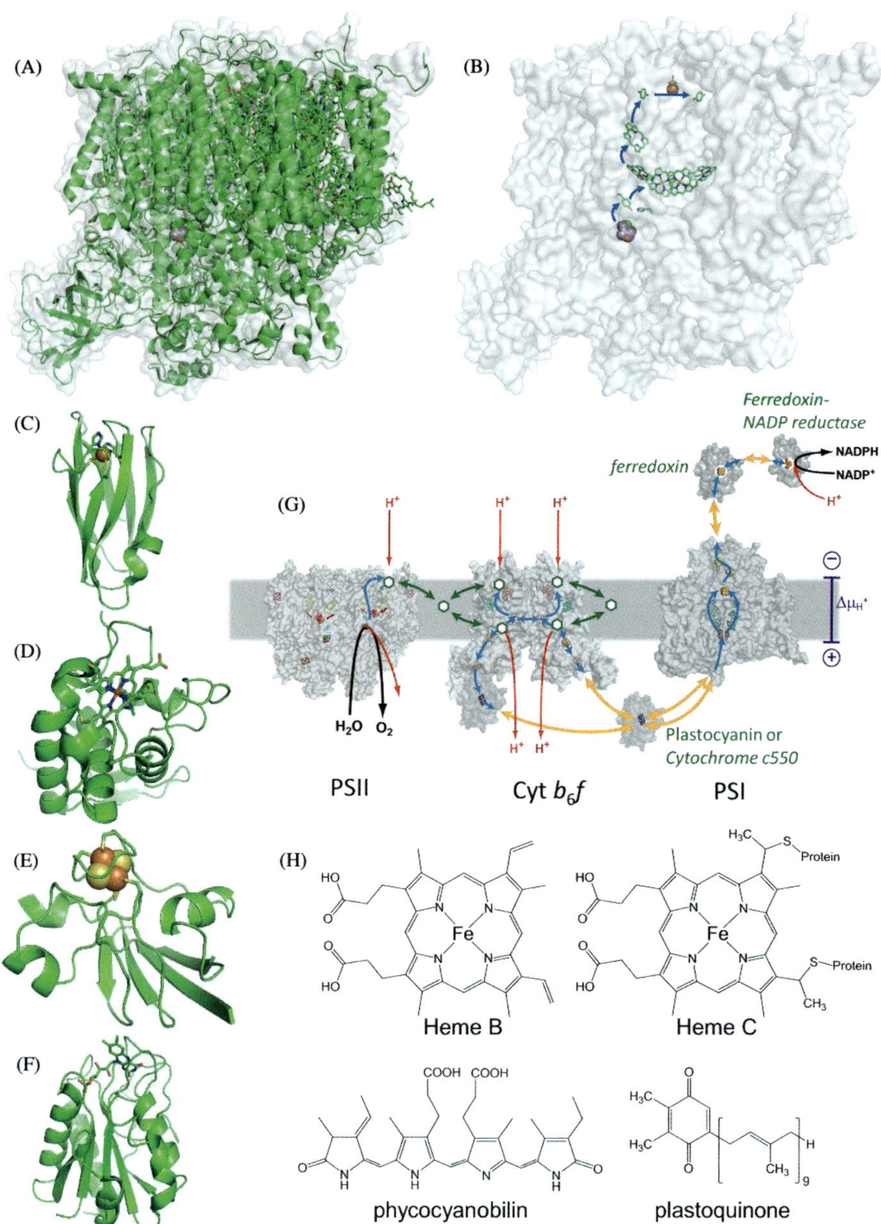

Figure 5. PSII ribbon diagram (A) and reaction center cofactors with blue arrows indicating electron transfer pathway (B) (PDB ID: 1S5L) [Ferreira *et al.*, 2004]. Photosynthetic electron transport proteins plastocyanin (C) (PDB ID: 7PCY) [Collyer *et al.*, 1990], cytochrome c_{550} (D) (PDB ID: 1E29) [Frazão *et al.*, 2001], ferrodoxin (E) (PDB ID: 1FXA) [Rypniewski *et al.*, 1991], and flavodoxin (F) (PDB ID: 1FLV) [Rao *et al.*, 1992] are shown as ribbon diagrams. (G) Photosynthetic membrane in gray with crystal structures of photosynthetic proteins shown to scale [Rypniewski *et al.*, 1991; Frazão *et al.*, 2001; Ben-Shem *et al.*, 2003; Kurisu *et al.*, 2003; Ferreira *et al.*, 2004; Anderson *et al.*, 2008]. (H) Chemical structures of some cofactors involved in photosynthesis.

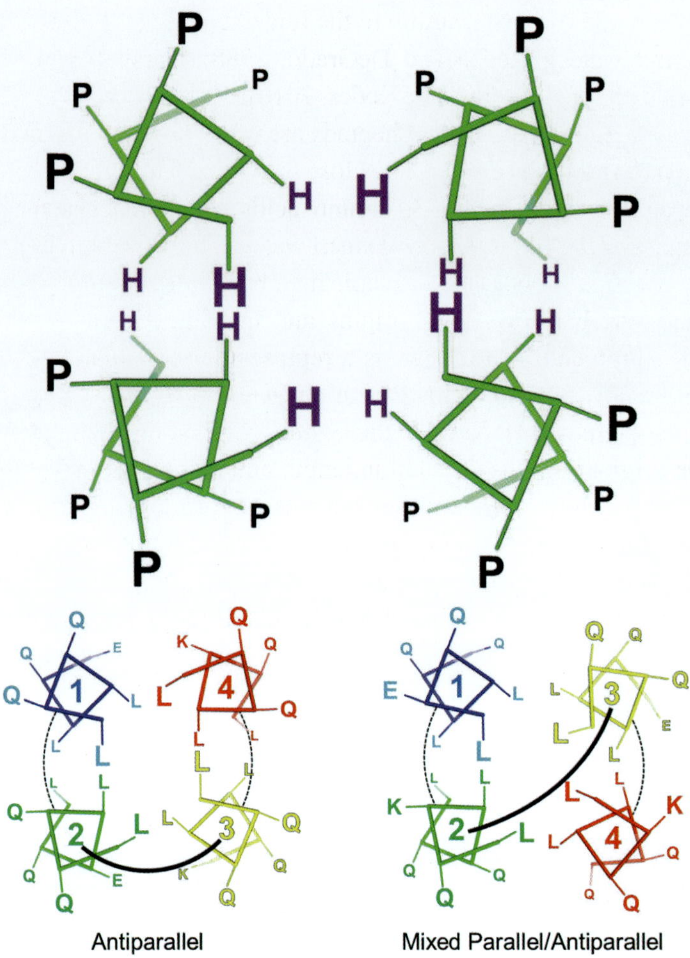

Figure 6. Top: Helical wheel diagram showing positions of amino acid α-carbons (connected by thick green lines) and β-carbons (connected to α-carbons by thin green lines) in an idealized heptad of a four-helix bundle maquette with hydrophobic (H) residues in the hydrophobic core and polar (P) residues on the exterior. Bottom: Single chain maquettes composed of leucine (L), glutamine (Q), positively charged lysine (K), and negatively charged glutamate (E) amino acids to create bundles with anti-parallel interfaces (left) and a mixture of parallel and anti-parallel interfaces (right). Helices are connected by loops represented by thick black lines for loops nearest the viewer and thin dotted black lines for loops on the far end of the bundle. Threading of the helices is controlled by the nature of the loops, charge patterning and hydrophobic core packing.

and K is lysine) can be designed to fold into an α-helix that tetramerizes to form a bundle. A four-α-helix maquette is comprised of four longitudinally aligned helices with a hydrophobic interior (*e.g.* leucine or phenylalanine) and a hydrophilic exterior of mixed positively charged (*e.g.* lysine or arginine) and negatively charged (*e.g.* glutamate or aspartate) amino acids positioned to optimize charge interactions and add thermal stability to the folded bundle (*e.g.* resisting unfolding in near boiling water) [Regan and DeGrado, 1988; Marshall and Mayo, 2001; Huang *et al.*, 2014; Brisendine and Koder, 2016].

These seven amino acid helical heptads are well known to form two complete turns to extend the length of a helix by close to 10.8 Å. Thus, a maquette with four heptad repeats provided by 25–30 amino acids easily reaches the length and molecular mass of helices in a typical small water-soluble protein found in nature.

Figure 7 (top) introduces the principle members of the family of *apo*-maquette proteins, and briefly shows their origins and how they became related during maquette developments. On the left is a representation of the two-heptad repeat peptides of Regan and DeGrado [Regan and DeGrado, 1988] assembled into a short four-helix bundle (**1**). Next, to the right is our first maquette (**2**) which compared to the originating **1** is doubled in length and comprises two pairs of chemically synthesized peptides disulfide linked at the N-terminal cysteines; these form a homo-dimer in either *syn-* (**2a**) or *anti-* (**2b**) configuration [Robertson *et al.*, 1994]. Helices in the *apo*-assembly as first designed adopt an *anti*-configuration [Huang *et al.*, 2003, 2004]. Pairs of helices of this *anti*-configured helical pair were repositioned and joined by a loop; these helix pairs were linked in turn by a di-cystenyl disulfide link either between the N-terminus of each di-helix peptide [Chen *et al.*, 1999] or as shown in the figure, between the loops to produce the "candelabra" maquette **3** [Koder *et al.*, 2009]. The next step involved eliminating the disulfide link and adding a third loop to **3** to produce a native-like single chain four-α-helix maquette protein **5** reflecting size and a common motif of natural proteins enabling, importantly, one-step expression in *E. coli* (see [Farid *et al.*, 2013; Solomon *et al.*, 2014]). This water-soluble single-chain maquette has since been patterned externally with hydrophobic residues to produce the amphiphilic structure that incorporates into and spans a membrane bilayer **6** [Goparaju *et al.*,

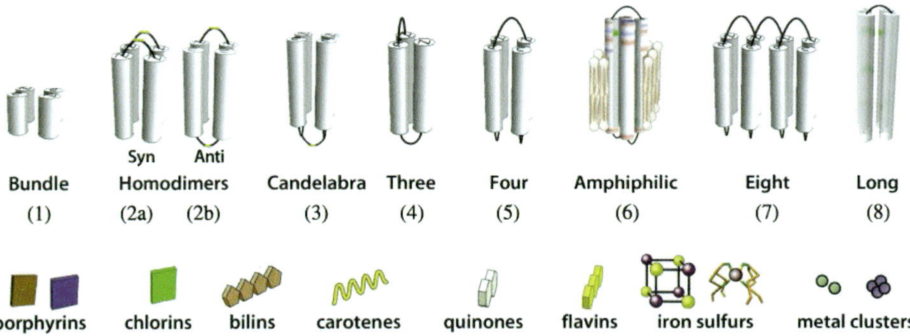

Figure 7. Use of various helical bundle maquettes (top row) and cofactors (bottom row) yields holo-maquettes equipped for various functions.

2016]. Further flexibility has been demonstrated by expression of water-soluble single-chain maquettes comprising eight-α-helices **7** [D. Auman and G. Kodali unpublished] or helices lengthened from about 45 Å to 62 Å (**8**) for which crystals are revealing a complete structure [Ennist, 2017].

The bottom row of Figure 7 likewise introduces cofactors that are common in natural oxidoreductases and that have become the functionalizing partners to the structural *apo*-maquettes. On the left are cyclic tetrapyrroles (brown, purple and green tablets). When these are decorated with an array of peripheral substitutions, equipped with a range of central metals and provided with different ligations from the supporting protein, they form the large class of porphyrins and chlorins well-known in nature and textbooks. With a central iron, the porphyrins become the hemes that are central to cytochromes for electron transfer, myoglobin and hemoglobin for dioxygen transport and many oxidoreductases involved in photosynthetic and respiratory energy conversion and intermediary metabolism. With a central magnesium, the chlorins become the light-harvesting and redox charge-separation units of photosynthesis. For maquette design and engineering, this cofactor class offers many of light- and redox-active cofactor possibilities from natural sources (for recent work see [Farid *et al.*, 2013; Solomon *et al.*, 2014; Goparaju *et al.*, 2016]) and an enormous resource of synthetic variants (see for example [Kodali *et al.*, 2017]) that on satisfying simple binding requirements, ligate to all the *apo*-maquettes except **1** and **4**. However while a great deal is learned from synthetic tetrapyrroles, the very significant advantage regarding the use of natural heme cofactors such as hemes A, B and C in functionalizing maquettes, as will be described later, is their potential for co-production *in vivo* with the single-chain maquette proteins such as **5** [Anderson *et al.*, 2014; Watkins *et al.*, 2016]). And although less developed, chlorins are showing promise for co-synthesis and ligation with maquette proteins in cyanobacteria [J. A. Mancini, A. Nagarajan, H. Pakrasi unpublished].

Next in Figure 7 is the linear tetrapyrrole class of bilin cofactors that in nature dominate the central part of the spectrum in harvesting light in photosynthesis and in light signaling (*e.g.* phytochrome); see Figure 5F. Like the hemes, various bilins have been shown to bind with sufficient affinity to the interior of maquettes and to promote a covalent link *via* their terminal vinyl group to cysteine in the loop and helical domains of single chain maquettes **5** [Mancini *et al.*, in preparation].

More challenging has been the incorporation of the highly hydrophobic cofactors. Carotenes have been persuaded to partition (one or two molecules) into **5** under non-biological conditions; nevertheless, these carotenoid demonstrated photo-protection of Zn chlorins ligated to the same maquette [G. Kodali unpublished]. Similarly flavins, well-known as catalytic elements for dehydrogenase activity in oxidoreductases but less known as the photoactive unit of cryptochromes, has to-date been chemically ligated to **2b** [Sharp *et al.*, 1998] and **5** [Bialas *et al.*, 2016] *via* cysteine sulfur displacement of flavin bromine substituents. Quinone as a cofactor is best known as proton-linked electron transporters diffusing in the membrane bilayer between sites on respiratory complexes of photosynthesis and respiration (Figure 5G). Our aim is to co-produce a specific quinone and corresponding specific maquette binding site. In the meantime we have taken advantage of the flexibility of maquette frame to incorporate the quinone as a synthetic naphthoquinone amino acid; this was synthesized and stitched into a precursor of **5** *via* intein chemistry [Lichtenstein *et al.*, 2012, 2015].

Iron-sulfur clusters are well-known as critical redox cofactors in many oxidoreductases they transport electrons through protein embedded chains over distances up to 100 Å. In small proteins (for instance, ferredoxins; see also Figure 5B) they serve as electron transporters diffusing in the aqueous compartments between the larger oxidoreductases of respiratory and photosynthetic energy coupling (Figure 5C). Early demonstrations showed iron sulfur clusters spontaneously assemble in maquettes **2** singly or in pairs, with or without heme(s); later single-chain variants were designed [Gibney *et al.*, 1996; Mulholland *et al.*, 1998; Musgrave *et al.*, 2002; Nanda *et al.*, 2005; Grzyb *et al.*, 2010; Nanda *et al.*, 2016]. Maquette frames of the kind in Figure 7 have also proved to offer a creative environment for metals alone and in clusters (Figure 5C) [Dieckmann *et al.*, 1997; Farrer *et al.*, 2000; Calhoun *et al.*, 2005; Geremia *et al.*, 2005; Touw *et al.*, 2007; Calhoun *et al.*, 2008; Tegoni *et al.*, 2012; Zastrow *et al.*, 2012; Roy *et al.*, 2014].

Redox active amino acids tyrosine and tryptophan perform as components of electron-transport chains that operate at high oxidizing potentials in natural enzymes such as PSII (tyrosine) and ribonucleotide reductase (tyrosine and tryptophans) or cryptochromes (3 or 4 tryptophans). Because these amino acids are a common part of natural protein sequences, the first indication of their involvement as an active redox component in a protein is a demonstration of their oxidation and reduction. For maquettes, this has been done by applying cyclic voltammetric methods the three-helix maquette (**4**) [Tommos *et al.*, 1999; Dai *et al.*, 2002; Hay *et al.*, 2005, 2007; Westerlund *et al.*, 2008; Berry *et al.*, 2012] and later in photoactive maquettes (**5,8**) [Conlan *et al.*, 2009; Glover *et al.*, 2014; Ennist, 2017]. [Moser *et al.*, 2016] has further general details regarding these maquettes that are on the threshold of performing useful activities and functions.

This set of α-helical structures and their assembled cofactors is proving to offer an adaptable framework for assembly in solution, on surfaces, spanning membranes or to enhancing function inside living cells (Figure 8). For such pursuits, the working frame is proving reliable in guiding selection of the distances between two or more cofactors during development of maquettes supporting different light and redox-driven actions exemplified in the natural proteins of

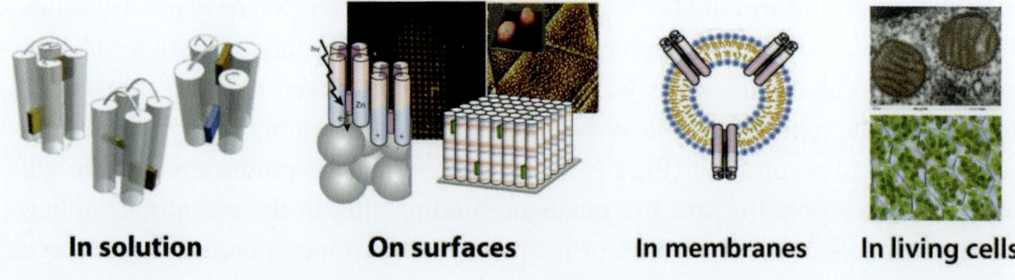

Figure 8. Many possible uses of maquettes in different environments arise from the numerous protein scaffolds and cofactors available to the experimenter.

Figure 5. Maquettes are designed to reproduce the function of natural cofactor chains in transferring electrons over distance through protein interiors, grouping cofactors in clusters for multi-electron catalysis, and acting as singular diffusing electron transporters to link separated sources of oxidants and reductants in cells.

2.4. *Electron transfer in protein understood for engineering*

The third development in maquette engineering is a practical understanding of electron tunneling in protein. Over one third of classified enzymes belong to the oxidoreductase family; many resolved molecular structures contribute critical electrochemical information on the single-electron oxidations and reductions. The oxidoreductase family continues to serve as a test bed for identifying what parameters drawn from general electron-transfer theory have been naturally selected to govern the engineering of electron transfer in both light- and redox-driven reactions. This third development has yielded a set of empirical expressions drawn from semi-classical Marcus theory (Marcus/Hopfield) as shown in Figure 9 [Moser *et al.*, 1992; Page *et al.*, 1999; Moser *et al.*, 2005; Moser and Dutton, 2006; Moser *et al.*, 2006b, 2010]. Exhaustive testing of electron transfers in many redox proteins by many laboratories over more than 25 years has proven the expressions sufficient to quantify electron-tunneling processes in biological energy conversions generally. The expressions confirm Eyring's maximum rate constant ($\sim 10^{13}$ s^{-1} or log 13) observed at van der Waals contact and demonstrate that the log of the tunneling rate constant through protein slows linearly over 12 orders of magnitude as the distance of tunneling through protein from the closest edge of one cofactor to the edge of another increases up to at least distances of 25 Å. They applied to downhill exergonic electron tunneling and linked by the Boltzmann term, to uphill endergonic electron-tunneling steps [Woodbury *et al.*, 1986; Page *et al.*, 1999]. The expressions have been demonstrated to apply equally to electron-tunneling events between two cofactors in proteins that are physiologically beneficial, deleterious or are not physiological at all. The expressions

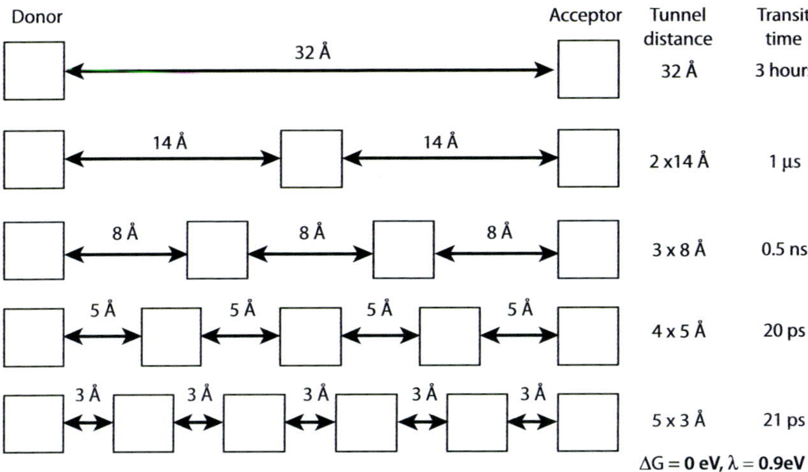

$$\log k_{et}^{exer} = 13 - 0.6\,(R\text{-}3.6) - 3.1\,(\Delta G^{O}\text{-}\lambda)^2/\lambda$$

$$\log k_{et}^{ender} = 13 - 0.6\,(R\text{-}3.6) - 3.1\,(\Delta G^{O}\text{-}\lambda)^2/\lambda - \Delta G^{O}/0.06$$

Figure 9. First principles electron tunneling in proteins. Empirically derived electron tunneling rate expressions guide the placement of redox centers in maquettes during design. The electron tunneling rate, k_{et}, is given for exergonic (exer) and endergonic (ender) reactions.

Figure 10. Electron-tunneling rate has an exponential dependence on the edge-to-edge distance; inserting redox cofactors between an electron donor and acceptor can lower the lifetime of a 32 Å electron tunneling reaction from hours to picoseconds.

provide the understanding of electron-transfer processes in natural proteins generally and hence provide the bases for the design and assembly of two or more cofactors arranged in a maquette protein to promote and suppress light- and redox-driven electron transfer. Figure 10 demonstrates in general terms the enormous engineering contributions that can be made by cofactors arranged as pairs or more in chains for extraordinary control over tunneling rates over very long distances, or in clusters sufficiently closely spaced to facilitate ultrafast tunneling or scaling of the substantial endergonic barriers familiar in catalysis.

3. Practical Strategies for Development of Functional Light- and Redox-active Proteins

3.1. *Lessons learned from the first heme protein maquettes*

Heme B and its interactions in the natural proteins of cytochromes, globins and oxidoreductases is understood well enough to play a dominant role in testing the viability of the maquette strategy. Despite the enormous acts of simplification made to inspiring natural multi-helical transmembrane and water soluble proteins bearing one or more hemes, their translation into elementary *de novo* designed scaffolds using the maquette strategy has been accomplished in a straightforward manner.

As introduced in Figure 7, early maquettes **1** and **2** were produced on a commercial synthesizer with its practical restrictions regarding the length of the sequence generated. Thus in **2a/2b** each designed helix employed an N-terminus cysteine with a view to link two identical helical sequences together. The helix pairs spontaneously associated to form a homodimeric four-α-helical maquette protein with an approximate molecular mass of 15 kDa, typical of many small diffusing electron-transporting proteins [Robertson *et al.*, 1994].

Not surprisingly the homodimer in the four helices adopted one of two dominant structural alignments, either anti-parallel (*anti*) or parallel (*syn*) as shown schematically in Figures 11A and 11B. The intrinsic properties of this first maquette protein have proven extraordinary. As designed, it bound one or two hemes per monomer up to a total of four with practical micro-molar or tighter dissociation constants. The oxidized and reduced hemes assumed the characteristics of individual hemes with absorbance peaks varying by a few nanometers and redox midpoint potentials (E_m values) ranging from −80 to −250 mV, as do bis-His ligated hemes B in natural cytochromes. When hemes were ligated in adjacent positions and oxidized to assume a net positive charge, the observed splitting of heme redox midpoint potentials demonstrated conspicuous heme–heme electrostatic interactions. The patterns of altered redox midpoint potentials and binding affinities of one two or more ferric hemes showed that the two homodimeric units adopted a *syn*-topology (Figure 11B; see Section 3.2.1). Corresponding charge–charge coupling is observed between ferric hemes and the acid-base equilibria of interfacial amino acids; for instance pK_a values of structurally nearby glutamates change from near 4 when the heme iron ferrous to above 7 shifts of higher than 7

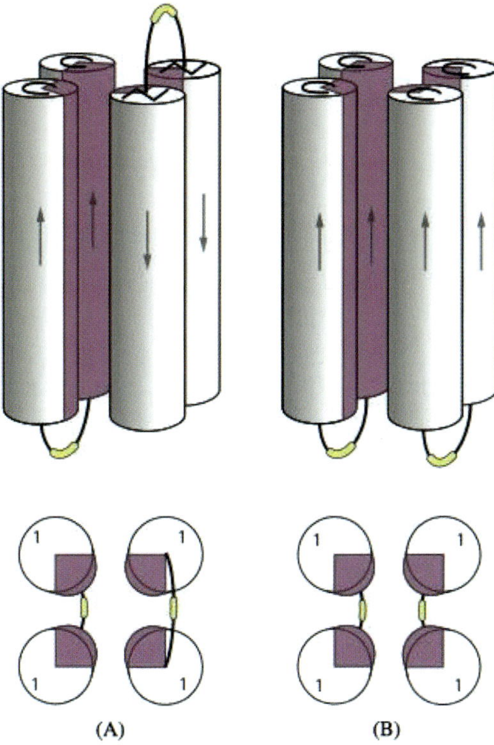

Figure 11. Homodimeric four-α-helical maquette protein in the (A) *anti-* or (B) *syn*-topology.

when the heme is ferric [Shifman *et al.*, 1998, 2000]. To a lesser degree, E_m values are altered to higher, more oxidizing potentials or lower, more reducing potentials as the exterior net charge of the protein is rendered more positive or negative. The evident internal electric fields also affected binding affinities of adjacent ligated ferric hemes in the maquette.

It is remarkable that physical-chemical and internal electric field effects functionally associated with natural light- and redox-energy-converting oxidoreductases can be readily achieved in such small, simple rudimentary proteins. On another level it is equally remarkable that these simple maquette structures are found to be so stable. The resistance of this homodimeric two-heme maquette to pH change became routinely recognized; ferrous hemes dissociate from the maquette at pH values below 4 and above 12, while the ferric heme dissociates at below 2.5 and 10. Furthermore, heme maquettes (beyond this first maquette) resist unfolding and heme dissociation not just at elevated temperatures already mentioned [Regan and DeGrado, 1988; Marshall and Mayo, 2001; Huang *et al.*, 2014; Brisendine and Koder, 2016] but at autoclave temperatures above 100°C [Ennist, 2017].

Another demonstration of the controlled maquette stability was seen when assembled on surfaces. The first maquette and variants were readily tolerant to altered net charge and charge patterns and cysteine substitutions on their exterior, opening the door to directed electrostatic and covalent assembly on a range of different surfaces [Chen *et al.*, 1998, 1999, 2002; Ye *et al.*, 2004].

The most telling characteristics of maquette flexibility and stability are evident in their properties on the Langmuir–Blodgett trough. At low applied lateral surface pressure, the area covered approximated the two-dimensional equivalent of the ideal gas laws with the *apo*-maquette bundle helices lying flat on the surface. At a characteristic area and pressure they associated into the size of a folded four-helix bundle lying on its side. At the highest pressure the area per bundle adopted was that of a stable upright orientation with an area per molecule expected for the bundle cross section normal to the long axis of the bundle. With one or more hemes ligated, the maquette spread on the trough surface at low surface pressures as a four-α-helix bundle on its side and with increasing pressure the bundle it reorganized to become upright. The maquettes retained the heme(s) in each arrangement and proved readily transferable onto a solid surface for further analysis and application [Chen *et al.*, 1998].

3.2. *Characterization of one maquette promotes development of others*

One of the "In extended use" OED definitions in Figure 1 describes "his art as a prolonged maquette for some ultimate synthesis or other." While this implies perhaps that he was never satisfied, we also have found that the development of a maquette to a satisfactory working product yields insights that tempt continued maquette development toward new products. Below as one example we track the progression that emerged from the heme B binding maquette in Section 3.1.

3.2.1. *Charge-activated conformational switch in a homodimeric four-α-helix heme B protein [Grosset et al., 2001]*

X-ray crystallographic structure determination of maquette **2** described above revealed that an *anti*-topology [Huang *et al.*, 2003, 2004] (Figure 11A) was favored. However as already mentioned, after ligation of two ferric hemes B the topology switched to favor the *syn*-topology (Figure 11D) despite expected repulsive charges on the adjacent hemes. To pursue this, an optical probe, coproporphyrin, was attached to the loop region of each helix pair to report the balance between *syn*- and *anti*-topology (Figure 12). Selected alteration of one alanine per helix in the core of the homodimeric four-α-helix structure to a more polar serine

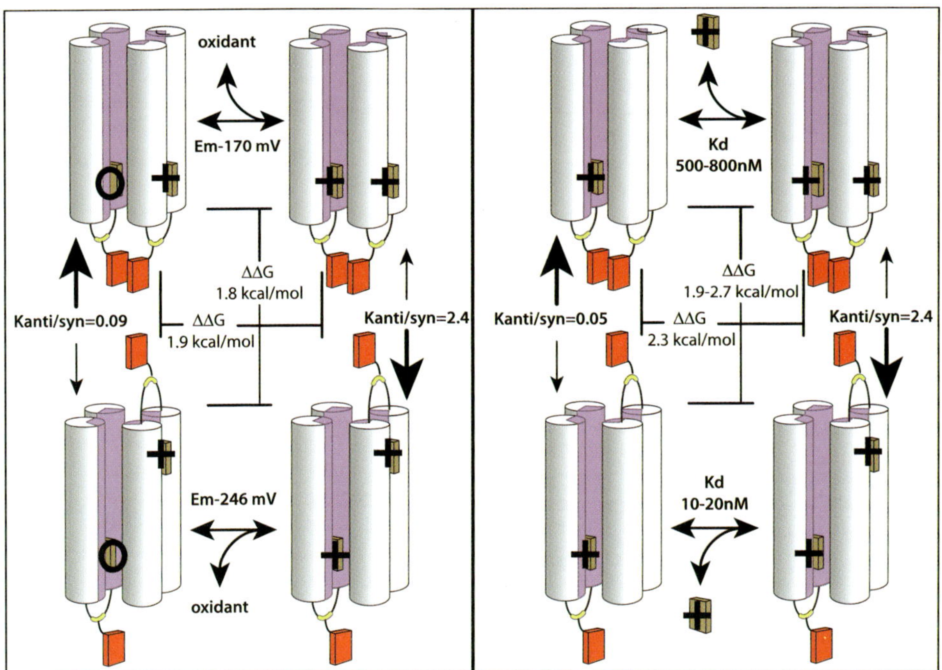

Figure 12. Thermodynamic squares relating the interactions between the *syn-* and *anti-*topologies described by heme B redox midpoint potentials (E$_m$ values; left). The + sign signifies the cationic charged ferric-heme while the 0 sign indicates the uncharged ferrous-heme and oxidized and reduced dissociation constants (K$_d$ values; right) and reported by a coproporphyrin optical probe (red).

promoted *anti*-topology on addition of ferric hemes B, but now on removal of the repulsive charges by reduction of the hemes shifted the balance to the *syn*-topology. Thus, the basic physical-chemical characterizations of **2** described in Section 3.1 (E$_m$ values, binding affinities of oxidized and reduced hemes), together with the introduction of a probe to assess the *syn-anti* balance, allowed the effects of oxidation and reduction of two adjacent hemes to be quantified. As shown in Figure 12, this turned maquette **2** into an optically monitored redox-driven switch.

3.2.2. *Electron tunneling between monolayers of linked-homodimeric heme protein and gold electrodes [Chen et al., 1999, 2002]*

Confidence in the characteristics of the homodimeric heme B maquette **2** drawn from references [Robertson *et al.*, 1994] and [Grosset *et al.*, 2001] and analyses of many other related maquettes, led to the maquette developments shown in Figure 13. Part A shows the restructuring of **2** to comprise two synthesized 68 residue helix-turn-helix peptides. These were linked together with a disulfide link

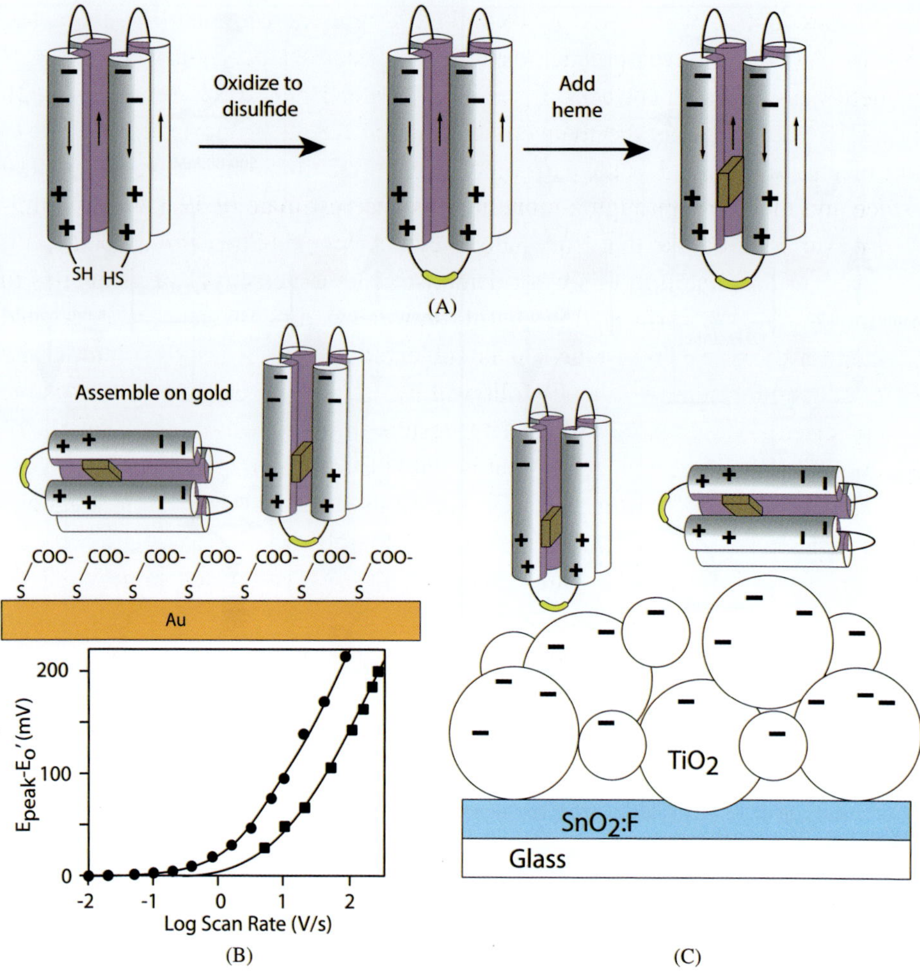

Figure 13. (A) Steps taken from the homodimeric four-helix maquette to a disulfide-linked maquette. At neutral pH, binding of the charge-patterned maquette occurs by electrostatic interaction between the net positive residues (lysine) of the maquette and the negatively charged surfaces. (B) Maquette assembled on the anionic undecanoate-coated gold surface. The half values of peak-to-peak (E_{peak} — E_o') separations from cyclic voltammetry traces as a function of scan rate. The solid circles represent the heme maquette at 120 s^{-1} and the solid squares are for cytochromes *c* at 30 s^{-1} [Chen *et al.*, 2002]. (C) Maquette assembled on nano-structured TiO$_2$ with binding to negatively charged oxygen moieties on the surface of the metal oxide electrode [Topoglidis *et al.*, 2003].

provided by cysteines on the N-terminus of each di-helix peptide to form a linked homodimeric four-α-helix bundle capable of *bis*-His ligation of one heme B. The exterior was decorated with a rather extreme charge patterning to assist in binding and orientation of the maquette on the prepared surfaces. Nevertheless, the new maquette carried forth similar heme E_m values and ~3 unit pK shifts on

oxidation-reduction that herald redox-linked proton-exchange as first characterized in **2**. These developments established electrostatically-driven heme B
maquettes assembly on surfaces [Chen *et al.*, 1998, 1999, 2002; Ye *et al.*, 2002].
Figure 13 part B shows the maquette assembled on undecanoate thio-linked to
gold electrode. Applied voltages promoted electron tunneling between the gold
surface and the heme through a tunneling distance estimate of 12.3 Å in the mil-
lisecond time, much like that from natural cytochrome *c* [Chen *et al.*, 2002].

A natural extension to this work demonstrates the versatility of maquettes to
function on range of surfaces. Figure 13C shows the same maquette is assembled
on titanium dioxide (TiO_2) where it is subjected to pulsed laser excitation and
heme reduction-oxidation kinetics followed by absorbance changes in the visible
range. Pulsed UV excitation of the TiO_2 results in microsecond or faster photoreduction of heme in the maquette that is stable for milliseconds before re-oxidation and return to the starting state. On substitution of the heme in the maquette
for an analogous photoactive Zn porphyrin, visible light-activation injects an
electron from the singlet excited state of the Zn-porphyrin into the TiO_2
conduction band in nanoseconds again to produce a long-lived transient absorption signal. These demonstrations compare well with earlier work with Zn-porphyrin
substituted natural cytochrome *c*.

3.2.3. *Linked-homodimeric heme protein oxygen transporter [Koder et al., 2009]*

Another maquette developed from homodimeric maquette **2**, follows a similar
process of re-directed loops and chemical linkage to form the four-α-helix bundle as shown on the left of Figure 14 with a view to suppress heme redox reactions and to stably bind molecular oxygen to the ferrous heme akin to natural
globins. The redesign also introduced phenylalanines into the helical hydrophobic interior for improved packing stability and heme binding, as detailed in
Figure 1 of Koder *et al.* [2009]. Without the disulfide link, the introduction of
oxygen to the ferrous-heme-ligating maquette oxidized the heme(s) in milliseconds to produce superoxide and peroxide. With the loop linkage in place, an
oxyferrous state was stably formed on the same timescale with oxygen affinities
and exchange timescales matching those of the natural globin family. It has
become clear that this simple link introduced to constrain inter-helical dynamics
does not impede access of small molecules like dioxygen and carbon monoxide
into the maquette interior but raises a barrier that slows the access of water by
orders of magnitude to suppress electron transfer with the loss of the oxyferrous
bond to produce ferric heme and superoxide [Anderson *et al.*, 2008]. As was
concluded by Koder *et al.* [2009],

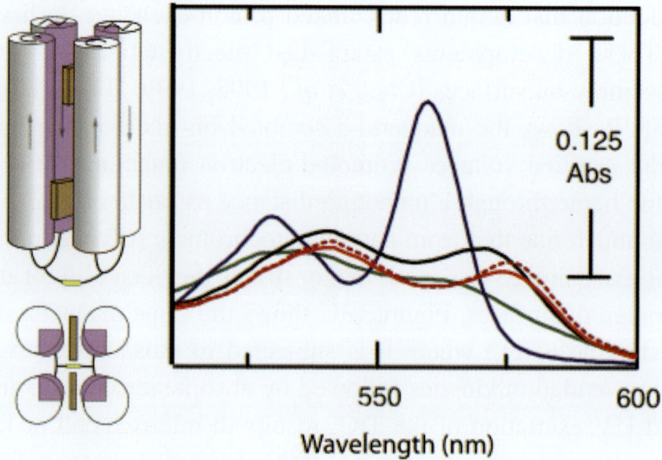

Figure 14. Heme maquette spectra showing oxidized/ferric (green), reduced/ferrous (blue), carboxyferrous (black) and oxyferrous (red)

"… the ease with which globin-like properties can be reproduced in a completely unrelated and simply engineered maquette indicates that the relatively complex globin fold is for the most part unremarkable, and may be common in nature not because of a uniquely capable design for oxygen binding, but simply because it is good enough."

4. Single-Chain Four-α-Helix Maquettes

Post-expression disulfide assembly confines the homodimeric maquette engineering to *in vitro* functions, while duplications and symmetries in the sequence severely restrict choices regarding diverse cofactor ligation and the positional freedom demanded for more sophisticated mechanistic functions. Despite these limitations, the first design and expression of a single chain four-α-helix protein ligating a cofactor appeared only a decade ago [Bender *et al.*, 2007]. Figure 15E shows the transformation into a single-chain four-α-helix to escape from these restrictions, emphasizing the transparency of historical relationships with maquette predecessors.

4.1. *From disulfide linked homodimeric bundles to single-chain four-α-helix structures*

The expression of the *apo*-single-chain four-α-helix has proven to be straightforward, quick with high yield in a pure form, typically 1–200 mg 4 hours after induction from a 2 L culture. The physical-chemical characteristics of

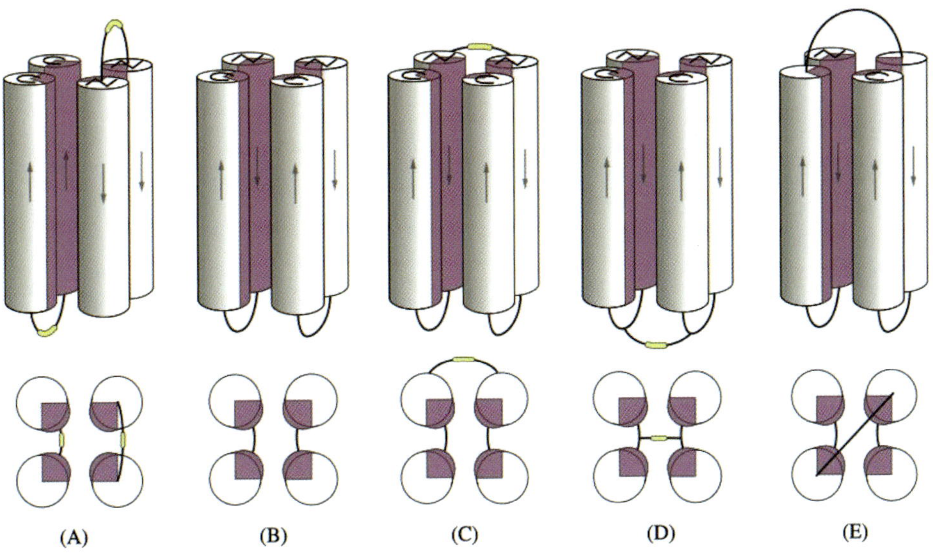

Figure 15. Stages in the development of single-chain four-α-helix maquettes.

disulfide-linked homodimeric predecessors were carried forward into the single chain structure without significant alterations [Farid *et al.*, 2013]. For example, the rotational strain on the histidine ligations to the heme iron promoted by locally acting interfacial glutamates in helical column [Koder *et al.*, 2009] continued to support the formation, dynamics and stability of the oxyferrous state. Similarly, the disulfide link to the homodimer essential for preventing water access and loss of the oxyferrous heme [Anderson *et al.*, 2008] was functionally replaced by the diagonal loop linkage in the single-chain structure. In maquette redesign, transfers between earlier structural forms to the single-chain four-α-helix structure are manageable and quantifiable.

Other properties of single-chain four-α-helix structures have opened new avenues for functional engineering of *de novo* designed proteins. The most prominent and promising is the identification of the presence of operationally distinct domains that lend themselves to separate engineering. As best stated in the original paper [Farid *et al.*, 2013],

"…these domains include the interior of paired helices 1 and 3 and the interior of paired helices 2 and 4 [see numbered helices in Figure 15 Part E], as well as the external charged surfaces; loop regions also operate as mostly independent domains. The functional evidence for independent domains is well complemented by CD monitored thermal stabilities of the secondary α-helices and NMR views of the tertiary structuring. Even with substantial conformational changes associated with helical rotation on heme binding and ligand exchange including a 3 unit ΔpK_a

glutamate-linked, proton coupling to heme oxidation-reduction [Shifman *et al.*, 1998] adjacent helical pairs are serviceably independent."

Likewise, analyses of light-activated charge separation and recombination between several different asymmetrically ligated cofactor dyads reflect behavior of earlier chemical constructs and adhere to established photochemistry and electron-tunneling guidelines.

Taken altogether these advances and their characterizations have provided confidence that insights derived from earlier homodimeric constructs are relevant to new applications in single-chain four-α-helix structures. It is also clear that the single-chain four-α-helix protein alone provides the maquette strategy with a highly creative resource for the pursuit of more ambitious functions both inside and outside of living cells.

4.2. *Compatibility of single-chain four-α-helix maquettes with natural proteins*

Our expressed *de novo* designed single-chain four-α-helix proteins interact effectively with natural proteins. This includes recognition, binding and inter-protein functions exemplified by *in vitro* electrostatic-driven interactions of the single-chain four-α-helix heme B maquette with natural cytochrome *c* [Fry *et al.*, 2016] (Figure 16), and by *in vivo* transmembrane transport and engagement with maturation machinery to covalently link a heme C inside the single-chain four-α-helix maquette [Anderson *et al.*, 2014] (Figure 17).

4.2.1. *Single-chain four-α-helix heme B maquette interactions with natural cytochrome c*

It has long been understood that successful electron-transfer contact between mobile redox protein partners can be guided by complementary surface electrostatics [Margoliash and Bosshard, 1971; Koppenol *et al.*, 1978; Dixon *et al.*, 1989]. The examples of cytochrome c_{550}, plastocyanin and ferredoxin electron transporters connecting larger catalytic redox complexes in the chloroplast were shown in Figure 5. These small diffusing transporters tolerate being replaced by other transporters of similar size and favorable redox properties but equipped with a different cofactor [Meschi *et al.*, 2011]. Similarly sized maquettes that tolerate dramatic sequence and charge patterning changes without compromising redox function, are good candidates for alternative, non-natural diffusing electron transporter proteins.

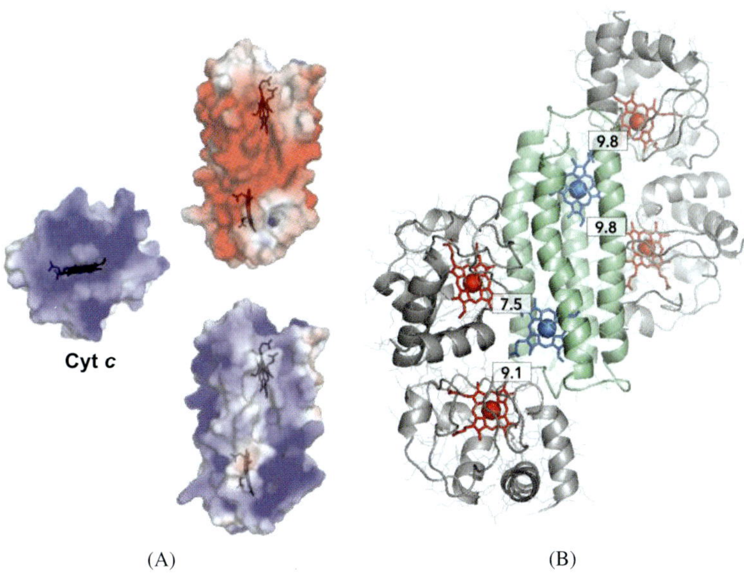

Figure 16. (A) Electrostatic surface estimates are shown for bovine cytochrome *c*, and two single-chain four-α-helix heme B maquettes created with different exterior charge patterning yielding net positive (red) and net negative (blue) charges; calculated with APBS [Baker *et al.*, 2001]. (B) Model to determine electrostatic surface-driven assembly of the net negatively-charged heme B maquette and natural cytochrome *c* within practical tunneling distances; with two hemes present in the maquette, the conjugated edges of the hemes B and heme C porphyrins hemes of four cytochrome *c* come within the functional distances indicated.

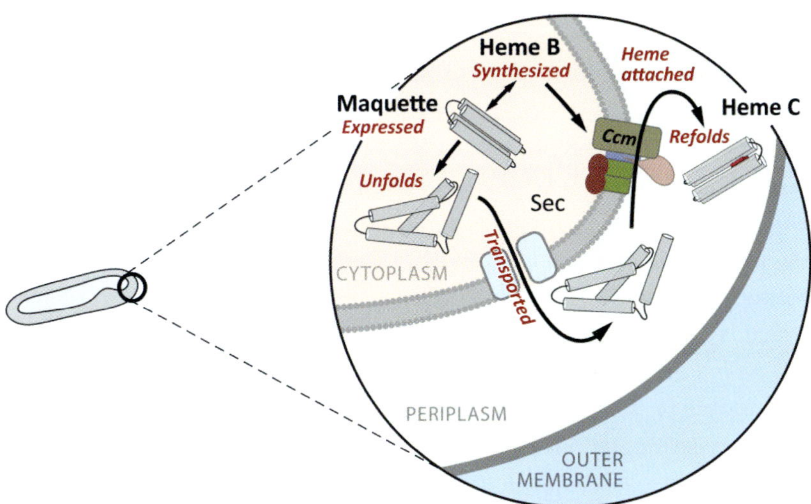

Figure 17. Schematic of the assembly process of the single-chain four-α-helix heme C maquette in *E coli.*

A series of heme B maquettes were engineered to offer complementary or clashing electrostatic interaction surfaces to the positive charge pattern around the heme binding site of cytochrome *c*. Electron transfer from the ferrous-heme B of the maquette to ferric-heme C of cytochrome *c* in aqueous solution reproduces the timescales and charge complementing modulation observed in natural systems. The ionic strength dependence of inter-protein electron transfer from 8.4×10^6 $M^{-1}s^{-1}$ to $1.3 \times 10^9\, M^{-1}s^{-1}$ follows a simple Debye–Hückel model for attraction between +8 net charged oxidized cytochrome *c* and −19 net charged heme maquette, with no indication of significant protein dipole moment steering.

The demonstrated compatibility of expressed *de novo* designed single-chain four-α-helix heme B protein with natural cytochrome *c* is essential to engineer future maquette insertion into natural electron-transfer networks. Indeed, each of the natural diffusive electron-transfer proteins has a *de novo* maquette counter part incorporating appropriate redox cofactors: iron-sulfur cluster [Gibney *et al.*, 1996]; flavin [Sharp *et al.*, 1998]; quinone [Lichtenstein *et al.*, 2012]; copper [Tegoni, 2014] and iron [Faiella *et al.*, 2009]. Successful creation of small *de novo* electron transporting proteins designed to redirect electrons in energy conversion systems or in metabolism holds promise for *in vivo* clinical intervention and for the production of useful molecular products.

4.2.2. *In vivo single-chain four-α-helix protein covalently linking heme C: A novel cytochrome c*

For any promise or potential to be turned into reality for clinical usefulness requires confidence that the genetic system of the cell can go much further than simple expression of our *de novo* designed proteins done so far in prokaryotes, not to mention eukaryotic systems. To start to address this challenge we considered the redesign of the single-chain four-α-helix protein as a heme C protein in *E. coli* [Anderson *et al.*, 2014] (Figure 5H). As indicated in Figure 17, the assembly of a covalently linked heme C involves multiple steps requiring the introduction of plasmids and precursors to promote or enhance several expressions of different contributing machinery. The first step involves separate expression of the (*apo*) protein and synthesis of cofactor heme B in the cytoplasm. The *apo*-protein equipped with a natural N-terminal signal sequence for the Sec translocon is sufficiently unstable to unfold (required by Sec) and be translocated across the membrane into the periplasm. Prompt removal of the signal sequence after transport renders the protein available to interact with the membrane-asociated cytochrome *c* maturation machinery (Ccm) from the periplasmic side of the membrane. Heme B moves from the cytoplasm through the Ccm maturation machinery (CcmA-H) to meet the *apo*-maquette protein. The maquette, equipped with a cytochrome *c* consensus CXXCH, binds to the Ccm where the heme B is covalently attached to the protein backbone *via* thioether linkages between the peripheral vinyl substituents on the porphyrin and two cysteine sidechains within the consensus CXXCH motif. The attachment of the heme C confers practical stability to the maquette protein. Efficiencies were very high with yields typically about 100 mg of purified single chain four-a-helix heme C protein from a 2 L culture.

The expression of the single-chain four-α-helix heme C protein in *E. coli* equipped with plasmids for the Sec translocon and the Ccm was successful. This demonstration could not have been clearer regarding the smooth working of the sequence of events that characterized the process and the demonstrated compatibility between the natural cellular machinery and the *de novo* designed sequences in the maquette frame. The exceptionally high efficiency and yields obtained for the expressed stable single-chain four-α-helix heme C is the result of accompanying trials directed to optimize the overall train of events in the serial process. The process readily accommodates two hemes C in the four-α-helix maquette, and remarkably, four hemes C in the eight-α-helix maquette (**7** in Figure 7). The C-type hemes, like the previous heme B occupants of the maquette, also readily interacted with molecular oxygen to form a stable oxyferrous heme C state, demonstrating that the pair of thioether links between the protein cysteines and heme vinyls do not interfere with this process. Thus, the heme C protein offers a potential high capacity oxygen transporter for clinical use without the risk of the highly detrimental loss of hemes during transfusions.

5. Prospects and Previews

The above elementary four-α-helix heme B and C proteins demonstrate key compatibility requirements for multiple expression and function as biological transporters of electrons and of molecular oxygen. These demonstrations have set the stage for cellular expression and processing of other cofactors as introduced in Figure 7 to be developed in maquettes for useful functions in a range of situations as suggested in Figure 8.

Recent work that has turned from the cyclic tetrapyrroles of hemes and chlorins to the natural linear tetrapyrroles of the bilin pigments (see Figure 5H) common in mid-visible range photosynthetic light harvesting serve to further demonstrate the power of the maquette approach. Maquettes have been designed that bind two Zn-chlorins that have been fused *in vivo* with natural biliprotein subunits containing phycocyanobilin or phycoerythrobilin to demonstrate efficient downhill multistep energy transfer from the bilin through the two Zn-chlorins [Mancini *et al.*, 2017]. Also following in the footsteps of the covalent *in vivo* expression of heme C in *E. coli*, a set of supporting plasmids for heme oxygenase,

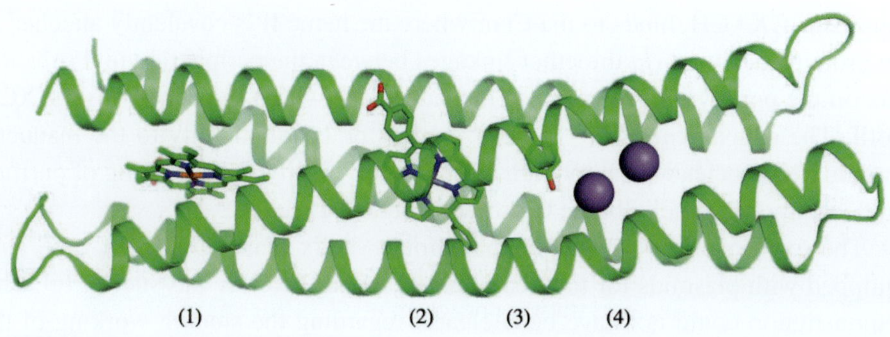

Figure 18. Crystal structure of a PSII maquette under development for light-driven water splitting and generation of useful fuels and chemicals. In the figure (1) is heme B (2) is a synthetic Zn-porphyrin (synthesized by T. Esipova), (3) is a tyrosine (H-bonded to a histidine not shown) and (4) is a metal binding site containing two manganese.

bilin reductases and lyases (provided by collaborator Donald A. Bryant in Mancini *et al.* in preparation) effectively co-express with the maquette protein to covalently link the terminal vinyl groups of bilins such as phycoerythrobilin, phycocyanobilin and biliverdin to cysteines for interior binding. Steady progress is being made to establish rigid binding of the bilins to optimize light-excitation energy transfer for light-harvesting and fluorescent yield for cellular imaging [Mancini *et al.* in preparation; and Iwanicki *et al* in preparation].

Finally after several years of work we now have in hand high-resolution structures of maquettes ligating cofactors [Ennist, 2017]. Figure 18 is a preview of a developing working "prototype" PSII maquette introduced as a "long" maquette (**8**) in Figure 7. The structure shown includes four different cofactors: (1) a heme B that is co-expressible; (2) a synthetic Zn-porphyrin (synthesized by Tatiana Esipova); (3) a tyrosine H-bonded to a histidine expressed to manage redox-linked proton exchange; and (4) a metal binding site (Mn) inspired by the bacterioferritin iron binding site (see [Conlan *et al.*, 2009]). Light-activation of the Zn chlorin delivers an electron to the heme to produce a ferrous heme and an oxidized Zn porphyrin. There is still a large amount of work to be done on this PSII maquette directed to increasing the quantum yield, energy conversion efficiency, tyrosine oxidation rate and effective multi-turnover Mn oxidation. These are necessary for oxidative water-splitting water catalysis as a unique renewable resource for multi-electron reductive generation of fuels and chemical useful to mankind. But in comparing the PSII maquette in Figure 18 with the structure of PSII in Figure 5B we can say at this point that we have made useful steps toward the reproduction of one of the most significant biological electron transfer chains on Earth.

Acknowledgements

PLD, CCM and JAM acknowledge PARC Photosynthetic Antenna Research Center (PARC), an Energy Frontier Research Center funded by the U.S. Department of Energy, Office of Science, Office of Basic Energy Sciences, under Award DESC0001035. BMD acknowledges NIH R21EY027562 entitled Biophysical Design Strategies for Next-Generation Maquette-based Genetically Encoded Voltage Indicators (GEVIs). DBA, CB and MJI acknowledge NIH GM008275T32 predoctoral training program in the "structural biology and molecular biophysics.

References

Anderson, J.L.R., Armstrong, C.T., Kodali, G., Lichtenstein, B.R., Watkins, D.W., Mancini, J.A., Boyle, A.L., Crump, M.P., Moser, C.C. and Dutton, P.L. (2014). Constructing a man-made c-type cytochrome maquette *in vivo*: electron transfer, oxygen transport and conversion to a photoactive light harvesting maquette, *Chem. Sci.*, 5, 507–514.

Anderson, J.L.R., Koder, R.L., Moser, C.C. and Dutton, P.L. (2008). Controlling complexity and water penetration in functional *de novo* protein design, *Biochem. Soc. Trans.*, 36, 1106–1111.

Baker, N.A., Sept, D., Joseph, S., Holst, M.J. and McCammon, J.A. (2001). Electrostatics of nanosystems: application to microtubules and the ribosome, *Proc. Natl. Acad. Sci. USA*, 98, 10037–10041.

Baradaran, R., Berrisford, J.M., Minhas, G.S. and Sazanov, L.A. (2017). Crystal structure of the entire respiratory complex I, *Nature*, 494, 443–448.

Bender, G.M. *et al.* (2007). *De novo* design of a single-chain diphenylporphyrin metalloprotein, *J. Am. Chem. Soc.*, 129, 10732–10740.

Ben-Shem, A., Frolow, F. and Nelson, N. (2003). Crystal structure of plant photosystem I, *Nature*, 426, 630–635.

Berry, B.W., Martínez-Rivera, M.C. and Tommos, C. (2012). Reversible voltammograms and a Pourbaix diagram for a protein tyrosine radical, *Proc. Nat. Acad. Sci.*, 109, 9739–9743.

Bialas, C., Jarocha, L.E. and Henbest, K.B. (2016). Engineering an artificial flavoprotein magnetosensor, *J. Am. Chem. Soc.*, 138, 16584–16587.

Bridgham, J.T., Ortlund, E.A. and Thornton, J.W. (2009). An epistatic ratchet constrains the direction of glucocorticoid receptor evolution, *Nature*, 461, 515–519.

Brisendine, J.M. and Koder, R.L. (2016). Fast, cheap and out of control — insights into thermodynamic and informatic constraints on natural protein sequences from *de novo* protein design, *Biochim. Biophys. Acta*, 1857, 485–492.

Calhoun, J.R., Liu, W., Spiegel, K., Dal Peraro, M., Klein, M.L., Valentine, K.G., Wand, A.J. and DeGrado, W.F. (2008). Solution NMR structure of a designed metalloprotein and complementary molecular dynamics refinement, *Structure*, 16, 210–215.

Calhoun, J.R., Nastri, F., Maglio, O., Pavone, V., Lombardi, A. and DeGrado, W.F. (2005). Artificial diiron proteins: from structure to function, *Biopolymers*, 80, 264–278.

Chen, X., Discher, B.M., Pilloud, D.L., Gibney, B.R., Moser, C.C. and Dutton, P.L. (2002). *De novo* design of a cytochrome *b* maquette for electron transfer and coupled reactions on electrodes, *J. Phys. Chem. B*, 106, 617–624.

Chen, X.X., Moser, C.C., Pilloud, D.L. and Dutton, P.L. (1998). Molecular orientation of Langmuir–Blodgett films of designed heme protein and lipoprotein maquettes, *J. Phys. Chem. B*, 102, 6425–6432.

Chen, X., Moser, C.C., Pilloud, D.L., Gibney, B.R. and Dutton, P.L. (1999). Engineering oriented heme protein maquette monolayers through surface residue charge distribution patterns, *J. Phys. Chem. B*, 103, 9029–9037.

Collyer, C.A., Guss, J.M., Sugimura, Y., Yoshizaki, F. and Freeman, H.C. (1990). Crystal structure of plastocyanin from a green alga, *Enteromorpha prolifera*, *J. Mol. Biol.*, 211, 617–632.

Conlan, B., Cox, N., Su, J.H., Hillier, W., Messinger, J., Lubitz, W., Dutton, P.L. and Wydrzynski, T. (2009). Photo-catalytic oxidation of a di-nuclear manganese centre in an engineered bacterioferritin "reaction centre," *Biochim. Biophys. Acta*, 1787, 1112–1121.

Dai, Q.-H., Tommos, C., Fuentes, E.J., Blomberg, M.R.A., Dutton, P.L. and Wand, A.J. (2002). Structure of a *de novo* designed protein model of radical enzymes, *J. Am. Chem. Soc.*, 124, 10952–10953.

Darwin, C. (1872). *The Origin of Species by Means of Natural Selection* (Earlton House, New York).

Deisenhofer, J., Epp, O., Sinning, I. and Michel, H. (1995). Crystallographic refinement at 2.3 Å resolution and refined model of the photosynthetic reaction centre from *Rhodopseudomonas viridis*, *J. Mol. Biol.*, 246, 429–457.

Dieckmann, G.R., McRorie, D.K., Tierney, D.L., Utschig, L.M., Singer, C.P., O'Halloran, T.V., Penner-Hahn, J.E., DeGrado, W.F. and Pecoraro, V.L. (1997). *De Novo* design of mercury-binding two- and three-helical bundles, *J. Am. Chem. Soc.*, 119, 6195–6196.

Dixon, D.W., Hong, X. and Woehler, S.E. (1989). Electrostatic and steric control of electron self-exchange in cytochromes c, c_{551}, and b_5, *Biophys. J.*, 56, 339–351.

Ennist, N.M. (2017). Design, structure, and action of an artificial photosynthetic reaction center, PhD Thesis, University of Pennsylvania, Philadelphia.

Faiella, M., Andreozzi, C., de Rosales, R.T.M., Pavone, V., Maglio, O., Nastri, F., DeGrado, W.F. and Lombardi, A. (2009). An artificial di-iron oxo-protein with phenol oxidase activity, *Nat. Chem. Biol.*, 5, 882–884.

Farid, T.A., Kodali, G., Solomon, L.A., Lichtenstein, B.R., Sheehan, M.M., Fry, B.A., Bialas, C., Ennist, N.M., Siedlecki, J.A., Zhao, Z., Stetz, M.A., Valentine, K.G., Anderson, J.L.R., Wand, A.J., Discher, B.M., Moser, C.C. and Dutton, P.L. (2013). Elementary tetrahelical protein design for diverse oxidoreductase functions, *Nat. Chem. Biol.*, 9826–833.

Farrer, B.T., McClure, C.P., Penner-Hahn, J.P. and Pecoraro, V.L. (2000). Arsenic(III)–cysteine interactions stabilize three-helix bundles in aqueous solution, *Inorg. Chem.*, 39, 5422–5423.

Ferreira, K.N., Iverson, T.M., Maghlaoui, K., Barber, J., Barber, J. and Iwata, S. (2004). Architecture of the photosynthetic oxygen-evolving center, *Science*, 303, 1831–1838.

Finnigan, G.C., Hanson-Smith, V., Stevens, T.H. and Thornton, J.W. (2012). Evolution of increased complexity in a molecular machine, *Nature*, 481, 394–398.

Frazão, C., Enguita, F.J., Coelho, R., Sheldrick, G.M., Navarro, J.A., Hervás, M., De la Rosa, M.A. and Carrondo, M.A. (2001). Crystal structure of low-potential cytochrome c_{549} from *Synechocystis* sp. PCC 6803 at 1.21 Å resolution, *J. Biol. Inorg. Chem.*, 6, 324–332.

Fry, B.A., Solomon, L.A., Dutton, P.L. and Moser, C.C. (2016). Design and engineering of a man-made diffusive electron-transport protein, *Biochim. Biophys. Acta*, 1857, 513–521.

Gao, X., Wen, X., Esser, L., Quinn, B., Yu, L., Yu, C.-A. and Xia, D. (2003). Structural basis for the quinone reduction in the bc_1 complex: a comparative analysis of crystal structures of mitochondrial cytochrome bc_1 with bound substrate and inhibitors at the Qi site, *Biochemistry*, 42, 9067–9080.

Geremia, S., Di Costanzo, L., Randaccio, L., Engel, D.E., Lombardi, A., Nastri, F. and DeGrado, W.F. (2005). Response of a designed metalloprotein to changes in metal ion coordination, exogenous ligands, and active site volume determined by x-ray crystallography, *J. Am. Chem. Soc.*, 127, 17266–17276.

Gibney, B.R., Mulholland, S.E., Rabanal, F. and Dutton, P.L. (1996). Ferredoxin and ferredoxin-heme maquettes, *Proc. Natl. Acad. Sci. USA*, 93, 15041–15046.

Glover, S.D., Jorge, C., Liang, L., Valentine, K.G., Hammarstrom, L. and Tommos, C. (2014). Photochemical tyrosine oxidation in the structurally well-defined alpha Y-3 protein: proton-coupled electron transfer and a long-lived tyrosine radical, *J. Am. Chem. Soc.*, 136, 14039–14051.

Goparaju, G., Fry, B.A., Chobot, S.E., Wiedman, G., Moser, C.C., Dutton, P.L. and Discher, B.M. (2016). First principles design of a core bioenergetic transmembrane electron-transfer protein, *Biochim. Biophys. Acta*, 1857, 503–512.

Grosset, A.M., Gibney, B.R., Rabanal, F., Moser, C.C. and Dutton, P.L. (2001). Proof of principle in a *de novo* designed protein maquette: an allosterically regulated, charge-activated conformational switch in a tetra-alpha-helix bundle, *Biochemistry*, 40, 5474–5487.

Grzyb, J., Xu, F., Weiner, L., Reijerse, E.J., Lubitz, W., Nanda, V. and Noy, D. (2010). *De novo* design of a non-natural fold for an iron-sulfur protein: alpha-helical coiled-coil with a four-iron four-sulfur cluster binding site in its central core, *Biochim. Biophys. Acta*, 1797, 406–413.

Hay, S., Westerlund, K. and Tommos, C. (2005). Moving a phenol hydroxyl group from the surface to the interior of a protein: effects on the phenol potential and pK_A, *Biochemistry*, 44, 11891–11902.

Hay, S., Westerlund, K. and Tommos, C. (2007). Redox characteristics of a *de novo* quinone protein, *J. Phys. Chem. B*, 111, 3488–3495.

Huang, S.S., Gibney, B.R., Stayrook, S.E., Dutton, P.L. and Lewis, M. (2003). X-ray structure of a maquette scaffold, *J. Mol. Biol.*, 326, 1219–1225.

Huang, S.S., Koder, R.L., Lewis, M., Wand, A.J. and Dutton, P.L. (2004). The HP-1 maquette: from an apoprotein structure to a structured hemoprotein designed to promote redox-coupled proton exchange, *Proc. Nat. Acad. Sci.*, 101, 5536–5541.

Huang, P.S., Oberdorfer, G., Xu, C., Pei, X.Y., Nannenga, B.L., Rogers, J.M., DiMaio, F., Gonen, T., Luisi, B. and Baker, D. (2014). High thermodynamic stability of parametrically designed helical bundles, *Science*, 346, 481–485.

Iwata, S., Ostermeier, C., Ludwig, B. and Michel, H. (1995). Structure at 2.8 Å resolution of cytochrome *c* oxidase from *Paracoccus denitrificans*, *Nature*, 376, 660–669.

Kodali, G., Mancini, J.A., Solomon, L.A., Espisova, T.V., Roach, N., Hobbs, C.J., Wagner, P., Mass, O.A., Aravindu, K., Barnsley, J.E., Gordon, K.C., Officer, D.L., Dutton, P.L. and Moser, C.C. (2017). Design and engineering of water-soluble light-harvesting protein maquettes, *Chem. Sci.*, 8, 316–324.

Koder, R.L., Anderson, J.L.R., Solomon, L.A., Reddy, K.S., Moser, C.C. and Dutton, P.L. (2009). Design and engineering of an O_2 transport protein, *Nature*, 458, 305–309.

Koppenol, W.H., Vroonland, C.A.J. and Braams, R. (1978). The electric potential field around cytochrome *c* and the effect of ionic strength on reaction rates of horse cytochrome *c*, *Biochim. Biophy. Acta*, 503, 499–508.

Kurisu, G., Zhang, H., Smith, J.L. and Cramer, W.A. (2003). Structure of the cytochrome $b_6 f$ complex of oxygenic photosynthesis: tuning the cavity, *Science*, 302, 1009–1014.

Lichtenstein, B.R., Bialas, C., Cerda, J.F., Fry, B.A., Dutton, P.L. and Moser, C.C. (2015). Designing light-activated charge-separating proteins with a naphthoquinone amino acid, *Angew. Chem. Int. Ed.*, 127, 13830–13833.

Lichtenstein, B.R., Moorman, V.R., Cerda, J.F., Wand, A.J. and Dutton, P.L. (2012). Electrochemical and structural coupling of the naphthoquinone amino acid, *Chem. Commun.*, 48, 1997–1999.

Mancini, J.A., Kodali, G., Jiang, J., Reddy, K.R., Lindsey, J.S., Bryant, D.A., Dutton, P.L. and Moser, C.C. (2017). Multi-step excitation energy transfer engineered in genetic fusions of natural and synthetic light-harvesting proteins, *J. R. Soc. Interface*, 14, 20160896.

Margoliash, E. and Bosshard, H.R. (1971). Guided by electrostatics, a textbook protein comes of age, *Trends Biochem. Sci.*, 8, 316–320.

Marshall, S.A. and Mayo, S.L. (2001). Achieving stability and conformational specificity in designed proteins *via* binary patterning, *J. Mol. Biol.*, 305, 619–631.

Prince, S.M., Freer, A.A., Hawthornthwaite-Lawless, A.M., Papiz, M.Z., Cogdell, R.J. and Isaacs, N.W. (1995). Crystal structure of an integral membrane light-harvesting complex from photosynthetic bacteria, *Nature*, 374, 517–521.

Meschi, F., Wiertz, F., Klauss, L., Blok, A., Ludwig, B., Merli, A., Heering, H.A., Rossi, G.L. and Ubbink, M. (2011). Efficient electron transfer in a protein network lacking specific interactions, *J. Am. Chem. Soc.*, 133, 16861–16867.

Moser, C.C., Anderson, J.L.R. and Dutton, P.L. (2010). Guidelines for tunneling in enzymes, *Biochim. Biophys. Acta*, 1797, 1573–1586.

Moser, C.C. and Dutton, P.L. (2006). Photosystem I; the light driven plastocyanin: ferredoxin oxidoreductase in photosynthesis. In *Application of Marcus Theory to*

Photosystem I Electron Transfer, Golbeck, J.H., ed. vol. 24, Chap. 34 (Dordrecht: Springer), pp. 583–594.

Moser, C.C., Farid, T.A., Chobot, S.E. and Dutton, P.L. (2006b). Electron-tunneling chains of mitochondria, *Biochim. Biophy. Acta*, 1757, 1096–1109.

Moser, C.C., Keske, J.M., Warncke, K., Farid, R.S. and Dutton, P.L. (1992). Nature of biological electron-transfer, *Nature*, 355, 796–802.

Moser, C.C., Page, C.C. and Dutton, P.L. (2005). Tunneling in PSII, *Photochem. Photobiol. Sci.*, 4933–4939.

Moser, C.C., Page, C.C. and Dutton, P.L. (2006a). Darwin at the molecular scale: selection and variance in electron tunnelling proteins including cytochrome *c* oxidase, *Philos. Trans. R. Soc. Lond. B: Biol. Sci.*, 361, 1295–1305.

Moser, C.C., Sheehan, M.M., Ennist, N.M., Kodali, G., Bialas, C., Englander, M.T., Discher, B.M. and Dutton, P.L. (2016). *De novo* construction of redox active proteins, *Methods Enzymol.*, 580, 365–388.

Mulholland, S.E., Gibney, B.R., Rabanal, F. and Dutton, P.L. (1998). Characterization of the fundamental protein ligand requirements of [4Fe-4S](2+/+) clusters with sixteen amino acid maquettes, *J. Am. Chem. Soc.*, 120, 10296–10302.

Muller, H.J. (1918). Genetic variability, twin hybrids and constant hybrids, in a case of balanced lethal factors, *Genetics*, 3, 422–499.

Muller, H.J. (1964). The relation of recombination to mutational advance, *Mutat. Res.*, 1, 2–9.

Musgrave, K.B., Laplaza, C.E., Holm, R.H., Hedman, B. and Hodgson, K.O. (2002). Structural characterization of metallopeptides designed as scaffolds for the stabilization of nickel(II)-Fe(4)S(4) bridged assemblies by X-ray absorption spectroscopy, *J. Am. Chem. Soc.*, 124, 3083–3092.

Nanda, V., Rosenblatt, M.M., Osyczka, A., Kono, H., Getahun, Z., Dutton, P.L., Saven, J.G. and DeGrado, W.F. (2005). *De novo* design of a redox-active minimal rubredoxin mimic, *J. Am. Chem. Soc.*, 127, 5804–5805.

Nanda, V., Senn, S., Pike, D.H., Rodriguez-Granillo, A., Hansen, W.A., Khare, S.D. and Noy, D. (2016). Structural principles for computational and *de novo* design of 4Fe–4S metalloproteins, *Biochim. Biophys. Acta*, 1857, 531–538.

OED Online (2017). "maquette, n." Oxford University Press, March 2017.

Page, C.C., Moser, C.C., Chen, X. and Dutton, P.L. (1999), Natural engineering principles of electron tunnelling in biological oxidation-reduction, *Nature*, 402, 47–52.

Paz, R. (1989). Feynman's office: the last blackboards, *Phys. Today*, 42, p. 88.

Rao, S.T., Shaffie, F., Yu, C., Satyshur, K.A., Stockman, B.J., Markley, J.L. and Sundaralingam, M. (1992). Structure of the oxidized long-chain flavodoxin from *Anabaena*-7120 at 2-Angstrom resolution, *Protein Sci.*, 1, 1413–1427.

Regan, L. and DeGrado, W.F. (1988). Characterization of a helical protein designed from first principles, *Science*, 241, 976–978.

Robertson, D.E., Farid, R.S., Moser, C.C., Mulholland, S.E., Urbauer, J.L., Pidikiti, R., Lear, J.D., Wand, A.J., DeGrado, W.F. and Dutton, P.L. (1994). Design and synthesis of multi-heme proteins, *Nature*, 368, 425–431.

Roy, A., Sommer, D.J., Schmitz, R.A., Brown, C.L., Gust, D., Astashkin, A. and Ghirlanda, G. (2014). A *de novo* designed 2[4Fe-4S] ferredoxin mimic mediates electron transfer, *J. Am. Chem. Soc.*, 136, 17343–17349.

Rypniewski, W.R., Breiter, D.R., Benning, M.M., Wesenberg, G., Oh, B.-H., Markley, J.L., Rayment, I. and Holden, H.M. (1991). Crystallization and structure determination of 2.5-.Ang. resolution of the oxidized iron-sulfur [2Fe-2S] ferredoxin isolated from *Anabaena* 7120, *Biochemistry*, 30, 4126–4131.

Sharp, R.E., Moser, C.C., Rabanal, F. and Dutton, P.L. (1998). Design, synthesis, and characterization of a photoactivatable flavocytochrome molecular maquette, *Proc. Natl. Acad. Sci. USA*, 95, 10465–10470.

Shifman, J.M., Gibney, B.R., Sharp, R.E. and Dutton, P.L. (2000). Heme redox potential control in *de novo* designed four-alpha-helix bundle proteins, *Biochemistry*, 39, 14813–14821.

Shifman, J.M., Moser, C.C., Kalsbeck, W.A., Bocian, D.F. and Dutton, P.L. (1998). Functionalized *de novo* designed proteins: mechanism of proton coupling to oxidation/reduction in heme protein maquettes, *Biochemistry*, 37, 16815–16827.

Solomon, L.A., Kodali, G., Moser, C.C. and Dutton, P.L. (2014). Engineering the assembly of heme cofactors in man-made proteins, *J. Am. Chem. Soc.*, 136, 3192–3199.

Sun, F., Huo, X., Zhai, Y., Wang, A., Xu, J., Su, D., Bartlam, M. and Rao, Z. (2005). Crystal structure of mitochondrial respiratory membrane protein complex II, *Cell*, 121, 1043–1057.

Tegoni, M. (2014). *De novo* designed copper α-helical peptides: from design to function, *Eur. J. Inorg. Chem.*, 2014, 2177–2193.

Tegoni, M., Yu, F., Bersellini, M., Penner-Hahn J.E. and Pecoraro, V.L. (2012). Designing a functional type 2 copper center that has nitrite reductase activity within α-helical coiled coils, *Proc. Nat. Acad. Sci.*, 109, 21234–21239.

Tommos, C., Skalicky, J.J., Pilloud, D.L., Wand, A.J. and Dutton, P.L. (1999). *De novo* proteins as models of radical enzymes, *Biochemistry*, 38, 9495–9507.

Topoglidis, E., Moser, C.C., Discher, B.M., Dutton, P.L. and Durrant, J.R. (2003). Functionalizing nanocrystalline metal oxide electrodes with robust synthetic redox proteins, *Chembiochem*, 4, 1332–1339.

Touw, D.S., Nordman, C.E., Stuckey, J.A. and Pecoraro, V.L. (2007). Identifying important structural characteristics of arsenic resistance proteins by using designed three-stranded coiled coils, *Proc. Natl. Acad. Sci. USA*, 104, 11969–11974.

Tsukihara, T., Aoyama, H., Yamashita, E., Tomizaki, T., Yamaguchi, H., Shinzawa-Itoh, K., Nakashima, R., Yaono, R. and Yoshikawa, S. (1995). Structures of metal sites of oxidized bovine heart cytochrome *c* oxidase at 2.8 A, *Science*, 269, 1069–1074.

Umena, Y., Kawakami, K., Shen, J.R. and Kamiya, N. (2011). Crystal structure of oxygen-evolving photosystem II at a resolution of 1.9 angstrom, *Nature*, 473, 55–61.

Watkins, D.W., Armstrong, C.T., Beesley, J.L., Marsh, J.E., Jenkins, J.M.X., Sessions, R.B., Mann, S. and Anderson, J.L.R. (2016). A suite of *de novo* c-type cytochromes for functional oxidoreductase engineering, *Biochim. Biophy. Acta*, 1857, 493–502.

Westerlund, K., Moran, S.D., Privett, H.K., Hay, S., Jarvet, J., Gibney, B.R. and Tommos, C. (2008). Making a single-chain four-helix bundle for redox chemistry studies, *Protein Eng. Des. Sel.*, 21, 645–652.

Woodbury, N.W., Parson, W.W., Gunner, M.R., Prince, R.C. and Dutton, P.L. (1986). Radical-pair energetics and decay mechanisms in reaction centers containing anthraquinones, naphthoquinones or benzoquinones in place of ubiquinone, *Biochim. Biophys. Acta*, 851, 6–22.

Ye, S., Strzalka, J., Chen, X., Moser, C.C., Dutton, P.L. and Blasie, J.K. (2002). Assembly of a vectorially oriented four-helix bundle at the air/water interface *via* directed electrostatic interactions, *Langmuir*, 19, 1515–1521.

Ye, S., Strzalka, J.W., Discher, B.M., Noy, D., Zheng, S., Dutton, P.L. and Blasie, J.K. (2004). Amphiphilic 4-helix bundles designed for biomolecular materials applications, *Langmuir*, 20, 5897–5904.

Zastrow, M.L., Peacock, A.F.A., Stuckey, J.A. and Pecoraro, V.L. (2012). Hydrolytic catalysis and structural stabilization in a designed metalloprotein, *Nat. Chem.*, 4, 118–123.

Zhu, J., Vinothkumar, K.R. and Hirst, J. (2016). Structure of mammalian respiratory complex I, *Nature*, 536, 354–358.

Chapter 2

Free, Stalled, and Controlled Rotation Single Molecule Experiments on F_1-ATPase and their Relationships

Sándor Volkán-Kacsó* and Rudolph A. Marcus[†]

*Noyes Laboratory of Chemical Physics,
1200 E. California Blvd., Pasadena, CA 91125, USA
*svk@caltech.edu,
[†]ram@caltech.edu

We show how an elastic group transfer theory can be used to interpret and treat free, stalling, and controlled rotation single molecule experiments on F_1-ATPase. It is shown how predictions can be made and tested within this class of experiments using our recent theoretical treatment of rotor angle dependent rate constants for this biological motor. The theory is also used to suggest an additional type of analysis of single molecule trajectories involving dwell angles.

1. Introduction

It is a pleasure to participate in this volume honoring our colleagues John Walker and Les Dutton. This chapter includes some of the results that we presented at their joint 75th birthday symposium. It touches on both their works, since it is on studies of F_1-ATPase, single molecule studies that rely heavily on John's structural discoveries, and it uses the functional form of a rate equation that expresses the activation free energy of each reaction step in terms of the standard free energy and "reorganization energy" λ for that step. The equation was originally derived for electron transfers, a field in which Les has made many contributions, some of them using this equation. This equation has now been adapted, as in earlier work [*e.g.* Marcus, 1968], to treat transfer reactions [Sutin, 1968; Murdoch, 1983; Lewis and Hu, 1984] that are unrelated to electron transfers.

The chapter is organized as follows: In Section 2 some of the experimental background in single molecule experiments is summarized, in particular, the free rotation, stalling and controlled rotation experiments. In Section 3 the functional form of the free energy of activation *versus* standard free energy of reaction for the various processes is described and used to obtain an equation for the various rotor angle dependent rate constants. Experimental results are described in Section 4. A theory of the free rotation experiments used by us [Volkán-Kacsó and Marcus, 2015] is described in Section 4, leading to Eq. (13), that was used to extract a key quantity $\left(\Delta G_0^0\right)$ that is used in predicting angle dependent rate constants in other experiments. The derivation is based on fluctuations from a "dwell" angle and angle dependent rate constants.

In Section 4.3 the dwell angles and substeps are discussed. For concreteness and self consistency we use in this article the dwell angles estimated in single molecule experiments, but since the equations are expressed in terms of the symbols θ_i and θ_f other values can be used instead. It is shown in Section 4.3 how an approximate difference in dwell angles for two successive reaction steps can be predicted from kinetic studies. It is applied to the ATP binding step and to the hydrolysis step. The results illustrate how incomplete or sparse data, in this case for the hydrolysis step, can be used and is particularly helpful when the existing kinetic data are very sparse, as they are for the hydrolysis step.

The biological function of the $F_O F_1$-ATP synthase is to regenerate, from ADP and Pi, "spent" ATP molecules used as a "fuel" in a multitude of cellular processes. [Boyer, 1993] ATP synthesis being energetically uphill, it is achieved by coupling a unidirectional mechanical rotation of a central γ shaft in the F_O ring, prompted by a local proton transfer between offset ion channels in the F_O (*E.g.*, [Junge and Nelson, 2015]), and the chemistry in the F_1 ring. Single-molecule studies have been thus far focused on the study of the water-soluble F_1-ATPase, [*e.g.*, Noji *et al.*, 1997; Yasuda *et al.*, 2001; Adachi *et al.*, 2012; Sielaff *et al.*, 2008; Spetzler *et al.*, 2009; Watanabe *et al.*, 2010] a complex formed by the F_1 ring and the γ shaft, which functions in a reverse direction and converts the free energy of the inverse process, ATP hydrolysis, to unidirectional rotation in the opposite direction.

Single molecule techniques have thus been providing quantitative data on the dynamics of the component F_1 of $F_O F_1$-ATP synthase, complementing ensemble biochemical and structural experiments [Braig *et al.*, 2000; Boyer, 1993; Senior, 2007; Weber, 2010]. In treating research data, typically in the hydrolysis direction of the F_1-ATPase motor, the goal is that the mechanism for the enzyme ATP synthase can be elucidated, assuming reversibility in the chemo-mechanical rotation Scheme [Adachi *et al.*, 2007, 2012].

In the present article we analyze three types of single molecule studies: stalling, controlled rotation and free rotation experiments using a chemo-mechanical theory. Controlled rotation data were predicted using the stalling data [Volkan-Kacso and Marcus, 2016]. We derive in Section 3.1 an equation, Eq. 2, that we assumed previously in treating the free rotation experiments. This equation is key to relating the properties of the latter system to the quasistatic rate constants extracted from the stalling and controlled rotation experiments.

We review in Section 3 a chemical-mechanical theory [Volkán-Kacsó and Marcus, 2015] that we formulated to treat several types of single molecule experiments on a biomolecular motor [Adachi *et al.*, 2007, 2012]. The theory is adapted from a theory of electron and other transfers in solution [Marcus, 1968; Marcus and Sutin, 1985; Marcus, 1993] and in addition contains the torsional elasticity effect [Sielaff *et al.*, 2008] in the motor in the various physical and chemical processes, such as ATP binding and ATP hydrolysis.

2. Description of Three Single-Molecule Experiments

2.1. *Free rotation experiments: Stepping rotation and concerted kinetics*

By attaching the F_1 "stator" ring subunit to a microscope slide the rotation of the "rotor" shaft has been monitored by dark field and fluorescent single molecule microscopy. In one version of this experiment a microfilament was attached to the rotor which rotates under heavy viscous load due to friction in the solution [Noji *et al.*, 1997]. When the viscous drag was reduced by replacing the probe with a submicron-sized bead, a counter-clockwise stepping rotation (when viewed from the F_O side) with dwell angles 120° apart was resolved [Adaci *et al.*, 2000]. Later, substeps of 40° and 80° were resolved using a lower ATP concentration [Yasuda *et al.*, 2001].

A chemo-mechanical scheme shown in Figure 1 has been established by a series of single-molecule experiments, in which ATP binding and ADP release (from a different β subunit) are concerted events (80° substep) and ATP hydrolysis and Pi release (again, from a different β subunit) occur during a 40° substep [Nishizaka *et al.*, 2004; Adachi *et al.*, 2007; Senior, 2007; Watanabe *et al.*, 2010]. In the scheme, when following the events in the same subunit, ATP binding occurs during the 80° step following a binding dwell set to 0° [Yasuda *et al.*, 2001]. The ATP does not undergo hydrolysis until much later in the rotation, during the 200° to 240° step, a 40° step. ADP release follows during the subsequent 80° step, from 240° to 320°. Finally, Pi release occurs during the last 40° step, from 320° to 360°.

The timescale of rotation of the γ subunit the biochemically active ATP synthase is comparable to that occurring in the free rotation experiments (ms).

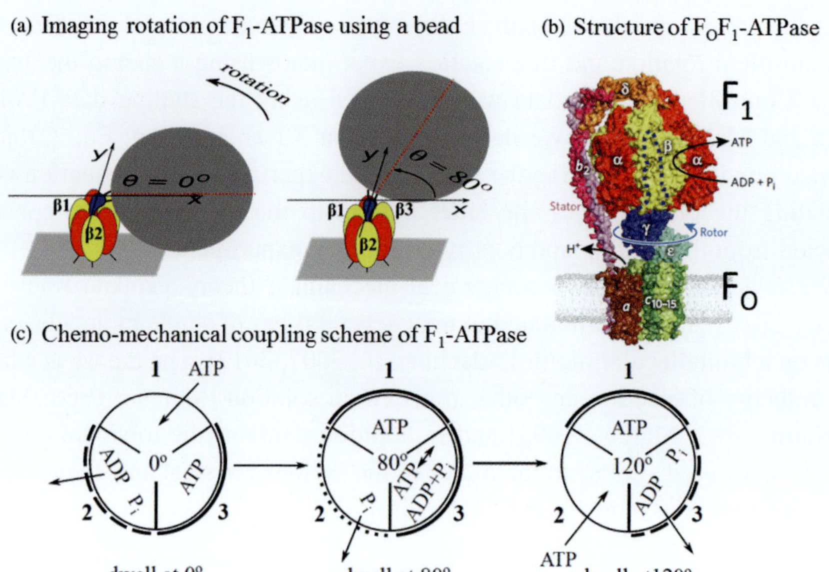

Figure 1. (a) Imaging rotation of F_1-ATPase with rotor fixed to the microscope slide and imaging bead attached to the rotor. The example shows a step of 80°. (b) Structure of the complete membrane embedded F_1F_o-ATPase, seen upside-down relative to the F_1-ATPase in a, reproduced from Weber [2010] with journal permission. (c) Scheme of coupled processes in F_1-ATPase during free rotation. The dwell angle increases in the counter clockwise direction. The species occupying the pockets of ring β subunits 1, 2 and 3.

We note that a high-speed rotation assay using gold nanorods as probes attached to the shaft [Spetzler *et al.*, 2009] permitted the study of the angular velocity of the rotor during the transition between dwells (timescales of μs). In this article we focus instead on free rotation experiments (ms) that display the stepping kinetics and yield trajectories from which the angle-dependent rate constant data were extracted from the dwell times. More will undoubtedly be learned from the short timescale fast rotation on the physiological timescale.

2.2. *Stalling experiments*

In these experiments the imaging beads were replaced by double beads possessing a permanent magnetic dipole moment [Adachi *et al.*, 2012]. When these micro-magnets were attached to the rotor (Figure 2a) and an external magnetic field was applied, the rotor aligned itself to the direction of the controlling external magnetic field. The setup, termed *magnetic tweezers,* can be used to set an arbitrary rotor angle θ.

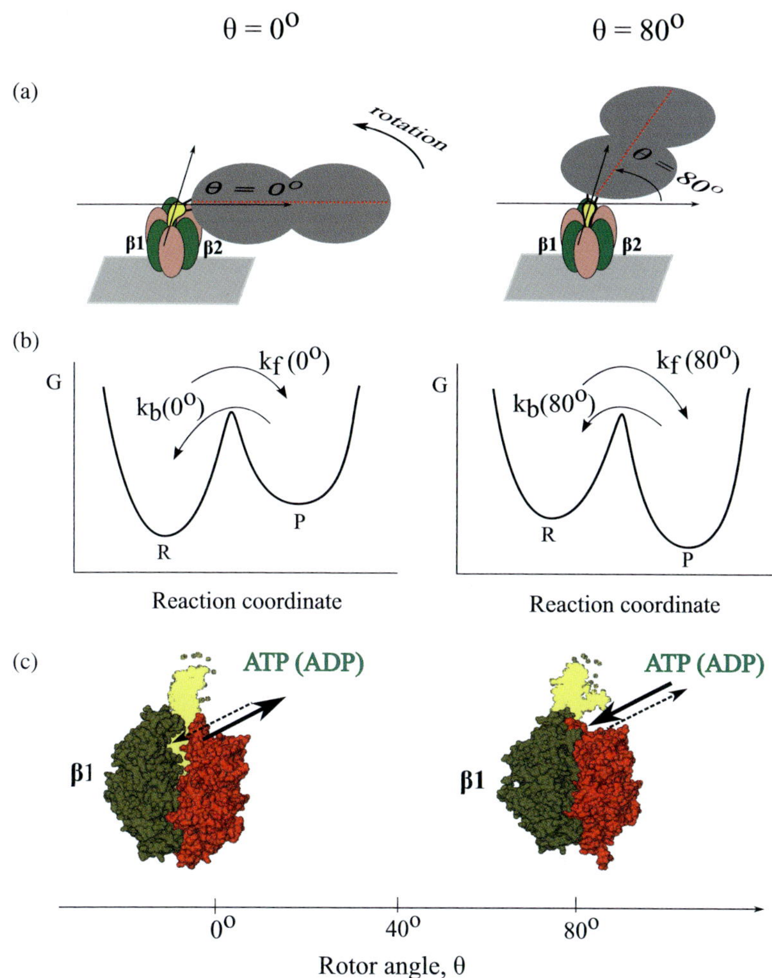

Figure 2. (a) F_1-ATPase in single molecule imaging and controlled rotation experiments at two rotor angles: 0° and 80°. A double-bead is attached to the γ rotor shaft (in yellow) and rotated against the stator ring (active subunits β_1-β_3). (b) Free energy profile for nucleotide binding (k_f) and release (k_b) rate constants at the two angles. (c) Open-to-close changes in the nucleotide binding β_1 subunit as a function of rotor angle.

Noji and coworkers have devised an ingenious stalling technique for using the magnetic tweezers to extract rate constant data [Watanabe *et al.*, 2010]. In these stalling experiments, a freely rotating shaft was allowed to reach a dwell angle (*e.g.*, the binding dwell), then the tweezers were immediately activated to set the rotor to an arbitrary angle θ. The rotor was then stalled at θ for a given time, after which it was released and it either moved back to the original dwell or forward to the next dwell.

After repeating these experiments for a series of dwell times, Adachi *et al.* [2012] interpreted the statistics of forward and back rotations in terms of a two-state model. They assumed that (1) during the stall time, the system can undergo several forward and backward processes, and (2) the rotations (forward or back) upon release depended on the state of the underlying process (*e.g.*, for ATP binding either ATP-bound or empty) at the moment of release. The observed relative number of forward and back events as a function of stall time yielded the forward (k_f) and backward (k_b) rate constants of ATP binding, or other processes. The model of elastic transfer of nucleotides was proposed to explain this rate constants *versus* θ data in terms of the effect of the elastic response of the rotor structure on the free energy surface (Figure 2b) [Volkán-Kacsó and Marcus, 2015].

2.3. *Controlled rotation experiments*

By replacing the "wild-type" nucleotides, ATP or ADP, by a modified nucleotide that fluoresces, single-molecule fluorescence microscopy was used to monitor individual binding and release events [Adachi *et al.*, 2007]. In particular, a fluorescent Cy3 moiety was attached to the nucleotides *via* a flexible alkane tether. In the experiments, when the Cy3-nucleotide bound to the F_1-ATPase the Cy3 moiety emitted a fluorescence signal but when it was in the solution the fluorescence was quenched.

Adachi *et al.* [2012] employed the fluorescence Cy3-nucleotide in conjunction with the rotation microscopy and magnetic tweezers using the micro-bead probes. To do so, the shaft was rotated at a constant rate and the individual binding and release events of fluorescent nucleotides, and so the site occupancy, were directly monitored [Adachi *et al.*, 2012]. By keeping a low ATP concentration, events of occupancy changing between 0 and 1 were determined and analyzed (Figure 2c). Using a grouping criterion of binding and release times, the number of $(0 \rightarrow 1)$ and $(1 \rightarrow 0)$ events state yielded forward and reverse rate constants as a function of rotor angle. The feature that is especially informative is comparing the results of such low occupancy experiments with the physiologically relevant experimental results where the occupancy is 2 or 3.

3. Group Transfer Theory in Stalling and Controlled Rotation Experiments

We give in this section an overview of the group transfer theory first proposed to treat the stalling experiments [Volkán-Kacsó and Marcus, 2015] and then used to predict controlled rotation experiment data [Volkán-Kacsó and Marcus, 2016], and later including composite events to extract hydrolysis and synthesis data from the trajectories [Volkán-Kacsó and Marcus, 2017]. Then, using the theoretical framework, the average rate constants extracted from dwell distributions in the free rotation experiments were then calculated.

3.1. *Angle dependent rate constants and free energies*

In a transition-state picture, the activation free energy barrier ΔG^* for a reaction determines the rate constant of the process. The rate constant k is given by

$$k = A \exp(-\Delta G^*/kT) \tag{1}$$

where A depends on the reaction order and the details of the transition state of the process. For a bimolecular process the A can be expressed in terms of a collision theory [Volkán-Kacsó and Marcus, 2015] in treating ATP binding that involves a weak binding site of an empty β subunit. The pre-exponential term A is a collision frequency Z, and a term related to an ATP binding outside of the β subunit. For the binding and release of ATP there are extensive single molecule data obtained by different methods [Yasuda *et al.*, 2001; Spetzler *et al.*, 2009; Watanabe *et al.*, 2010; Braig *et al.*, 2000; Watanabe *et al.*, 2012; Adachi *et al.*, 2012]. In the ATP binding an ATP first forms a collision complex with the F_1-ATPase, followed by an ATP binding in an 80° step in the overall hydrolysis direction [Oster and Wang, 2000].

To treat the free energy barrier in Eq. (1), we note that the equations of the "weak-overlap" theory of electron transfer reactions were adapted to treat "strong overlap" reactions, such as the transfer of an atom in a reaction [Marcus, 1968; Marcus and Sutin, 1985; Marcus, 1993]. We have applied this idea to treat the transfer of an even larger species, such as a nucleotide, ATP or ADP, during binding to the F_1-ATPase, while the rotor angle θ is fixed, *e.g.*, as in the stalling experiments. Taking into account a work term W^r for attaching the ATP from solution to the exterior of the F_1-ATPase (a weak binding site), to the binding pocket (strong binding site) the well-known quadratic equation is written as a function of the rotor angle θ.

$$\Delta G^*(\theta) = W^r + [\lambda + \Delta G^0(\theta)]^2/4\lambda. \tag{2}$$

In Eq. (2) ΔG^0 denotes the standard free energy reaction of the transfer process, a function of θ, and λ denotes the "reorganization energy" for the reaction. The quadratic formalism has been tested experimentally in its adaptation to strong overlap processes [Sutin, 1968; Lewis and Hu, 1984] and theoretically [Murdoch, 1983]. A related approach leading to a related equation is based on a bond energy-bond order model (BEBO) [Marcus, 1968 and reference cited therein] in which the sum of the bond orders of the reactant and product state bonds is assumed constant during the reaction. It served to explain why the quadratic equation given in Eq. (2) derived initially for weak overlap electron transfer reactions, can be

$$\Delta G^0(\theta)$$

$$G_r^0(\theta) \quad \rightarrow \quad G_p^0(\theta)$$

$$w^r \uparrow \qquad\qquad \downarrow -w^p$$

$$G_r^0(\theta_i) \quad \rightarrow \quad G_p^0(\theta_f)$$

$$\Delta G_0^0$$

Scheme 1.

applied to other transfer reactions. In the BEBO model the activation energy of the transfer is roughly 10% of the total bond energy. It is easier to break a chemical bond if at the same time a new chemical bond is also forming, in this case the process is 10 times easier in terms of energetics. This cooperation is assumed when applying Eq. (2) to the physical processes of binding a nucleotide and release, which involve the breaking and forming of bonds such as hydrogen bonds and "hydrophobic" bonds.

To calculate the θ dependence of the standard free energy of reaction $\Delta G^0(\theta)$ a thermodynamic cycle (Scheme 1) was used [Volkán-Kacsó and Marcus, 2015]. The energy balance around the cycle provides a relationship between the free energies of a ligand binding in free rotation, ΔG_0^0, and the binding free energy $\Delta G^0(\theta)$ at a constant rotor angle θ.

In Scheme 1, $\Delta G_r^0(\theta)$ and $\Delta G_p^0(\theta)$ are the free energies of the system in the reactant and product states when the rotor is at an angle θ, while $\Delta G_r^0(\theta_i)$ and $\Delta G_p^0(\theta_f)$ are the reactant and product free energies in the relaxed system at dwell angles θ_i and θ_f, respectively. We assume an elastic coupling first suggested by Junge and coworkers [Panke *et al.*, 2001; Sielaff *et al.*, 2008] between the rotor angle and the stator subunit conformation [Volkán-Kacsó and Marcus, 2015] involved in each of these steps. Accordingly, the system, while in the reactant state, can be rotated from a dwell angle θ_i to an arbitrary angle θ by elastically distorting the structure, by doing a work $w_r(\theta)$ provided externally by the external magnetic field of the magnetic tweezers. To achieve a rotation from θ_i to θ, parts of the structure (mainly the rotor and to some extent the lever arm of the β subunit) undergo an elastic twisting deformation of stiffness κ.

Analogously, the F_1-ATPase in the product state, can be rotated from an angle θ to the relaxed state at θ_f by doing a work $w_p(\theta)$, a work done by the system. To achieve this rotation an elastic twisting deformation of stiffness κ must occur. The elastic work terms then are

$$w^r = \frac{\kappa}{2}(\theta - \theta_i)^2 \quad \text{and} \quad w^p = \frac{\kappa}{2}(\theta - \theta_f)^2. \tag{3}$$

The energy balance in Scheme 1 requires that

$$\Delta G^0(\theta) = \Delta G_0^0 + w^p(\theta) - w^r(\theta) = \Delta G_0^0 - \kappa\,(\theta_f - \theta_i)(\theta - \theta_c), \tag{4}$$

where the notation $\theta_c = (\theta_f + \theta_i)/2$ was applied and ΔG_0^0 is the free energy change corresponding to the difference between the free energy minima in the final dwell state *versus* that in the initial dwell state. Using Eqs. (1)–(5) we have shown [Volkán-Kacsó and Marcus (2017)] that the theory provides a symmetric expression for the rate constant *versus* rotor angle dependence,

$$kT \ln k_f(\theta) = kTk_f(\theta_c) + (1/2 + \Delta G_0^0/2\lambda)\Delta\theta_r - (\Delta\theta_r)^2/4\lambda,$$
$$kT \ln k_b(\theta) = kTk_b(\theta_c) + (-1/2 + \Delta G_0^0/2\lambda)\Delta\theta_r - (\Delta\theta_r)^2/4\lambda, \tag{5}$$

where $\Delta\theta_r = \kappa(\theta_f - \theta_i)(\theta - \theta_c)$ and $\theta_c = (\theta_i + \theta_f)/2$, as defined in Volkán-Kacsó and Marcus [2015]. The quadratic term is small in the case of the ATP binding step (less than 8% contribution). We note that for the equilibrium constant this dependence is predicted to be exponential,

$$kT \ln K(\theta) = kT \ln K(\theta_c) - \kappa\,(\theta_f - \theta_i)(\theta - \theta_c). \tag{6}$$

3.2. *Application to ATP binding in the overlapping θ-range*

The stalling and controlled rotation experiments are complementary in several respects. Both experiments provide the θ-dependent rate constants and equilibrium constants for the steps in the overall process. There is a region of angles θ roughly $-45°$ to $+45°$ where the controlled rotation and stalling data overlap. Important differences are that in the former a fluorescent species, Cy3-ATP and Cy3-ADP, is used instead of ATP and ADP. Further, the nucleotide concentration in the controlled rotation experiments is kept low such that the occupancy is at most 1, while in the stalling experiments the system has the physiologically relevant occupancy 2 to 3 by nucleotides. Due to the finite time resolution of the single molecule fluorescence signal, in the controlled rotation experiments, there are certain missed events when the rate of nucleotide release is very fast. These missed events were calculated using the theory embodied in the previous equations [Volkán-Kacsó and Marcus, 2016].

Turning now to the quantitative analysis of the rate constant *versus* rotor angle data in these experiments, we first note that the ln $K(\theta)$ seen in Eq. (7) to be linear in θ, in agreement with the experimental data [Volkán-Kacsó and Marcus, 2015]. With data from other ensemble [Boyer, 1993; Weber and Senior, 1997] and single molecule [Yasuda *et al.*, 2001, Adachi *et al.*, 2012; Spetzler *et al.*, 2009] experiments Eqs. (5)–(6) were used to predict [Volkán-Kacsó and Marcus, 2015] for the stalling experiments on ATP binding the Bronsted slope $\alpha = \partial \ln k_f / \partial \ln K$ of 0.47,

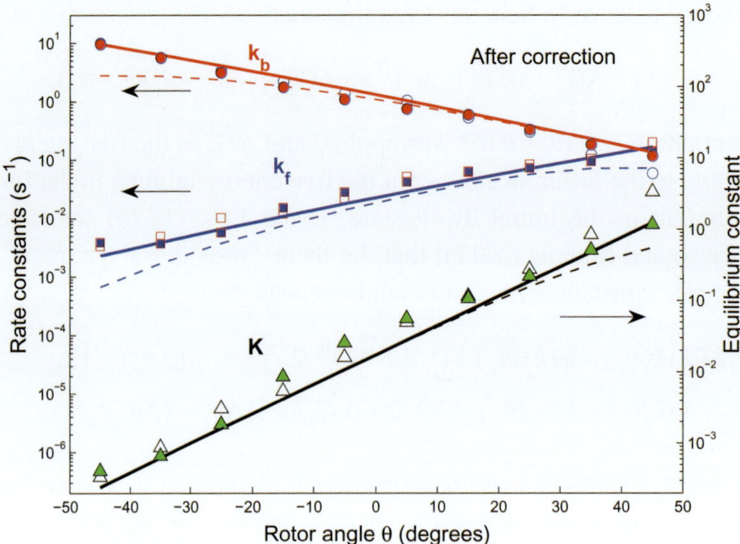

Figure 3. Corrected binding and release rate and equilibrium rate constants *versus* θ angle for Cy3-ATP in the presence (solid squares, circles and triangles) and absence of Pi (open symbols) in solution adapted from Volkán-Kacsó and Marcus [2016]. The experimental data of Adachi *et al.* [2012] corrected for missed events (and an error due to replacing the time spent in the empty state by total time of a trajectory) are compared with their theoretical counterparts (solid lines). Dashed lines show the data without corrections.

which compares with the value of 0.48 in the stalling experiment for the rate of ATP binding and release over the θ range studied. For the ATP binding rate k_f the predicted slope of the $\ln k_f(\theta)$ *versus* θ was found to be in reasonable agreement (~within 10%) with experiment [Volkán-Kacsó and Marcus, 2015]. For the spring constant κ of the rotor that appears in Eq. (3) the value of $\kappa = 16$ pN nm rad^{-2} obtained from the stalling experiments was used.

The stalling data, supplemented by ensemble data, was used to predict, using Eq. (5), the rate constants in the controlled rotation experiments with no adjustable parameters. The comparison is given in Figure 3, in the θ-range of overlap. Good agreement was found between these experiments and our calculations, seen there, where the points (symbols) are experimental and the solid curves theoretical data.

3.3. *Turnover, near symmetry and long binding events in the controlled rotation experiments*

To explain a turnover in binding seen in controlled rotation data seen in Figure 4, it was suggested [Volkán-Kacsó and Marcus, 2016] that a slow diffusion process occurs after the binding to the outside of the enzyme: the diffusion of ATP in a channel at the interface of an α and β subunit, is slowed by the partially closed subunit.

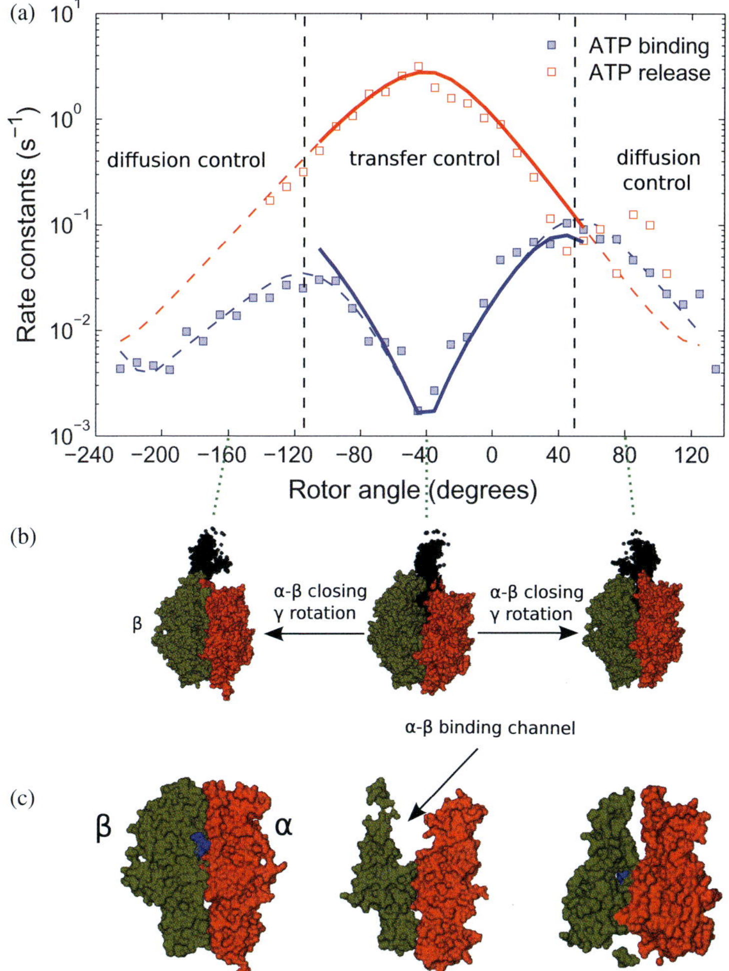

Figure 4. (a) Reported binding and release rate constants *versus* controlled rotation angle for fluorescent ATP in the presence of Pi in solution. The reported uncorrected experimental data (squares) are compared with theoretical counterparts (solid lines) by calculating missed events and also correcting for an error due to replacing the time in the empty state T_0 by the total time T. Dashed lines show a fit to the experimental data. (b) F_1-ATPase structure at three different rotor angles with β subunits in green, α subunits in red and the γ subunit in black. (c) Cutaway of the three structures revealing the binding channel at the α–β interface and its narrowing as the rotor angle is changed.

Structurally, the turnover indicates that the closing of the binding channel continues as the angle is rotated beyond 80°. The turnover is consistent with structural data showing a continuum of open-to-close states between –40° and 140°. A closure of the binding channel (Figure 4c) as the rotor is moved beyond 80° (Figure 4b) retards the diffusion process and it becomes the bottleneck during binding. In a diffusion-reaction scheme [Marcus, 1960; Noyes, 1961], the transfer reaction is the bottleneck in the θ-region from –120° to +45°.

In Figure 4c the β subunit is the most open at $-40°$ and the most closed presumably at $+140°$ ($0°$ is defined as the binding dwell angle at which the β subunit in question is empty and is about to undergo ATP binding associated with a subsequent $80°$ substep). The minimum at $-40°$ does not correspond to a specific dwell angle. Diffusion reaction treatments are given in Noyes [1961] and Marcus [1960].

Recently, this theory of elastic group was further extended [Volkán-Kacsó and Marcus, 2017] to treat the experimentally observed lifetime distribution of long binding events in the controlled rotation. Using these distributions the long binding events in the experiments can be calculated and the rate constants for the hydrolysis and synthesis reactions occurring during these events extracted. As discussed in a later section a near symmetry of the data about the angle of $-40°$ and a "turnover" in the binding rate data *versus* rotor angle for angles greater than about $40°$ was interpreted using diffusion-reaction kinetics.

4. Application of the Theory to Free Rotation

4.1. *Rate constant of a free rotation experiment*

In this subsection we derive a key equation Eq. (10), used to obtain some of the quantities, including ΔG_0^0, used [Volkán-Kacsó and Marcus, 2015] to predict the controlled rotation results from the stalling data. In a free rotation experiment we suppose that the system jumps from θ_i to θ_f during some fluctuation from its initial value θ_i. The probability of a fluctuation in θ is proportional to $\exp\left[-C(\theta)/kT\right]$ where

$$C(\theta) = \kappa(\theta - \theta_i)^2/2. \tag{6}$$

At this θ the free energy of activation for binding for an ATP attached to the outside of an ATPase β-subunit is (apart from a W^r term in Volkán-Kacsó and Marcus [2015] that we include later as part of a factor A). Using Eq. (2) for the forward rate constant $k_f(\theta)$ one finds that

$$k_f(\theta) = A\exp(-\Delta G^*(\theta)/kT), \tag{7}$$

where A is given in Volkán-Kacsó and Marcus [2015] and $\Delta G^*(\theta)$ is given by Eq. (2). The average value of the forward rate constant in a free rotation is then given by

$$\left\langle k_f \right\rangle = \int_{\theta_L}^{\theta_H} A e^{-[\Delta G^*(\theta) + C(\theta)]/kT} d\theta \Big/ \int_{\theta_L}^{\theta_H} e^{-C(\theta)/kT} d\theta. \tag{8}$$

We note that the effective integration limits enclose a peak angle θ^* of the integrand, so $\theta_L < \theta^* < \theta_H$, and are determined by the steepness of the integrand about its maximum. If one neglects the term quadratic in $\Delta G°(\theta)$ in Eq. (4) (valid, in the case of ATP binding, as seen in the closeness of the Bronsted slope α to the value

0.5) experiment one finds from Eqs. (4)–(7) that the minimum value of exponent in the numerator occurs when θ is given by

$$\theta^* - \theta_i \approx (\theta_f - \theta_i)/2. \tag{9}$$

The integrand of the exponent in the numerator of Eq. 8 is peaked at this angle $\theta = \theta^*$. Using the saddle point method to evaluate the integral, in the first approximation we find that

$$\left\langle k_f \right\rangle \approx k_f(\theta^*), \tag{10}$$

a result used in Volkán-Kacsó and Marcus [2015] to obtain the value of the ΔG_0^0 from the experimental data.

The analysis of the reverse process, ATP release, can be performed in an analogous manner. An important difference is that the initial state for the release, the dwell at 80°, is beyond the interval covered in the stalling experiments. Controlled rotation data can be used to predict the step size and dwell angles, by taking into account the effect of the Cy3 moiety attached to the ATP on binding and release in both controlled and free rotation. Using a functional form for the release rate constant $k_b = A \exp(-\Delta G^b(\theta)/kT)$, where $\Delta G^b(\theta)$ is the barrier for the ATP release step that can be extracted from the controlled rotation data, the analogue of Eq. (8) is

$$\left\langle k_b \right\rangle = \int_{\theta_L}^{\theta_H} A e^{-[\Delta G^b(\theta)+C(\theta)]/kT} \, d\theta \bigg/ \int_{\theta_L}^{\theta_H} e^{-C(\theta)/kT} \, d\theta. \tag{12}$$

According to Eq. (5) and as described for Pi release in Volkán-Kacsó and Marcus [2015], and for nucleotide release in Volkán-Kacsó and Marcus [2016] the $\Delta G^b(\theta)$ is an approximately linear function of the rotor angle. If the saddle point method is used, Eq. (12) yields in first approximation the analogue of Eq. (10),

$$\left\langle k_b \right\rangle \approx k_b(\theta^*), \tag{13}$$

and the maximum will be at $\theta^* = \theta_c = 40°$.

4.2. *Relation between controlled and free rotation experiments*

The structural changes of opening and closing are consistent with the trends seen in the slopes of the binding and release rate *versus* rotor angle data and with spontaneous stepping rotation. How can the features revealed in stalling and controlled rotation data in Figures 3 and 4 be connected with the rotational kinetics

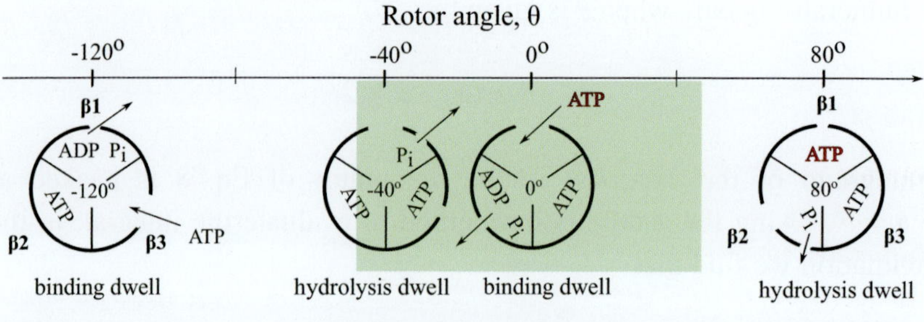

Figure 5.   Shaft rotation, nucleotide binding/release activity and conformational changes of a ring subunit in single molecule experiments of stalling (a), controlled rotation (b) and free stepping rotation (c). In (a–c) the system is represented at four different values of the rotor angle, corresponding to dwell angles at −120°, −40°, 0° and 80°. θ is defined relative to subunit 1, *i.e.* θ = 0 for the binding dwell with an empty β subunits.

(Figure 1c) and structural changes? The correspondence between the results of these three experiments, given in the present analysis, based on the theoretical treatment, are summarized in Figure 5.

On the one hand, displacing the rotor to the left or right starting from –40° both accelerate the rate of binding in the controlled rotation experiments. Such displacements of the rotor invariably induce a closing of the β subunit, as seen in the structure in Figure 4c. These effects are consistent with the spontaneous stepping forward rotation induced by binding of ATP in free rotation experiments occurring in a θ-range to the right of –40° (from 0° to 80°).

On the other hand, rotating the rotor towards –40° from the left or from the right both induce an opening of the β subunit which accelerates nucleotide release. This effect, in turn, is consistent with the spontaneous forward rotation induced by the release of ADP in free rotation experiments occurring in a θ-range to the left of –40° (from −120° to –40°).

4.3. *Use of the rate constant versus rotor angle data to predict the step size and dwell angles in free rotation*

Here we show how the single molecule free rotation kinetics, including step size $(\theta_f - \theta_i)$ and dwell angles θ_i and θ_f of the free system, which cannot be resolved in free rotation assays, can be predicted in the rare case that $\ln K(\theta)$ *versus* θ is linear over the entire range $(\theta_f - \theta_i)$. It also presents an example of combining different types of sparse data.

The theory has provided simple mathematical relations between the dwell angles, the spring constant and $\ln k_f$, k_b or K *versus* rotor angle slopes [Volkán-Kacsó and Marcus, 2015; Volkán-Kacsó and Marcus, 2016, 2017]. To predict dwell angles, an independent estimate of the spring constant κ is needed, *e.g.*, from fluctuation experiments.[Sielaff *et al.*, 2008] Using the expression from Eq. (6) for the slope $kT\partial \ln k(\theta) / \partial\theta = K(\theta_f - \theta_i)$ of angle-dependent rate constant, obtained from controlled rotation experiments, the predicted step size in free rotation experiments, $(\theta_f - \theta_i)$ is given by

$$\theta_f - \theta_i = \frac{kT}{\kappa} \frac{\partial \ln K(\theta)}{\partial\theta}. \tag{14}$$

In the present application of Eq. (14) for the hydrolysis step the linearity spans an angular range of $(150°, 250°)$ [Watanabe *et al.*, 2012] with $\theta_i = 200$. From these data and Eq. (14) one obtains $\theta_f = \theta_i + 15° = 215°$.

The experimental slope for the equilibrium constant $\partial \ln K(\theta) / \partial\theta = 1.9$ deg^{-1} and a $\kappa = 16pN\ nm/rad^2$ yield a predicted step size of $\theta_f - \theta_i = 15°$. We note that the stiffness $\kappa = 16pN\ nm/rad^2$ corresponds to the ATP binding step, and we use this value in the present analysis to approximate the κ for the hydrolysis step. Different steps likely have somewhat different κ values, *e.g.* in Volkán-Kacsó and Marcus [2016] we found for ADP binding a $\kappa = 12pN\ nm/rad^2$ which is smaller than the $\kappa = 16pN\ nm/rad^2$ of ATP binding.

A separate hydrolysis and a Pi release step was not directly resolved in single molecule trajectories, but a difference of about $10°$ in the stepping size in the $90°/30°$ and $80°/40°$ stepping ratio seen in different experiments lead Adachi *et al.* [2007] to speculate that this $10°$ difference is due to the hydrolysis step. The present calculated value of $\theta_f - \theta_i = 15°$ is in reasonable agreement with this assertion, considering the sparsity of the data of Adachi *et al.* [2012]. These data also suggest a Bronsted slope close to unity for the hydrolysis step which we speculate could indicate the presence of two or more substeps during hydrolysis [Shweins and Warshel, 1996]. We note that Eq. (14) does not apply for binding since $\ln K(\theta)$ *versus* θ is not linear over the whole range of (θ_i, θ_f) for binding (*e.g*, as seen in the lack of the linearity over the *complete* range in Figure 4a for binding).

In a crystallographic method of studying the kinetic states (dwells) in active rotation Walker and coworkers prepared stable states by using substitutions in the binding species [Bason *et al.*, 2015]. By resolving such the X-ray crystallography structures of mammalian (human and bovine) F_1-ATPase, three stable states were found about 65°, 25° and 30° apart and it was suggested that these states correspond to dwell angles that define the stepping spontaneous rotation of the F_1-ATPase. The 65°, 25° and 30° division is somewhat similar to the 80°, 10° and 30° division proposed from single molecule experiments by Kinosita and coworkers for the thermophilic bacillus F_1-ATPase [Adachi *et al.*, 2007].

5. Concluding Remarks

In the present chapter, we have illustrated how an elastic group transfer theory can be used to treat three types of single molecule experiments: stalling, controlled rotation and free rotation experiments. To do so, we first reviewed the relations for the rate constants of substeps such as ATP binding or ADP release provided using the theory to treat and predict the stalling and controlled rotation data in previous studies [Volkán-Kacsó and Marcus, 2015, 2016, 2017]. A new feature in the present book chapter is the calculation of average rate constant that can be extracted from the dwell time distributions in the free rotation experiments, Eq. (10), used in an earlier work. We also demonstrated how the difference in dwell angles for two successive reaction steps in the free rotation experiments can be predicted from rate constant *versus* rotor angle data from stalling and/or controlled rotation experiments. These predictions were applied to the ATP binding 80° substep, and the hydrolysis substep, supporting an earlier contention that the latter is a small (10°) step.

Acknowledgements

This work was supported by the Office of the Naval Research, the Army Research Office, and the James W. Glanville Foundation.

Appendix

Evaluation of Eq. (8)

We explore here the approximation for ATP binding in more detail by evaluating the numerator and the denominator. For ATP binding $\theta_i = 0°$ and $\theta_f = 80°$ and we found using the present Eqs. (9) and (10) [Volkán-Kacsó and Marcus, 2015] that $\Delta G_0^0 = 6\,kcal\,/\,mol \ll 2\lambda = 136\,kcal\,/\,mol$ and $\kappa = 16pN\,nm/rad^2$.

In the numerator in Eq. (10), $\theta^* = \theta_c = 40°$ and Eq. 7 is strictly applicable in the θ range covered by stalling experiments of Adachi *et al.* [2012], roughly $-45° < \theta < 45°$. The integrand at the lower bound of $\theta_L \approx -45°$ is smaller than the maximal value by a factor of $\exp\left(1/8\,\beta\kappa\,c^2 - \alpha\lambda^2\right)/\exp\left(-3/4\,\beta\kappa\,c^2 - \alpha\lambda^2\right) \approx \exp(5.3) \approx 200$. The angle $\theta^* \approx 40°$ is close to a 45° upper bound, and so an empirical functional form can be used as described in Volkán-Kacsó and Marcus [2017], to extend Eq. (7) beyond the turn-over in controlled rotation experiments. In this case, for an upper bound at $\theta_H \approx 60°$, the integrand is estimated to have decayed roughly by a factor of $\exp\left(1/8\,\beta\kappa\,c^2\right)/\exp\left(-1/8\,\beta\kappa\,c^2\right) \approx \exp(2.2) \approx 9$. Using the saddle-point approximation that treats the integrand as a Gaussian in a $(-\infty, +\infty)$ interval of θ, we have

$$\int_{\theta_L}^{\theta_H} A\,e^{-[\Delta G^*(\theta)+C(\theta)]/kT}\,d\theta \approx \sqrt{2\pi/\kappa}\,A\,e^{-[\Delta G^*(\theta^*)+C(\theta^*)]/kT}. \qquad (11)$$

In the denominator of Eq. (8) the Gaussian is centered at $\theta^* = 0°$, the elastic approximation in Eq. (6) has been also found to apply within the range of the stalling experiments [Adachi *et al.*, 2012], although it likely applies beyond it [Panke *et al.*, 2001]. For a $\theta_L \approx -45°$ and $\theta_H \approx -45°$ the integrand has decayed relative to the peak by a factor of about $\exp(1.34) \approx 4$ so the Laplace approximation yields $\sqrt{2\pi/\kappa}\,\exp\left(-C(\theta^*)/kT\right)$.

References

Adachi, K., Oiwa, K., Nishizaka, T., Furuike, S., Noji, H., Itoh, H., Yoshida, M. and Kinosita, K., Jr. (2007). Coupling of rotation and catalysis in F1-ATPase revealed by single molecule imaging and manipulation, *Cell*, 130, 309–321.

Adachi, K., Oiwa, K., Yoshida, M., Nishizaka, T. and Kinosita, K., Jr. (2012). Controlled rotation of the F1-ATPase reveals differential and continuous binding changes for ATP synthesis, *Nat. Commun.*, 3, 1022.

Adachi, K., Yasuda, R., Noji, H., Itoh, H., Harada, Y.S., Yoshida, M. and Kinosita, K., Jr. (2000). Stepping rotation of F1-ATPase visualized through angle-resolved single-fluorophore imaging, *Proc. Natl. Acad. Sci. USA*, 97, 7243–7247.

Boyer, P.D. (1993). The binding change mechanism for ATP synthase—some probabilities and possibilities, *Biochim. Biophys. Acta*, 1140, 215–250.

Braig, K., Menz, R.I., Montgomery, M.G., Leslie, A.G. and Walker, J.E. (2000). Structure of bovine mitochondrial F1-ATPase inhibited by $Mg^{2+}ADP$ and aluminium fluoride, *Structure*, 8, 567–573.

Chidsey, C.E.D. (1991). Free energy and temperature dependence of electron transfer at the metal-electrolyte interface, *Science*, 251, 919–922.

Junge, W. and Nelson, N. (2015). ATP synthase, *Annu. Rev. Biochem.*, 84, 631–657.

Lewis, E.S. and Hu, D.D. (1984). Methyl transfers. 8. The Marcus equation and transfers between aerosulfonates, *J. Am. Chem. Soc.*, 106, 3292–3296.

Marcus, R.A. (1960). Discussion comment on mixed reaction-diffusion controlled rates, *Discuss. Faraday Soc.*, 29, 129.

Marcus, R.A. (1968). Theoretical relations among rate constants, barriers, and Brønsted slopes of chemical reactions, *J. Phys. Chem.*, 72, 891–899.

Marcus, R.A. (1993). Electron transfer reactions in chemistry: theory and experiment, including biographical sketch. In *Les Prix Nobel*, Frangsmyr, T., ed., Nobel Foundation (Almqvist & Wiksell, Stockholm, Sweden), p. 63.

Marcus, R.A. and Sutin, N. (1985). Electron transfers in chemistry and biology, *Biochim. Biophys. Acta*, 811, 265–322 (and references cited therein).

Masaike, T., Koyama-Horibe, F., Oiwa, K., Yoshida, M. and Nishizaka, T. (2008). Cooperative three-step motions in catalytic subunits of F1-ATPase correlate with 80 degrees and 40 degrees substep rotations, *Nat. Struct. Mol. Biol.*, 15, 1326–1333.

Murdoch, J.R. (1983). A simple relationship between empirical theories for predicting barrier heights of electron-transfer, proton-transfer, atom-transfer, and group transfer reactions, *J. Am. Chem. Soc.*, 105, 2159–2164.

Nishizaka, T., *et al.* (2004). Chemomechanical coupling in F1-ATPase revealed by simultaneous observation of nucleotide kinetics and rotation, *Nat. Struct. Mol. Biol.*, 11, 142–148.

Noji, H., Yasuda, R., Yoshida, M. and Kinosita, K., Jr. (1997). Direct observation of the rotation of F1-ATPase. *Nature*, 386, 299.

Noyes, R.M. (1961). Effects of diffusion rates on chemical kinetics, *Prog. React. Kinet.*, 1, 129–160.

Oster, G. and Wang, H. (2000). Reverse engineering a protein: the mechanochemistry of ATP synthase, *Biochim. Biophys. Acta*, 1458, 482–510.

Panke, O., Cherepanov, D.A., Gumbiowski, K., Engelbrecht, S. and Junge, W. (2001). Viscoelastic dynamics of actin filaments coupled to rotary F-ATPase: angular torque profile of the enzyme. *Biophys. J.*, 81, 1220–1233.

Sakaki, N., Shimo-Kon, R., Adachi, K., Itoh, H., Furuike, S., Muneyuki, E., Yoshida, M. and Kinosita, K., Jr. (2005). One rotary mechanism for F1-ATPase over ATP concentrations from millimolar down to nanomolar, *Biophys. J.*, 88, 2047–2056.

Schweins, T. and Warshel, A. (1996). Mechanistic analysis of the observed linear free energy relationships in p21[ras] and related systems, *Biochem.*, 35, 14232–14243.

Senior, A.E. (2007). ATP synthase: motoring to the finish line, *Cell*, 130, 220–221.

Sielaff, H., Rennekamp, H., Engelbrecht, S. and Junge, W. (2008) Functional halt positions of rotary F_OF_1-ATPase correlated with crystal structures, *Biophys. J.*, 95, 4979–4987.

Spetzler, D., Ishmukhametov, R., Hornung, T., Day, L.J., Martin, J. and Frasch, W.D. (2009). Single molecule measurements of F1-ATPase reveal an interdependence between the power stroke and the dwell duration. *Biochemistry* 48, 7979.

Sugawa, M., Okazaki, K.-I., Kobayashi, M., Matsui, T., Hummer, G., Masaike, T. and Nishizaka, T. (2016). F1-ATPase conformational cycle from simultaneous single-molecule FRET and rotation measurements, *Proc. Natl. Acad. Sci. USA*, 113, E2916–E2924.

Sutin, N. (1966). The kinetics of organic reactions in solution, *Annu. Rev. Phys. Chem.*, 17, 119–172.

Uchihashi, T., Iino, R., Ando, T. and Noji, H. (2011). High-speed atomic force microscopy reveals rotary catalysis of rotorless F1-ATPase, *Science*, 333, 755–758.

Volkán-Kacsó, S. and Marcus, R.A. (2015). Theory for rates, equilibrium constants, and Brønsted slopes in F1-ATPase single molecule imaging experiments, *Proc. Natl. Acad. Sci. USA*, 112, 14230–14235.

Volkán-Kacsó, S. and Marcus, R.A. (2016). Theory of controlled rotation experiments, predictions, tests and comparison with stalling experiments in F1-ATPase, *Proc. Natl. Acad. Sci. USA*, 113, 12029–12034.

Volkán-Kacsó, S. and Marcus, R.A. (2017). Theory of long binding events in single-molecule–controlled rotation experiments on F1-ATPase, *Proc. Natl. Acad. Sci. USA*, 114, 7272–7277.

Walker, J.E. (2013). The ATP synthase: the understood, the uncertain and the unknown, *Biochem. Soc. Trans.*, 41, 1–16.

Watanabe, R., Iino, R. and Noji, H. (2010). Phosphate release in F1-ATPase catalytic cycle follows ADP release, *Nat. Chem. Biol.*, 6, 814–820.

Watanabe, R., Okuno, D., Sakakihara, S., Shimabukuro, K., Iino, R., Yoshida, M. and Noji, H. (2012). Mechanical modulation of catalytic power on F1-ATPase, *Nat. Chem. Biol.*, 8, 86.

Watanabe, R. and Noji, H. (2014). Timing of inorganic phosphate release modulates the catalytic activity of ATP-driven rotary motor protein, *Nat. Commun.*, 5, 3486.

Weber, J. (2010). Structural biology: toward the ATP synthase mechanism, *Nat. Chem. Biol.*, 6, 794–795.

Weber, J. and Senior, A.E. (1997). Catalytic mechanism of F1-ATPase. *Biochim. Biophys. Acta*, 1319, 19–58.

Yasuda, R., Noji, H., Yoshida, M., Kinosita, K., Jr. and Itoh, H. (2001). Resolution of distinct rotational substeps by submillisecond kinetic analysis of F1-ATPase, *Nature*, 410, 898–904.

Zimmermann, B., Diez, M., Zarrabi, N., Gräber, P. and Börsch, M. (2005). Movements of the ε-subunit during catalysis and activation in single membrane-bound H⁺-ATP synthase, *EMBO J.*, 24, 2053–2063.

Chapter 3

The Role of the H-Channel in Cytochrome *c* Oxidase: A Commentary

Mårten Wikström

Institute of Biotechnology,
University of Helsinki, Helsinki, Finland
marten.wikstrom@helsinki.fi

The so-called K- and D-pathways of proton translocation have been identified in heme-copper oxidases by both functional and structural experimentation. In bacterial heme-copper oxidases these pathways have been shown to provide the proton uptake routes both for protons to be consumed in O_2 reduction to water, and for proton translocation across the membrane. A separate so-called H-channel has been claimed to be the unique pathway of the pumped protons in cytochrome *c* oxidase from mammalian mitochondria, and a corresponding mechanism of proton pumping has been proposed that differs substantially from the general mechanism proposed on the basis of time-resolved experiments with the homologous bacterial heme-copper oxidases of type A. In order to help to resolve this unfortunate discrepancy a brief summary of the experimental facts is presented here.

1. Introduction

Site-directed mutagenesis experiments and crystal structures have together identified the likely pathways of proton transfer in cytochrome *c* oxidase (Cco). The K-pathway is established as a pathway for uptake of "chemical protons in A-type heme-copper oxidases, and as the path for uptake of both "chemical" and "pumped" protons in heme-copper oxidases of type B and C [Fetter *et al.*, 1995;

Konstantinov *et al.*, 1997; Vygodina *et al.*, 1998; Ädelroth *et al.*, 1998]. The D-pathway is unique for the A-type heme-copper oxidases, which includes C*c*Os from both mitochondria and bacteria, as well as some bacterial quinol oxidases [Sousa *et al.*, 2012; Abramson *et al.*, 2000]. It is established by mutagenesis experiments with A-type C*c*O from bacteria that the D-pathway is used for uptake of all protons to be pumped across the membrane, as well as for uptake of 1–2 of the four protons consumed in the chemistry of O_2 reduction to water [see Brzezinski *et al.*, 2008; Belevich and Verkhovsky, 2008; Kaila *et al.*, 2010]. A separate H-pathway has been postulated to be specifically used for transfer of all the pumped protons in mammalian mitochondria [Tsukihara *et al.*, 2003; Shimokata *et al.*, 2007], and a specific proton-pumping mechanism has been proposed to explain how proton translocation *via* the H-channel is linked to the partial reactions of the catalytic cycle [Yoshikawa and Shimada, 2015; Yano *et al.*, 2016]. This mechanism differs substantially from the generally accepted mechanism of proton pumping that has been largely based on time-resolved experiments with bacterial A-type heme-copper oxidases [see *e.g.* Verkhovsky *et al.*, 1999; Wikström *et al.*, 2003; Popovic and Stuchebrukhov, 2004; Bloch *et al.*, 2004; Belevich *et al.*, 2007; Konstantinov, 2012; Rich and Maréchal, 2013; Wikström *et al.*, 2015].

The very different mechanisms of proton pumping proposed for the structurally closely related A-type C*c*Os from bacteria and mammals have created considerable confusion. For this reason I will briefly discuss key features here that distinguish the proposed mechanisms.

2. Conservation of the H-Channel Structure

The H-channel was first recognised from the crystal structure of mitochondrial C*c*O from bovine heart [Tsukihara *et al.*, 2003], and is to a large part conserved in the bacterial C*c*Os of type A, as well as in C*c*O from yeast mitochondria [Rich and Maréchal, 2013]. Extensive mutagenesis experiments with C*c*O from both *P. denitrificans* and *Rh. sphaeroides* as well as from *Saccharomyces cerevisiae* [Lee *et al.*, 2000; Salje *et al.*, 2005; Rich and Maréchal, 2013] show that it is not used for proton-pumping in those organisms. Most remarkably, the H-pathway is entirely absent from bacterial A-type heme-copper oxidases that use ubiquinol rather than cytochrome *c* as electron donor, such as cytochrome *bo*₃ from *E. coli* [Wikström *et al.*, 2015], even though these enzymes are proton pumps [Puustinen *et al.*, 1989] just like the C*c*os (Figure 1), and even though they are structurally closely related.

Cytochrome *bo*₃ also lacks the Ca^{2+}/Na^+ binding site [Abramson *et al.*, 2000] of heme-copper type A C*c*Os, which is associated with the H-channel structure [Vygodina *et al.*, 2014]. The presence of the H-pathway in A-type oxidases thus correlates with the nature of the electron donor (cytochrome *c versus* ubiquinol),

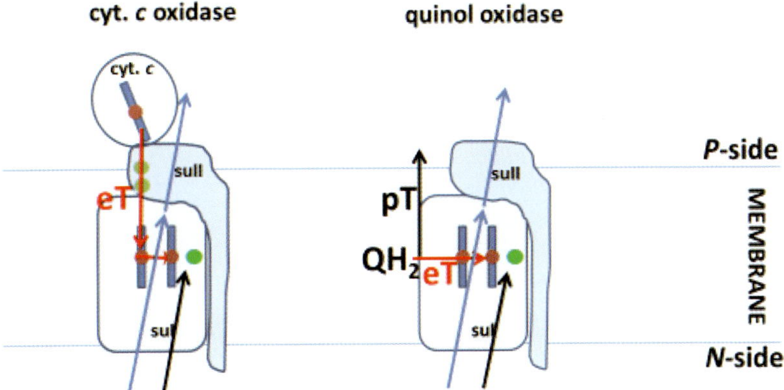

Figure 1. Proton-pumping by A-type heme-copper oxidases. C*c*Os to the left, quinol oxidases to the right. Note that subunit II (su II) has the pair of copper atoms (two green spheres) making up the Cu_A site in C*c*Os, which is missing from the structure of the A-type quinol oxidases. In both images the blue rectangulae represent the low-spin (left) and high-spin (right) haem groups, the green sphere on the right of a_3 being the Cu_B centre. eT = electron transfer; pT = proton transfer. Red arrows represents eT, black and blue arrows represent transfer of chemical and pumped protons, respectively.

but not with the function as a proton pump. For this reason it appears plausible that the H-channel may rather be a "dielectric well" that compensates electrically for the transmembrane electron transfer from cytochrome *c* to heme *a* [Rich and Maréchal, 2013; Wikström *et al.*, 2015].

3. The Linkage of Proton Pumping to Individual Steps of the Catalytic Cycle

Mitochondrial C*c*Os was discovered to be a proton pump by experiments showing both redox-linked proton release to the ***P***-side of the membrane and corresponding uptake from the ***N***-side, as well as translocation of two electrical charges per electron transferred [Wikström, 1977; Wikström and Saari, 1977]. The structures of the intermediates of the catalytic cycle are quite well-known today [see Wikström *et al.*, 2015] and are summarised in Figure 2.

Initially, thermodynamic equilibrium analysis of the P → F and F → O reaction steps in intact mitochondria led to the proposal [Wikström, 1989] that all proton-pumping was linked to these reactions, so that two protons would be pumped in each step in addition to the separation of charge due to electron transfer and proton uptake for the chemistry of water formation. Kinetic experiments showed, however, that this was incorrect, and that each of the four electron transfer reactions in the cycle is linked to pumping of one proton [Oliveberg *et al.*, 1991; Verkhovsky *et al.*,

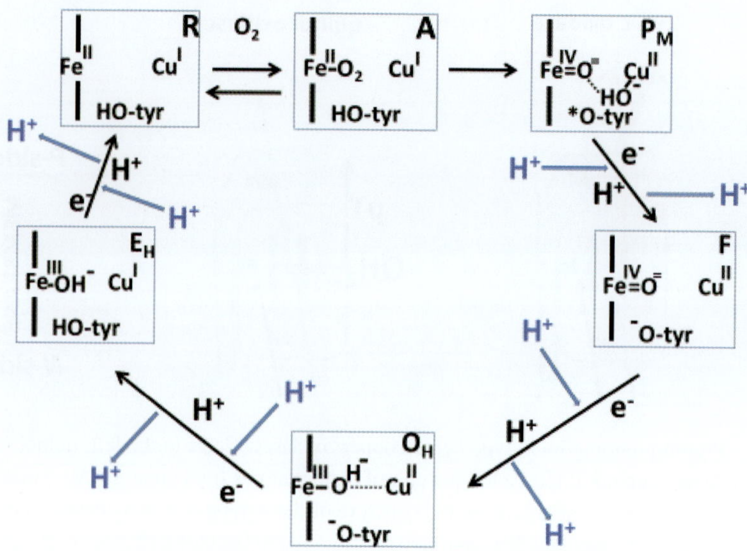

Figure 2.   Catalytic cycle of A-type heme copper oxidases. The rectangulae encompass the binuclear site (BNC), including haem a_3, Cu_B and the tyrosine covalently linked to one of the three histidine ligands of Cu_B (not shown). The name of the catalytic intermediate is given in the upper right corner. Blue arrows with protons (H^+) denote proton-pumping events; black H^+ denotes uptake of a "substrate proton." tyr-O* denotes a neutral tyrosine radical.

1999; Faxén *et al.*, 2005], as indicated in Figure 2. Moreover, time-resolved electrometric measurements of charge translocation showed unequivocally that it occurred to the same extent in each step, *viz.* to the extent of two electrical charges per electron transferred [see *e.g.* Belevich and Verkhovsky, 2008].

The recently proposed proton pumping mechanism involving the H-channel [Yoshikawa and Shimada, 2015; Yano *et al.*, 2016] can be schematically summarised as in Figure 3.

Here, all four protons to be pumped are taken up from the *N*-side of the membrane after full reduction of the binuclear centre, but before the binding of O_2, and are collected in the water cluster near the Mg^{2+} ion on the *P*-side of the heme groups. Remarkably, it is proposed that all four chemical protons are taken up (two in each step) in the P_{M-} > F and F– > O_H transitions, and that one pumped proton is released on the *P*-side in each one-electron reaction step. Although this yields the correct overall charge translocation stoichiometry, it is very difficult to reconcile with what is currently known about the structures of intermediates F and O_H (see Figure 2). For example, the FTIR data on the F intermediate suggest that the covalently bonded tyrosine residue in the binuclear site is deprotonated [Gorbikova *et al.*, 2008]. Uptake of all four chemical protons already at the O_H stage would also mean that the site has at that stage "used up" all the strongly basic

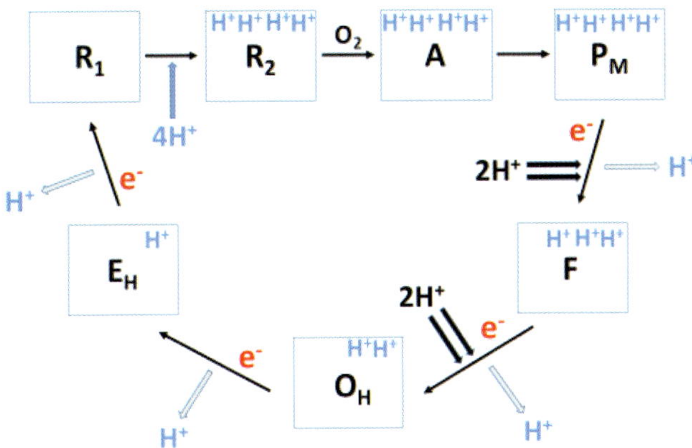

Figure 3.   Catalytic cycle according to Yano *et al.* [2016]. Blue H⁺ denote pumped protons, which are proposed to be collected near the Mg^{2+} site on the ***P***-side of the heme groups. Black H⁺ denote "chemical" protons consumed in the reduction of O_2 to water. The intermediate states are defined as in Figure 2.

groups the protonation of which has been thought to provide much of the driving force for proton pumping [Morgan *et al.*, 1994; Rich, 1995].

In the linkage scheme of Figure 3 the $O_H \rightarrow E_H$ and $E_H \rightarrow R$ reaction steps are each coupled only to release of a single proton from the proton loading site to the aqueous ***P***-phase. The electrogenicity of such a linkage can only be ca. 20–30% of the electrogenicity of translocating a single charge all across the membrane dielectric. It is therefore in direct contrast to the results of time-resolved electrometric electron injection experiments, which show that each of these reactions is coupled to overall translocation of two electrical charges (see above). However, one should note that the time-resolved experiments on the $O_H \rightarrow E_H$ transition were performed with A-type C*c*O from *P. denitrificans* [Bloch *et al.*, 2004; Belevich *et al.*, 2007]. Hence, if the linkage scheme of Figure 3 is correct, it can apply only to C*c*O from mammalian mitochondria, the mechanism of proton pumping of which must then be vastly different from that of A-type C*c*O from bacteria despite their strong structural and functional homology.

4. The Proton Pump Cycle

Egawa *et al.* [2013] have proposed another type of mechanism that utilises the H-channel, but otherwise bases proton translocation on coupling to heme *a* reduction and reoxidation. Loading the proton pump coupled to heme *a* reduction and releasing the proton to the other side of the membrane linked to reoxidation of heme *a* is actually a theme proposed as early as 1978 by Artzatbanov

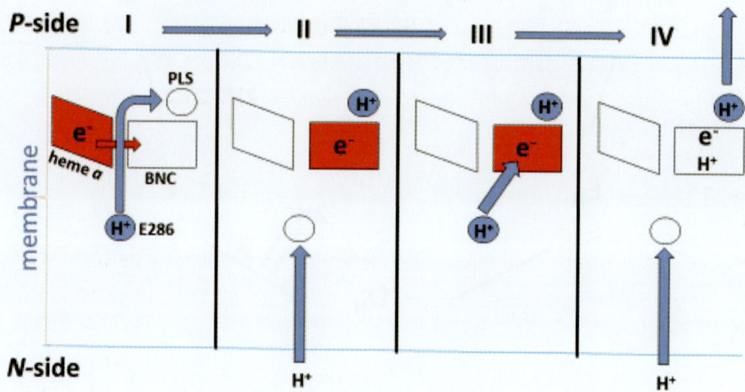

Figure 4. Reaction sequence of the proton pump. The two rectangulae denote heme *a* and the binuclear centre, respectively. Also shown schematically are the glutamic acid E286 at the end of the D-channel, and the proton loading site (PLS) near the propionate domain of heme a_3.

et al. [1978]. It was also entertained by Papa *et al.* [see Papa, 2005], and by the early work of Wikström and Krab [1979]. Although many of the ideas brought forward in the scheme of Artzatbanov *et al.* [1978] were amazingly accurate (such as *e.g.* the parallel location of the two heme groups in the membrane), especially considering that they were made almost 20 years before the 3D structure, the unique linkage of heme *a* oxidation-reduction to release and uptake of the pumped proton, respectively, has to be abandoned. As pointed out by Siletsky and Konstantinov [2012], H⁺ *release* to the *P*-phase is only secondary to uptake of the chemical proton (and is driven by it), and is not a result of heme *a* oxidation *per se*. Also, heme *a* reduction, as such, does not lead to loading of the PLS, neither in the A → F transition [Belevich *et al.*, 2006], nor in the O_H → E_H transition [Bloch *et al.*, 2004], but requires electron transfer from heme *a* to the binuclear site, as summarised in the scheme of Figure 4 [see Wikström *et al.*, 2015].

5. Conclusions

The H-channel structure is conserved in A-type heme-copper C*c*Os (but not in quinol oxidases) and has been proposed to catalyse translocation of all pumped protons in mammalian C*c*O. Mutagenesis data with A-type C*c*Os from bacteria and yeast shows that this proposal is not applicable to the C*c*Os from these organisms although they are structurally and functionally strongly related to those in mammals.

Acknowledgements

The author is indebted to the Nanyang Technical University, Singapore, for their hospitality, and to James Barber and Bertil Andersson for organising these conferences. The Magnus Ehrnrooth Foundation and Societas Scientiarum Fennica are acknowledged for financial support. I am grateful to Robert B. Gennis (Urbana, Illinois), Alexander A. Konstantinov (Moscow), and Peter R. Rich (London) for helpful discussions.

References

Abramson, J., Riistama, S., Larsson, G., Jasaitis, A., Svensson-Ek, M., Laakkonen, L., Puustinen, A., Iwata, S. and Wikström, M. (2000). The structure of the heme-copper oxidase from *Escherichia coli* and its binding site for ubiquinone, *Nat. Struct. Biol.*, 7, 910–917.

Ädelroth, P., Gennis, R.B. and Brzezinski, P. (1998). Role of the pathway through K(I-362) in proton transfer in cytochrome *c* oxidase from *R. sphaeroides*, *Biochemistry*, 37, 2470–2476.

Belevich, I., Bloch, D.A., Belevich, N., Wikström, M. and Verkhovsky, M.I. (2007). Exploring the proton pump mechanism of cytochrome *c* oxidase in real time, *Proc. Natl. Acad. Sci. USA*, 104, 2685–2690.

Belevich, I. and Verkhovsky, M.I. (2008). Molecular mechanism of proton translocation by cytochrome *c* oxidase, *Antioxid. Redox Signal.*, 10, 1–29.

Bloch, D., Belevich, I., Jasaitis, A., Ribacka, C., Puustinen, A., Verkhovsky, M.I. and Wikström, M. (2004). The catalytic cycle of cytochrome *c* oxidase is not the sum of its two halves, *Proc. Natl. Acad. Sci. USA*, 101, 529–533.

Brzezinski, P., Reimann, J. and Ädelroth, P. (2008). Molecular architecture of the proton diode of cytochrome *c* oxidase, *Biochem. Soc. Trans.*, 36, 1169–1174.

Egawa, T., Yeh, S.R. and Rousseau, D.L. (2013). Redox-controlled proton gating in bovine cytochrome *c* oxidase, *PLOS ONE*, 8, e63669.

Faxén, K., Gilderson, G., Ädelroth, P. and Brzezinski, P. (2005). A mechanistic principle for proton pumping by cytochrome *c* oxidase, *Nature*, 437, 286–289.

Fetter, J.R., Qian, J., Shapleigh, J., Thomas, J.W., Garcia-Horsman, A., Schmidt, E., Hosler, J., Babcock, G.T., Gennis, R.B. and Ferguson-Miller, S. (1995). Possible proton relay pathways in cytochrome *c* oxidase, *Proc. Natl. Acad. Sci. USA*, 92, 1604–1608.

Gorbikova, E.A., Wikström, M. and Verkhovsky, M.I. (2008). The protonation state of the cross-linked tyrosine during the catalytic cycle of cytochrome *c* oxidase, *J. Biol. Chem.*, 283, 34907–34912.

Kaila, V.R.I., Verkhovsky, M.I. and Wikström, M. (2010). Proton-coupled electron transfer in cytochrome oxidase, *Chem. Rev.*, 110, 7062–7081.

Konstantinov, A.A. (2012). Cytochrome *c* oxidase: intermediates of the catalytic cycle and their energy-coupled interconversion, *FEBS Lett.*, 586, 630–639.

Konstantinov, A.A., Siletsky, S., Mitchell, D., Kaulen, A. and Gennis R.B. (1997). The roles of the two proton input channels in cytochrome *c* oxidase from *Rhodobacter sphaeroides* probed by the effects of site-directed mutations on time-resolved electrogenic intraprotein proton transfer, *Proc. Natl. Acad. Sci. USA*, 94, 9085–9090.

Lee, H., Das, T.K., Rousseau, D.L., Mills, D., Ferguson-Miller, S. and Gennis, R.B. (2000). Mutations in the putative H-channel in the cytochrome *c* oxidase from *Rhodobacter sphaeroides* show that this channel is not important for proton conduction but reveal modulation of the properties of heme *a*, *Biochemistry*, 39, 2989–2996.

Morgan, J.E., Verkhovsky, M.I. and Wikström, M. (1994). The histidine cycle: a new model for proton translocation in the respiratory heme-copper oxidases, *J. Bioenerg. Biomembr*, 26, 599–608.

Oliveberg, M., Hallén, S. and Nilsson, T. (1991). Uptake and release of protons during the reaction between cytochrome *c* oxidase and molecular oxygen: a flow-flash investigation, *Biochemistry*, 30, 436–440.

Papa, S. (2005). Role of cooperative H^+/e^- linkage (redox Bohr effect) at heme *a*/CuA and heme *a*3/CuB in the proton pump of cytochrome *c* oxidase, *Biochemistry (Moscow)*, 70, 178–186.

Popović, D.M. and Stuchebrukhov, A.A. (2004). Proton pumping mechanism and catalytic cycle of cytochrome *c* oxidase: coulomb pump model with kinetic gating, *FEBS Lett.* 566, 126–130.

Puustinen, A., Finel, M., Virkki, M. and Wikström, M. (1989). Cytochrome o (bo) is a proton pump in *Paracoccus denitrificans* and *Escherichia coli*, *FEBS Lett.*, 249, 163–167.

Rich, P.R. (1995). Towards an understanding of the chemistry of oxygen reduction and proton translocation in the iron-copper respiratory oxidases, *Aust. J. Plant. Physiol.*, 22, 479–486.

Rich, P.R. and Maréchal, A. (2013). Functions of the hydrophilic channels in protonmotive cytochrome *c* oxidase, *J. R. Soc Interface*, 10(86), 20130183.

Salje, J., Ludwig, B. and Richter, O.M.H. (2005). Is a third proton-conducting pathway operative in bacterial cytochrome *c* oxidase? *Biochem. Soc. Trans.*, 33, 829–831.

Shimokata, K., Katayama, Y., Murayama, H., Suematsu, M., Tsukihara, T., Muramoto, K., Aoyama, H., Yoshikawa, S. and Shimada, H. (2007). The proton pumping pathway of bovine heart cytochrome *c* oxidase, *Proc. Natl. Acad. Sci. USA*, 104, 4200–4205.

Siletsky, S.A. and Konstantinov, A.A. (2012). Cytochrome *c* oxidase: charge translocation coupled to single-electron partial steps of the catalytic cycle, *Biochim. Biophys. Acta*, 1817, 476–488.

Sousa, F.L., Alves, R.J., Ribeiro, M.A., Pereira-Leal, J.B., Teixeira, M. and Pereira, M.M. (2012). The superfamily of heme-copper oxygen reductases: types and evolutionary considerations, *Biochim. Biophys. Acta*, 1817, 629–637.

Tsukihara, T., Shimokata, K., Katayama, Y., Shimada, H., Muramoto, K., Aoyama, H., Mochizuki, M., Shinzawa-Itoh, K., Yamashita, E., Yao, M., Ishimura, Y. and Yoshikawa, S. (2003). The low-spin heme of cytochrome *c* oxidase as the driving element of the proton-pumping process, *Proc. Natl. Acad. Sci. USA*, 100, 15304–15309.

Verkhovsky, M.I., Jasaitis, A., Verkhovskaya, M.L., Morgan, J.E. and Wikström, M. (1999). Proton translocation by cytochrome *c* oxidase, *Nature*, 400, 480–483.

Vygodina, T.V., Kirichenko, A. and Konstantinov, A.A. (2014). Cation binding site of cytochrome *c* oxidase: progress report. *Biochim, Biophys. Acta*, 1837, 1188–1195.

Vygodina, T.V., Pecoraro, C., Mitchell, D., Gennis, R. and Konstantinov, A.A. (1998). Mechanism of inhibition of electron transfer by amino acid replacement K362M in a proton channel of *Rhodobacter sphaeroides* cytochrome *c* oxidase, *Biochemistry*, 37, 3053–3061.

Wikström, M.K.F. (1977). Proton pump coupled to cytochrome *c* oxidase in mitochondria, *Nature* 266, 271–273.

Wikström, M. (1989). Identification of the electron transfers in cytochrome oxidase that are coupled to proton-pumping, *Nature*, 338, 776–778.

Wikström, M. and Krab, K. (1979). Proton-pumping cytochrome *c* oxidase, *Biochim. Biophys. Acta*, 549, 177–222.

Wikström, M.K.F. and Saari, H.T. (1977). The mechanism of energy transduction by mitochondrial cytochrome *c* oxidase, *Biochim. Biophys. Acta*, 462, 347–361.

Wikström, M., Sharma, V., Kaila, V.R.I., Hosler, J.P. and Hummer, G. (2015). New perspectives on proton pumping in cellular respiration, *Chem. Rev.*, 115, 2196–2221.

Wikström, M., Verkhovsky, M.I. and Hummer, G. (2003). Water-gated mechanism of proton translocation by cytochrome *c* oxidase, *Biochim. Biophys. Acta*, 1604, 61–65.

Yano, N., Muramoto, K., Shimada, A., Takemura, S., Baba, J., Fujisawa, H., Mochizuki, M., Shinzawa-Itoh, K., Yamashita, E., Tssukihara, T. and Yoshikawa, S. (2016). The Mg^{2+}-containing water cluster of mammalian cytochrome *c* oxidase collects four pumping proton equivalents in each catalytic cycle, *J. Biol. Chem.*, 291, 23882–23894.

Yoshikawa, S. and Shimada, A. (2015). Reaction mechanism of cytochrome *c* oxidase, *Chem. Rev.*, 115, 1936–1989.

Chapter 4

Cytochrome *c* Oxidase: Insight into Functions from Studies of the Yeast *S. cerevisiae* Homologue

Peter R. Rich

Glynn Laboratory of Bioenergetics,
Institute of Structural and Molecular Biology,
University College London, Gower Street, London WC1E 6BT, UK
prr@ucl.ac.uk

Different members of the superfamily of haem-copper oxidases have diverse substrates, cofactors and protein subunits. Nevertheless, all appear to catalyse oxygen reduction with a structurally-similar catalytic binuclear centre. This is housed in a common subunit I together with a low spin haem electron donor to the catalytic centre. All also appear to be coupled to additional proton translocation. However, the possible pathways for movement of protons through the protein structures are not universally conserved. Hence, although the basic mechanism of coupling is likely to follow similar principles, the detailed atomic mechanisms are likely to be different between the A and the B/C subgroups, and have even been suggested to differ between vertebrate mitochondrial cytochrome *c* oxidases and bacterial quinol and cytochrome *c* oxidases within the same A1 subgroup. Here, studies of the yeast mitochondrial homologue are reviewed that shed light on roles of different possible proton transfer pathways and some of the additional subunits found in mitochondrial enzymes.

1. Introduction: Catalysis, Coupling and Efficiency

Aerobic respiration in eukaryotic mitochondria requires an oxidase that reduces the oxygen which provides a terminal sink in order to maintain electron flow through the respiratory electron transfer chain. In mammals, the yeast *S. cerevisiae* (hereafter referred to only as yeast) and many other eukaryotes, this oxygen-reducing function is catalyzed solely by cytochrome *c* oxidase (C*c*O), a large multiprotein complex that spans the inner mitochondrial membrane. Electrons from reduced cytochrome *c* are transferred to a buried catalytic site, the binuclear centre (BNC), formed from a haem A (haem a_3) and a copper atom (Cu_B). It uses four electrons from cytochrome *c* in the intermembrane space (IMS), together with four protons from the mitochondrial matrix, to reduce molecular oxygen to two water molecules. As a result of the vectorial nature of the reaction, part of the energy released by the exergonic oxygen reduction reaction can be conserved in the transmembrane electrochemical proton gradient, as envisaged in the original chemiosmotic proposals of Mitchell [Mitchell, 1966].

The thermodynamic efficiency of energy conservation for this vectorial reaction may be estimated for typical physiological conditions. The standard potential E'_0 for O_2/H_2O at pH 7 is +815 mV *versus* SHE [Wood, 1988]. However, this refers to conditions with liquid water at 25°C and pH 7 and an O_2 fugacity of 1 atmosphere. Given Henry's Law, the 21% content of oxygen in air and the fact that its concentration can be far lower in tissues distant from the blood supply (perhaps as low as 2% [Carreau *et al.*, 2011]), this means that the operative E'_h is 10–25 mV lower. Assuming that the reductant source cytochrome c^{2+}/c^{3+} ratio in aerobic tissues is around 1, the reductant E'_h would correspond to its E_{m7} of +260 mV. Hence the overall $\Delta E'_h$ for the four-electron reduction of oxygen to water by cytochrome c^{2+} within cells at pH 7 is 545–530 mV ($\Delta G' \sim -50$ kcal.mol^{-1}). With a typical transmembrane PMF of ~200 mV (mostly $\Delta\psi$; positive outside), this means that only 35–40% of the available energy would be conserved from the four electrogenic PCET reactions due to their vectorial nature.

However, each full catalytic cycle also causes an additional four protons to be translocated from the matrix to the IMS [Wikström, 1977], hence increasing the fraction of energy that can be stored in the PMF to ~75% when operating against a large PMF. This means that, even with a high PMF of 200 mV, the overall operative $\Delta G'$ of the coupled energy-conserving reaction is ~ -12 kcal.mol^{-1}. Hence, the reaction is still operating far from equilibrium with 25% of the energy lost as heat. The reaction is essentially irreversible (for $\Delta G \sim -12$ kcal.mol^{-1}, ratio of forward/reverse reaction rate constants $\sim 10^9$) and it is not possible to drive oxygen generation by reversed electron flow, even with a high PMF, as has been confirmed experimentally [Wikström, 1988]. In environments such as exercising muscle, where the phosphate potential and the PMF may well drop considerably, the efficiency would be much lower, though the throughput rate and power output might increase [Rich, 1982].

The presence of this irreversible terminal step means that the upstream complexes of the respiratory chain can, provided that they have the necessary catalytic capacity, operate much closer to equilibrium and, therefore, operate closer to maximum efficiency. The fact that reversed electron transfer from reduced cytochrome *c* or ubiquinol to NAD^+ can occur readily in coupled mitochondria [Chance and Hollunger, 1961] suggests that this is the case. It also means that modulation of C*c*O activity, rather than activities of those complexes operating close to equilibrium, would most influence the overall throughput of the whole respiratory chain. This may well be one major reason why mitochondrial C*c*Os, in contrast to the other major respiratory complexes, possess a variety of isoforms of supernumerary subunits and ligand and phosphorylation sites that can influence activity allosterically [Kadenbach and Hüttemann, 2015], as discussed below.

2. Core Structural and Functional Variations in the HCO Superfamily

Mitochondrial C*c*Os are members of a haem-copper oxidase (HCO) superfamily, homologues of which are found throughout eukarya, bacteria and archaea [García-Horsman *et al.*, 1994; Sousa *et al.*, 2012; Hemp and Gennis, 2008]. The superfamily has been divided into A, B and C subgroups, and includes both C*c*Os and quinol oxidases (QOs). All have a homologous core subunit I which houses the catalytic BNC for oxygen reduction, formed by a high spin haem and a histidine-coordinated copper that has a catalytically-active tyrosine covalently linked to one of its histidine ligands [Tsukihara *et al.*, 1996]. Electron donation into the BNC is *via* a *bis*-histidine-coordinated low spin haem that is also housed in subunit I. Variations occur in haem types (A, B and O), the sequence location of the catalytic tyrosine and the immediate donor to the low spin haem (Cu and/or bound haem(s) C in the C*c*Os; direct from quinol substrate in the QOs). Further variations occur in the number and types of additional subunits and the distribution of domains between them. Despite these differences, it seems likely that the basic BNC-catalysed mechanism of oxygen reduction is essentially the same in all HCOs. This involves a four-electron transfer to dioxygen in a single step after binding that follows reduction of the two BNC metals [Wikström *et al.*, 2015; Rich, 2017]. This converts the dioxygen into two oxides without formation of kinetically-detectable, potentially damaging, reactive oxygen species. It seems that all HCOs share the same vectorial arrangement of delivery of reductant and substrate protons into the BNC described above that results in conservation of some of the energy of the exergonic oxygen reduction (in the case of QOs, the movement of the electron towards the N phase is replaced at least in part by expulsion of the positive quinol proton towards the P phase).

It also seems likely that all HCOs are coupled to the additional translocation of protons across the membrane that increases the efficiency of energy conservation, though coupling stoichiometry may be lower [Han *et al.*, 2011], or perhaps is more readily decreased by an opposing PMF [Wikström *et al.*, 2015]. However, different HCOs possess different possible pathways that could conduct protons within the protein structures [Hemp and Gennis, 2008]. Specifically, all A-type HCOs have two well conserved structures, the D and K channels [Konstantinov *et al.*, 1997]. The D channel connects the N phase to a buried glutamic acid (A1-type, which includes all eukaryotic mitochondrial CcOs) or tyrosine (A2-type) that is roughly 10 Å from both haem edges. The K channel connects the N phase to the catalytic tyrosine of the BNC. All A-type CcOs also have a third, H, channel, though with some variations (Figure 1) (and in A-type QOs the H structure is virtually absent [Wikström *et al.*, 2015]. The H channel potentially connects the N and P phases, running past the formyl and farnesyl substituents on the D- and A-pyrrole ring of the low spin haem [Tsukihara *et al.*, 1996]. There is ample evidence from kinetic analyses and mutation effects that the D channel conducts translocated protons in bacterial A-type CcOs [Wikström *et al.*, 2015]. However,

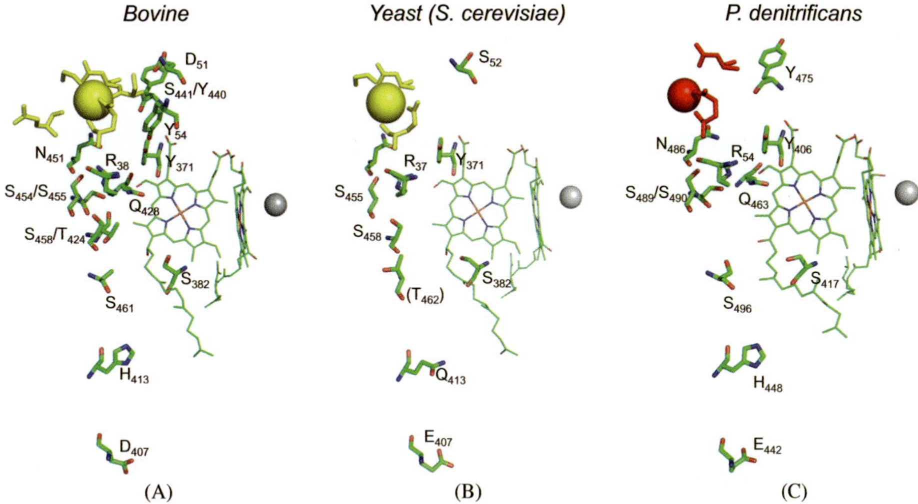

Figure 1. Comparison of residues within the H channel of bovine, yeast and *P. denitrificans* CcOs. Figures were constructed from PDB files of oxidised forms of CcOs: 5B1A (bovine, 1.8 Å [Yano *et al.*, 2016]); homology modelled yeast structure [Maréchal *et al.*, 2012] (with an assumed sodium rather than Ca^{++} ion); 3HB3 (*P. denitrificans*, 2.25 Å [Koepke *et al.*, 2009]. In panels B and C, only residues with sequence identity/similarity to bovine H channel residues are shown (yeast T$_{462}$ is one residue from the sequence-aligned residue A$_{461}$). Residues interacting with the sodium ion are shown in yellow (bovine and yeast); those interacting with the calcium ion (*P. denitrificans*) are shown in red.

it has been suggested from structural and functional data that the H channel fulfills this role in mammalian mitochondrial C*c*Os (reviewed in [Yoshikawa and Shimada, 2015]). Particularly surprising is the evident lack of either a functional D or H channel in the B- and C-type HCOs, despite their protonmotive ability. However, they do possess the equivalent of a K channel (though lacking the central lysine). These differences have led to the conclusion that protonmotive coupling in the B- and C-type HCOs should utilise the variant K channel for both substrate and translocated protons though, to date, no comprehensive model has emerged. Hence, a current issue is that, despite the likelihood that the oxygen chemistry has a common mechanism in all HCOs, coupling to proton translocations across the membrane may be achieved with quite different pathways and mechanisms in different subgroups of HCOs and even within the same A1 subgroup.

3. Yeast C*c*O — A Link Between Mammalian and A1 Bacterial C*c*Os

Yeast mitochondrial C*c*O offers an experimentally-amenable model system to explore possible functional variations in mitochondrial C*c*Os [Maréchal *et al.*, 2012]. Its core subunits have extensive sequence identity/similarity with mammalian subunits (for example, its subunit I has 59% sequence identity with bovine C*c*O subunit I). Homology modelling of structure confirms that it has D and K channels that are very similar to those present in other A1-type bacterial and mammalian C*c*Os. It also has a clear H channel, though with some differences. Figure 1 compares H channel residues in bovine and yeast mitochondrial C*c*Os together with those in the *P. denitrificans* A1-type C*c*O. All three have hydrophilic residues in the lower part of their structures that in bovine C*c*O is suggested to house water-accessible domains [Yoshikawa and Shimada, 2015], including an equivalent of residue S382, which undergoes potentially important conformational alterations [Yano *et al.*, 2016; Liu *et al.*, 2011]. Also conserved in all three are residues R38 and Y371 which interact with haem *a*. However, H413 that is in the centre of the lower part of the structure in mammalian and bacterial C*c*Os is replaced by a glutamine in yeast C*c*O.

Much more poorly conserved are residues that form the upper H-bonding network part of the bovine H channel. In particular, neither yeast nor *P. denitrificans* C*c*Os retain the proposed "gate" formed by S440–Y441 amide link, or the conformationally-flexible D51 at the P phase exit (though silent human polymorphisms D51/G and D51/Y have anyway been reported [mitomap, 2017]). The principal residues that form the bovine Na^+/Ca^{++} binding site are also only partly conserved in yeast and *P. denitrificans* C*c*Os, possibly explaining the closer similarity of yeast C*c*O with bacterial rather than bovine C*c*O in terms of ease of dissociation of bound Ca^{++} [Maréchal *et al.*, 2013]. Nevertheless, there is still a substantial potential H-bonding network of residues within this region in all three C*c*Os.

3.1. *The H channel and its possible roles*

Particularly useful in the yeast system is the availability of technology for site-directed mutagenesis of the core, mitochondrially-encoded subunits [Meunier *et al.*, 2012]. An extensive set of mutations have been introduced into the K, D and H channel regions of subunit I. Initial studies on effects on coupling ratio of a mutant with four mutations in the H channel were performed using vesicle-reconstituted enzyme. The data suggested that, like bacterial C*c*Os, coupling efficiency was not perturbed by such H channel interference [Dodia, 2014]. Further more precise studies of coupling efficiencies in enzyme with mutations of single residues in all three channels were made by measurements of "classical" H^+/O ratios determined from state 3/4 transitions induced by ADP additions to intact respiring mitochondria [Chance and Williams, 1955]. These investigations confirmed this finding and also indicated that mutations in the D channel have effects similar to those found in bacterial C*c*Os, including uncoupling of proton translocation from oxygen catalysis in a D channel mutant whose bacterial equivalent has the same effect [Maréchal *et al.*, 2014; Maréchal *et al.*, 2017]. As a result, an alternative possible "dielectric channel" role of the H channel in yeast mitochondrial and bacterial A1 C*c*Os has been suggested [Rich and Maréchal, 2013]. The structure of a dielectric channel should be similar to proton-conducting channels, being composed of charged or polarisable amino acids, water molecules [Nagle and Tristram-Nagle, 1983; DeCoursey, 2003] and other moieties. However, the structure is unable to support a continuous current of protons through the protein structure by the Grotthuss mechanism [Grotthuss, 2006]. This mechanism requires charge relocation along an H-bonded chain to a gated protonation sink, followed by molecular reorientations so that protons can be repetitively taken from the source and delivered to the sink. A dielectric channel can also relocate charge — in the case of the H channel this would occur in particular in response to charge change caused by reduction/oxidation of the low spin haem A — again by H-bonding changes as in the first part of the Grotthuss mechanism, or by charge/dipole movements more generally. However, unlike proton-conducting channels, the changes are not linked to the consumption of a proton at a source and delivery to a gated sink to create a net proton transfer with each redox cycle. Instead, any charge relocation (including any redox-linked protonation) occurs reversibly with each redox cycle, so that there is no resulting net transfer of a proton or other charged species. Hence, instead of facilitating a current of protons from source to sink, its function is to influence thermodynamic and kinetic properties of the buried electron transfer event by fully reversible charge relocations. The dielectric well effect could result in a reversible redox-linked protonation change within the channel if an appropriate protonatable group is present, akin to the redox Bohr proton effect [Papa *et al.*, 1979; Papa, 2005]. However, its physical basis can instead involve quite different groups whose positions or polarities change in concert to compensate the charge change. The induced changes can also be spatially extensive in that the charge rearrangements can be transmitted through the channel to one or both aqueous surfaces where remote protonation or ionic redistributions can occur. Whether the H channel fulfills such a function in mammalian CcOs in addition to, or instead of, providing a conduit for a current translocated protons remains to be established. It might be noted that a similar dielectric role has been suggested for the K channel in order to partially charge-compensate transiently-charged BNC intermediates [Jünemann *et al.*, 1997; Lepp *et al.*, 2008], hence facilitating their formation by lowering the free energy required. Interestingly, it has been noted that the H channel is absent in the A1-type QOs [Wikström *et al.*, 2015]. In this case, the electron donor is an electroneutral quinol from within the membrane and the system instead has a proton-conducting half channel to conduct protons released by quinol oxidation into the P phase [Abramson *et al.*, 2000], in addition to well-conserved D and K channels.

3.2. *Supernumerary and additional subunits*

The mitochondrial CcOs are considerably larger than their bacterial counterparts because they have additional supernumerary subunits. Human and bovine CcOs have 10 such additional subunits [Kadenbach and Hüttemann, 2015] and yeast CcO has eight, all of which are homologous to human/bovine subunits, though with low sequence identities in some cases [Maréchal *et al.*, 2012]. These additional subunits do not have direct roles in proton/electron transfer reactions. Some can occur as tissue- or environment-dependent isoforms. In yeast, only Cox5 (homologous to mammalian subunit IV) has isoforms but in mammalian CcOs isoforms of subunits IV, VIa, VIb, VIc, VIIa, VIIb and VIII have been found. Several subunits house sites that may become phosphorylated [Hüttemann *et al.*, 2012], acetylated [Liko *et al.*, 2016] or can bind nucleotides [Kadenbach and Hüttemann, 2015; Arnold, 2012]. Specific functions of some of these isoforms, their post-translational modifications and ligand binding sites have been suggested, though the supporting data in many cases remain inconsistent. Only one bacterial homologue of these supernumerary subunits has been reported, the *ctFm* gene product in the cyanobacterium *Synechocystis*, which has 50% and 20% sequence identity with yeast Cox5 and bovine subunit IV, respectively [Alge *et al.*, 1999].

In BN-PAGE gels of digitonin-extracted membranes, C*c*O can be found as monomers, dimers or in supercomplexes with complexes I and III [Wittig and Schägger, 2009; Stuart, 2008; Dudkina *et al.*, 2008; Cui *et al.*, 2014]. Bovine C*c*O crystallises as a homodimer with the dimer interface formed primarily by subunits VIb and III, with smaller contributions from Vb and VIa [Tsukihara *et al.*, 1996]. How important the dimer is in the natural membrane remains unclear. It has recently been suggested that propensity to form dimers may be influenced by iso-forms of subunits VIa and VIIa in mammalian cells types [Cogliati *et al.*, 2016]. In the known supercomplex structures, bovine and yeast C*c*Os are attached as monomers. However, even when incorporated into supercomplexes, the dimer-forming surface is sufficiently exposed that it could still allow dimerisation [Letts *et al.*, 2016] (Figure 2).

Several subunits are involved in interactions with partner complexes to form supercomplexes in mammalian systems (Figure 2). In the mammalian I/III$_2$/IV

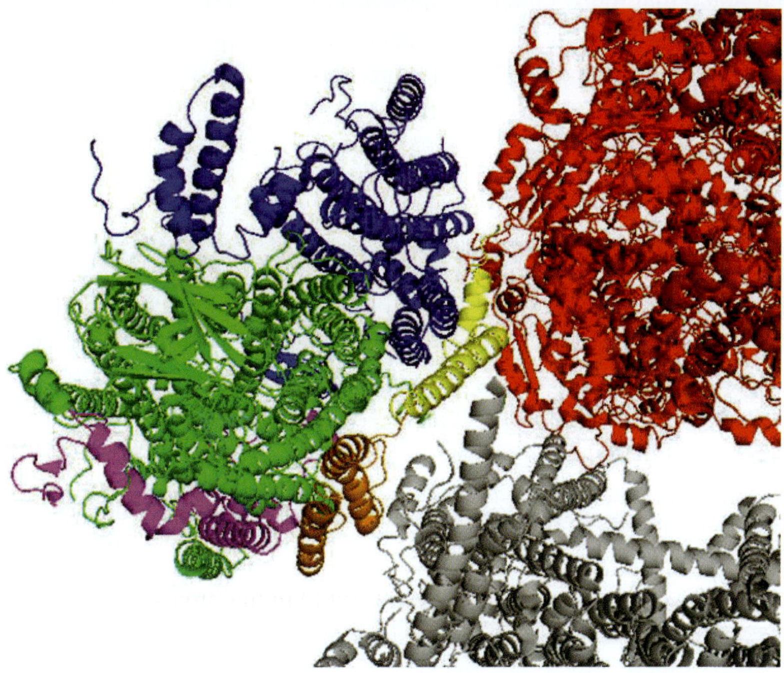

Figure 2. Relative positions of supernumerary subunits of ovine C*c*O involved in supercomplex interactions and dimerisation. The figure is drawn from coordinates of the tight form of the I/III$_2$/IV supercomplex in PDB 5J4Z [Letts *et al.*, 2016]. Interaction regions of complex I and complex III are shown in grey and red, respectively. Subunits of C*c*O are colour-coded as: brown, subunits VIIc, VIII interacting with complex I; yellow, subunit VIIa (or VIIa2l) interacting with complex III; blue, subunits VIb, III, Vb, VIa that contribute to the C*c*O dimer interface in crystal forms; magenta, subunit IV (homologue of yeast subunit Cox5); green, all other C*c*O subunits.

```
Yeast Cox7      1    .............................................MANKV      5
Bovine VIIa1    1    .........................MRALR.......VSQALVRSFSSTARNRFENRV     26
Bovine VIIa2    1    .......................MLRNLLA...... LRQIAKRTISTSSRRQFENKV     28
Bovine VIIa2l   1    MYYKFSGFTQKLAGAWASDAYSPQGLRPWSTEAPPIIFATPTKLSSGPTAYDYAGKNTV     60
                                                                            * *

Yeast Cox7      6    IQLQKIFQSSTK-PLWWRHPRSALYLYPFYAIFAVAWTP-LLYIPNAIRGIKAKKA      60
Bovine VIIa1   27    AEKQKLFQEDNGLPVHLKGGATDNILYRVTMTLCLGGTLYSLYCL-GWASFPHKK-      80
Bovine VIIa2   29    PEKQKLFQEDNGIPVHLKGGIADALLYRATLILTVGGTAYAMFEL-AVASFPKKQD      83
Bovine VIIa2l  61    PELQKFFQKSDGVPIHLKRGLPDQMLYRTTMALTVGGTIYCLIAL-YMASQPRNK-     114
                    : **:**..    *:                 : :.  .   :      .   ::
```

Figure 3. Clustal alignment of yeast Cox7 with homologous bovine subunits VIIa1, VIIa2 and with VIIa2l.

supercomplex [Letts *et al.*, 2016], subunit VIIa (or possibly the homologous subunit COX7a2l, also known as SCAF1 [Pérez-Pérez *et al.*, 2016; Cogliati *et al.*, 2016; Lapuente-Brun *et al.*, 2013; Ikeda *et al.*, 2013]) appears to interact with subunits UQCR11 and UQCR1 of complex III, and subunits VIIc and VIII interact with subunits ND5 and B12/NDUFB3 of complex I. However, it is not yet clear if the same interaction with complex III occurs in the yeast $III_2/IV_{1/2}$ supercomplex. Yeast Cox7 has weak (12–18%) sequence identity with all three isoforms of bovine subunit VIIa. It does not have the additional N-terminal extension characteristic of COX7a2l (Figure 3) that has been proposed to interact with complex III, and lacks an important histidine residue (H76 in the bovine VIIa2l sequence in Figure 3) [Cogliati *et al.*, 2016]. No other Cox7 homologues are evident in the *S. cerevisiae* genome. In addition, previous EM-derived models of the yeast III_2/IV_2 supercomplex have suggested that Cox5, which is spatially distant from Cox7, provides the interface with complex III [Mileykovskaya *et al.*, 2012].

Several additional proteins have been identified that are associated with digitonin-extracted mitochondrial C*c*Os but were not retained in the bovine C*c*O crystals produced with harsher detergents used to prepare samples for atomic structure determination (Table 1). In *S. cerevisiae* two proteins, Rcf1 and Rcf2, are associated with C*c*O, complex III_2 and III_2/IV supercomplexes [Vukotic *et al.*, 2012; Strogolova *et al.*, 2012] and may also regulate C*c*O activity independently of supercomplex formation [Rydström Lundin *et al.*, 2016; Hayashi *et al.*, 2015]. Two orthologues of Rcf1 are found in mammals and are also associated with C*c*O. However, surprisingly, neither was present in the recently solved ovine $I/III_2/IV$ supercomplex structure [Letts *et al.*, 2016] though again a direct effect of the Rcf1 orthologue (Higd1a) on catalytic activity has been suggested [Hayashi *et al.*, 2015].

Interestingly, Higd1a is induced in hypoxic conditions and is proposed to bind to a site close to subunit IV and the H channel of subunit I [Hayashi *et al.*, 2015]. This is in a relatively exposed region of C*c*O that is not involved in the dimer interface or in supercomplex interactions (Figure 2). Both bovine subunit IV and

Table 1.   Additional weakly-bound proteins associated with yeast and mammalian C*c*Os.

Yeast subunit	Mammalian subunit	Possible function
—	SCAF-1 (COX7a2l)	Forms III/IV interface
Rcf1	Rcf1a (Higd1a), Rcf1b	Required for supercomplex formation; regulatory function
Rcf2	—	Amplification of Rcf1 effects
YPR010C-A	NDUFA4	Regulation and biogenesis
COX26	—	Associated with III/IV supercomplex
AAC2		ADP/ATP exchange across IMM

its yeast homologue Cox5 have isoforms related to oxygen levels, raising the possibility that binding of Rcf/Higd1 isoforms are influenced by subunit IV/Cox5 isoforms. Interestingly, subunit IV/Cox5 interacts with two helices of core subunit I that provide part of the H channel structure [Rich and Maréchal, 2013]. Hence, it may be speculated that changes in subunit IV/Cox5 isoforms and/or binding of further protein partners or modifications such as phosphorylation might induces changes to catalysis parameters through allosteric effects of the nearby H channel by influencing its properties as a dielectric channel. An additional subunit, Cox26, has recently been found to be part of the yeast III$_2$/IV supercomplex [Levchenco *et al.*, 2016; Strecker *et al.*, 2016]. A further subunit, NDUFA4, originally thought to be a subunit of complex I, has now clearly been shown to be a component of mammalian complex IV [Balsa *et al.*, 2012] and also co-purifies with yeast III$_2$/IV supercomplex [Vukotic *et al.*, 2012], though its function is unknown. Association of the ADP/ATP carrier (Aac2) has also been found to be associated with the yeast supercomplex [Dienhart and Stuart, 2008]. For all of these more loosely-attached additional subunits, further work will be required to establish more firmly their roles in supramolecular structures, stability, allosteric control, assembly and/or interactions with as-yet unknown factors.

4. Concluding Remarks

In mitochondrial and bacterial forms of A1-type C*c*Os the pathways for protons through the protein must be made by one or more of the three identifiable hydrophilic networks of amino acids and associated waters. These may also provide a separate function as dielectric channels in some situations, a role which has not been widely recognised or tested to date. Further assessments of proton transfer *versus* dielectric roles of these channels may help to resolve the current discussions of whether conduction of translocated protons switches from the D channel in yeast mitochondrial and bacterial C*c*Os to the H channel in mammalian C*c*Os. Yeast C*c*O also offers a system to probe the possible control roles of its tightly bound "supernumerary" subunits and more weakly associated factors, many of which have mammalian homologues, and the effects of incorporation into supercomplexes.

Acknowledgements

Peter Rich is funded by the BBSRC with grants BB/K001094/1 and BB/L020165/1 that are directly relevant to this topic. I would like to thank Drs. Amandine Maréchal, Brigitte Meunier, Andrew Hartley and Tom Warelow for useful discussions relevant to this topic.

References

Abramson, J., Riistama, S., Larrson, G., Jasaitis, A., Svensson-Ek, M., Laakkonen, L., Puustinen, A., Iwata, S. and Wikström, M. (2000). The structure of the ubiquinol oxidase from *Escherichia coli* and its ubiquinone binding site, *Nat. Struct. Biol.*, 7, 910–917.

Alge, D., Wastyn, M., Mayer, C., Jungwirth, C., Zimmermann, U., Zoder, R., Fromwald, S. and Peschek, G.A. (1999). Allosteric properties of cyanobacterial cytochrome *c* oxidase: inhibition of the coupled enzyme by ATP and stimulation by ADP, *IUBMB Life*, 48, 187–197.

Arnold, S. (2012). The power of life-cytochrome *c* oxidase takes center stage in metabolic control, cell signalling and survival, *Mitochondrion*, 12, 46–56.

Balsa, E., Marco, R., Perales-Clemente, E., Szklarczyk, R., Calvo, E., Landàzuri, M.O. and Enríquez, J.A. (2012). NDUFA4 is a subunit of complex IV of the mammalian electron transport chain, *Cell Metab.*, 16, 378–386.

Carreau, A., el Hafny-Rahbi, B., Matejuk, A., Grillon, C. and Kieda, C. (2011). Why is the partial oxygen pressure of human tissues a crucial parameter? Small molecules and hypoxia, *J. Cell. Mol. Med.*, 15, 1239–1253.

Chance, B. and Hollunger, G. (1961). The interaction of energy and electron transfer reactions in mitochondria. IV. The pathway of electron transfer, *J. Biol. Chem.*, 236, 1562–1568.

Chance, B. and Williams, G.R. (1955). Respiratory enzymes in oxidative phosphorylation. III. The steady state, *J. Biol. Chem.*, 217, 409.

Cogliati, S., Calvo, E., Loureiro, M., Guaras, A.M., Nieto-Arellano, R., Garcia-Poyatos, C., Ezkurdia, N., Vázquez, J. and Enriquez, J.A. (2016). Mechanism of super-assembly of respiratory complexes III and IV, *Nature*, 539, 579–582.

Cui, T.-Z., Conte, A., Fox, J.L., Zara, V. and Winge, D.R. (2014). Modulation of the respiratory supercomplexes in yeast. Enhanced formation of cytochrome oxidase increases the stability and abundance of respiratory supercomplexes, *J. Biol. Chem.*, 289, 6133–6141.

DeCoursey, T.E. (2003). Voltage-gated proton channels and other proton transfer pathways. *Physiol. Rev.*, 83, 475–579.

Dienhart, M.K. and Stuart, R.A. (2008). The yeast Aac2 protein exists in physical association with the cytochrome bc_1-COX supercomplex and the TIM23 machinery. *Mol. Biol. Cell*, 19, 3934–3943.

Dodia, R.J. (2014). Structure-function relationship of mitochondrial cytochrome *c* oxidase: redox centres, proton pathways and isozymes. PhD Thesis, University College London.

Dudkina, N.V., Sunderhaus, S., Boekema, E.J. and Braun, H.-P. (2008). The higher level of organization of the oxidative phosphorylation system: mitochondrial supercomplexes, *J. Bioenerg. Biomembr.*, 40, 419–424.

García-Horsman, J.A., Barquera, B., Rumbley, J., Ma, J. and Gennis, R.B. (1994). The superfamily of heme-copper oxidases, *J. Bacteriol.*, 176, 5587–5600.

Grotthuss, C.J.T. (2006). (Translated version of) Memoir on the decomposition of water and of the bodies that it holds in solution by means of galvanic electricity, *Biochim. Biophys. Acta*, 1757, 871–875.

Han, H., Hemp, J., Pace, L.A., Ouyang, H., Ganesan, K., Roh, J.H., Daldal, F., Blanke, S.R. and Gennis, R.B. (2011). Adaptation of aerobic respiration to low O_2 environment, *Proc. Natl. Acad. Sci. USA*, 108, 14109–14114.

Hayashi, T., Asano, Y., Shintani, Y., Aoyama, H., Kioka, H., Tsukamoto, O., Hikita, M., Shinzawa-Itoh, K., Takafuji, K., Higo, S., Kato, H., Yamazaki, S., Matsuoka, K., Nakano, A., Asanuma, H., Asakura, M., Minamino, T., Goto, Y.-I., Ogura, T., Kitakaza, M., Komuro, I., Sakata, Y., Tsukihara, T., Yoshikawa, S. and Takashima, S. (2015). *Hig1a* is a positive regulator of cytochrome *c* oxidase, *Proc. Natl. Acad. Sci. USA*, 112, 1553–1558.

Hemp, J. and Gennis, R.B. (2008). Diversity of the heme-copper superfamily in archaea: insights from genomics and structural modelling. In *Results Problems in Cell Differentiation*, Schäfer, G. and Penefsky, H.S., eds. (New York: Springer-Verlag), pp. 1–31.

Hüttemann, M., Helling, S., Sanderson, T.H., Sinkler, C., Samavati, L., Mahapatra, G., Varughese, A., Lu, G., Liu, J., Ramzan, R., Vogt, S., Grossman, L.I., Doan, J.W. and Lee, I. (2012). Regulation of mitochondrial respiration and apoptosis through cell signaling: cytochrome *c* oxidase and cytochrome *c* in ischemia/reperfusion injury and inflammation, *Biochim. Biophys. Acta*, 1817, 598–609.

Ikeda, K., Shiba, S., Horie-Inoue, K., Shimokata, K. and Inoue, S. (2013). A stabilizing factor for mitochondrial respiratory supercomplex assembly regulates energy metabolism in muscle, *Nat. Commun.*, 4, 2147.

Jünemann, S., Meunier, B., Gennis, R.B. and Rich, P.R. (1997). Effects of mutation of the conserved lysine-362 in cytochrome *c* oxidase from *Rhodobacter sphaeroides*, *Biochemistry*, 36, 14456–14464.

Kadenbach, B. and Hüttemann, M. (2015). The subunit composition and function of mammalian cytochrome *c* oxidase, *Mitochondrion*, 24, 64–76.

Koepke, J., Olkhova, E., Angerer, H., Muller, H., Peng, G. and Michel, H. (2009). High resolution crystal structure of *Paracoccus denitrificans* cytochrome *c* oxidase: new insights into the active site and the proton transfer pathways. *Biochim. Biophys. Acta*, 1787, 635–645.

Konstantinov, A.A., Siletsky, S., Mitchell, D., Kaulen, A. and Gennis, R.B. (1997). The roles of the two proton input channels in cytochrome *c* oxidase from *Rhodobacter sphaeroides* probed by the effects of site-directed mutations on time-resolved electrogenic intraprotein proton transfer, *Proc. Natl. Acad. Sci. USA*, 94, 9085–9090.

Lapuente-Brun, E., Moreno-Loshuertos, R., Acín-Pérez, R., Latorre-Pellicer, A., Colás, C., Balsa, E., Perales-Clemente, E., Quirós, P.M., Calvo, E., Rodríguez-Hernández, M.A., Navas, P., Cruz, R., Carracedo, A., López-Otín, C., Pérez-Martos, A., Fernández-Silva, P., Fernández-Vizarra, E. and Enriquez, J.A. (2013). Supercomplex assembly determines electron flux in the mitochondrial electron transport chain, *Science*, 340, 1567–1570.

Lepp, H., Svahn, E., Faxén, K. and Brzezinski, P. (2008). Charge transfer in the K proton pathway linked to electron transfer to the catalytic site in cytochrome *c* oxidase, *Biochemistry*, 47, 4929–4935.

Letts, J.A., Fiedorczuk, K. and Sazanov, L.A. (2016). The architecture of respiratory supercomplexes, *Nature*, 537, 644–648.

Levchenco, M., Wuttke, J.-M., Rômpler, K., Schmidt, B., Neifer, K., Juris, L., Wissel, M., Rehling, P. and Deckers, M. (2016). Cox26 is a novel stoichiometric subunit of the yeast cytochrome *c* oxidase. *Biochim. Biophys. Acta*, 1863, 1624–1632.

Liko, I., Degiacomi, M.T., Mohammed, S., Yoshikawa, S., Schmidt, C. and Robinson, C. (2016). Dimer interface of bovine cytochrome *c* oxidase is influenced by local post-translational modifications and lipid binding, *Proc. Natl. Acad. Sci. USA*, 113, 8230–8235.

Liu, J., Qin, L. and Ferguson-Miller, S. (2011). Crystallographic and online spectral evidence for role of conformational change and conserved water in cytochrome oxidase proton pump, *Proc. Natl. Acad. Sci. USA*, 108, 1284–1289.

Maréchal, A., Haraux, F., Meunier, B. and Rich, P.R. (2014). Determination of H^+/e ratios in mitochondrial yeast cytochrome *c* oxidase, *Biochim. Biophys. Acta*, 1837, e100.

Maréchal, A., Iwaki, M. and Rich, P.R. (2013). Structural changes in cytochrome *c* oxidase induced by binding of sodium and calcium ions: an ATR-FTIR study, *J. Am. Chem. Soc.*, 135, 5802–5807.

Maréchal, A., Meunier, B., Lee, D., Orengo, C. and Rich, P.R. (2012). Yeast cytochrome *c* oxidase: a model system to study mitochondrial forms of the haem-copper oxidase superfamily, *Biochim. Biophys. Acta*, 1817, 620–628.

Maréchal, A., Xu, J.P., Genko, N., Meunier, B. and Rich, P.R. (2017). Identification of the proton pumping channel in yeast mitochondrial cytochrome *c* oxidase. In preparation.

Meunier, B., Maréchal, A. and Rich, P.R. (2012). Construction of histidine-tagged yeast mitochondrial cytochrome *c* oxidase for facile purification of mutant forms, *Biochem. J.*, 444, 199–204.

Mileykovskaya, E., Penczek, P.A., Fang, J., Mallampalli, V.K., Sparagna, G.C. and Dowhan, W. (2012). Arrangement of the respiratory chain complexes in *Saccharomyces cerevisiae* supercomplex III_2IV_2 revealed by single particle cryo-electron microscopy, *J. Biol. Chem.*, 287, 23095–23103.

Mitchell, P. 1966. *Chemiosmotic Coupling in Oxidative and Photosynthetic Phosphorylation.* (Glynn Research, Bodmin).

MITOMAP (2017). MITOMAP: mtDNA Coding Region & RNA Sequence Variants, http://www. mitomap.org/foswiki/bin/view/MITOMAP/PolymorphismsCoding, Last edited: Jan 30, 2017.

Nagle, J.F. and Tristram-Nagle, S. (1983). Hydrogen bonded chain mechanisms for proton conduction and proton pumping, *J. Membr. Biol.*, 74, 1–14.

Papa, S. (2005). Role of cooperative H^+/e linkage (redox Bohr effect) at heme a/Cu_A and heme a_3/Cu_B in the proton pump of cytochrome c oxidase, *Biochemistry (Moscow)*, 70, 178–186.

Papa, S., Guerrieri, F. and Izzo, G. (1979). Redox Bohr-effects in the cytochrome system of mitochondria, *FEBS Lett.*, 105, 213–216.

Pérez-Pérez, R., Lobo-Jame, T., Milenkovic, D., Mourier, A., Bratic, A., Garcia-Bartolomé, A., Fernández-Vizarra, E., Cadenas, S., Delmiro, A., Garcia-Consuegra, I., Arenas, J., Martin, M.A., Larsson, N.-G. and Ugalde, C. (2016). COX7A2L Is a mitochondrial complex III binding protein that stabilizes the III_2+IV supercomplex without affecting respirasome formation. *Cell Rep.*, 16, 2387–2398.

Rich, P.R. (1982). Electron and proton transfers in chemical and biological quinone systems, *Faraday Discuss. Chem. Soc.*, 74, 349–364.

Rich, P.R. (2017). Mitochondrial cytochrome c oxidase: catalysis, coupling and controversies, *Biochem. Soc. Trans.*, 45(3), 813–829.

Rich, P.R. and Maréchal, A. (2013). Functions of the hydrophilic channels in protonmotive cytochrome c oxidase, *J. R. Soc. Interface*, 10, 1–14.

Rydström Lundin, C., von Ballmoos, C., Ott, M., Ädelroth, P. and Brzezinski, P. (2016). Regulatory role of the respiratory supercomplex factors in *Saccharomyces cerevisiae*, *Proc. Natl. Acad. Sci. USA*, 113, 4476–4485.

Sousa, F.L., Alves, R.J., Ribeiro, M.A., Pereira-Leal, J.B., Teixeira, M. and Pereira, M.M. (2012). The superfamily of heme-copper oxygen reductases: types and evolutionary considerations, *Biochim. Biophys. Acta*, 1817, 629–637.

Strecker, V., Kadeer, Z., Heidler, J., Cruciat, C.-M., Angerer, H., Giese, H., Pfeiffer, K., Stuart, R.A. and Wittig, I. (2016). Supercomplex-associated Cox26 protein binds to cytochrome c oxidase, *Biochim. Biophys. Acta*, 1863, 1643–1652.

Strogolova, V., Furness, A., Robb-McGrath, M., Garlich, J. and Stuart, R.A. (2012). Rcf1 and Rcf2, members of the hypoxia-induced gene 1 protein family, are critical components of the mitochondrial cytochrome bc_1-cytochrome c oxidase supercomplex, *Mol. Cell. Biol.*, 32, 1363–1373.

Stuart, R.A. (2008). Supercomplex organisation of the oxidative phosphorylation enzymes in yeast mitochondria, *J. Bioenerg. Biomembr.*, 40, 411–417.

Tsukihara, T., Aoyama, H., Yamashita, E., Tomizaki, T., Yamaguchi, H., Shinzawa-Itoh, K., Nakashima, R., Yaono, R. and Yoshikawa, S. (1996). The whole structure of the 13-subunit oxidized cytochrome c oxidase at 2.8 Å, *Science*, 272, 1136–1144.

Vukotic, M., Oeljeklaus, S., Wiese, S., Vogtle, F., Meisinger, C., Meyer, H.E., Zieseniss, A., Katschinski, D.M., Jans, D.C., Jakobs, S., Warscheid, B., Rehling, P. and Deckers, M. (2012). Rcf1 mediates cytochrome oxidase assembly and respirasome formation, revealing heterogeneity of the enzyme complex, *Cell Metab.*, 15, 336–347.

Wikström, M.K.F. (1977). Proton pump coupled to cytochrome *c* oxidase in mitochondria, *Nature*, 266, 271–273.

Wikström, M. (1988). Mechanism of cell respiration. Properties of individual reaction steps in the catalysis of dioxygen reduction by cytochrome oxidase, *Chem. Scr.*, 28A, 71–74.

Wikström, M., Sharma, V., Kaila, V.R.I., Hosler, J.P. and Hummer, G. (2015). New perspectives on proton pumping in cellular respiration, *Chem. Rev.*, 115, 2196–2221.

Wittig, I. and Schägger, H. (2009). Supramolecular organization of ATP synthase and respiratory chain in mitochondrial membranes, *Biochim. Biophys. Acta*, 1787, 672–680.

Wood, P.M. (1988). The potential diagram for oxygen at pH 7, *Biochem. J.*, 253, 287–289.

Yano, N., Muramoto, K., Shimada, A., Takemura, S., Baba, J., Fujisawa, H., Mochizuki, M., Shinzawa-Itoh, K., Yamashita, E., Tsukihara, T. and Yoshikawa, S. (2016). The Mg^{2+}-containing water cluster of mammalian cytochrome *c* oxidase collects four pumping proton equivalents in each catalytic cycle, *J. Biol. Chem.*, 291, 2382–2394.

Yoshikawa, S. and Shimada, A. (2015). Reaction mechanism of cytochrome *c* oxidase, *Chem. Rev.*, 115, 1936–1989.

Chapter 5

Femtosecond Infrared Crystallography of Photosystem II Core Complexes: Watching Exciton Dynamics and Charge Separation in Real Space and Time

Marius Kaucikas*, James Barber*, Thomas Renger[+]
and Jasper J. van Thor*[,‡,]

*Imperial College London, South Kensington Campus,
Sir Ernst Chain Bld., SW7 2A2, UK
[+]Johannes Kepler University Linz, Institute of Theoretical Physics,
Altenberger Str. 69, A 4040 Linz, Austria
[‡]j.vanthor@imperial.ac.uk

We have shown structurally sensitive femtosecond time resolved infrared crystallography of oriented orthorhombic single crystals of photosystem II core complexes of *Synechococcus elongatus*, which allows a real-space analysis of exciton dynamics and charge separation. Directional selection of both the visible excitation and the infrared probes prepares and measures populations selectively which are analysed on the basis of X-ray crystallographic coordinates. Excited state chlorophyll *a* infrared absorption allows a real space probe of exciton dynamics in the local, single pigment, basis. Two contrasting models, known as the trap-limited and as transfer-to-the-trap limited model are critically tested. The dichroic amplitudes that result from photoselection are maintained on the ~60 ps time-scale that corresponds to the dominant energy transfer process which directly supports the transfer-to-the-trap limitation of the overall light-harvesting process. A detailed comparison with structure-based theory of exciton dynamics in photosystem II extends existing theory to the crystallographic case. On nanosecond time-scale non-crystallographic symmetry is developed for a structural measurement of the P_{680}^+/Pheo$^-$ charge separated state that supports the mode assignments for Pheo/Pheo$^-$ and P_{680}/P_{680}^+.

1. Introduction

Natural light harvesting and charge separation in photosystem II (PSII) core complexes involves dynamics of electronic excitations and electron transfer that occurs in 74 chlorin pigments in the case of *S. elongatus* cyanobacterial PSII core particles. A possible method to find structural information for the ultrafast processes retrieves real-space information from oriented and ordered materials such as molecular crystals with suitable symmetry properties. Even better, if information-rich spectrally dispersed spectroscopy information can be combined with structurally sensitive information, this will substantially aid modeling of complex data that by necessity has contributions from many origins. Photosynthetic light harvesting and charge separation represents a problem, where the added real space dimension is highly desirable to disentangle the very complex dynamics information.

Two contrasting models of photosynthetic light harvesting by PSII cores, known as the trap-limited [Holzwarth *et al.*, 2006; Schatz, Brock and Holzwarth, 1988] and as transfer-to-the-trap limited [Raszewski and Renger, 2008; Vasil'ev *et al.*, 2001] model are critically tested by VIS/IR spectroscopy on oriented single crystals. According to the trap-limited model exciton equilibration in the whole core complex occurs in less than 2 ps and reversible electron transfer reactions determine the slow decay time of fluorescence. In contrast, in the transfer-to-the-trap limited model this decay time reflects slow excitation energy transfer between the core antennae and the reaction center, where the excitation energy is irreversibly trapped. Here, the exciton equilibration that discriminates between these models is unambiguously revealed using structurally sensitive femtosecond time-resolved infrared crystallography of oriented single crystals of PSII core complexes of *Synechococcus elongatus*. Directional selection of both the visible excitation and the infrared probes prepares and measures populations which are analysed on the basis of X-ray crystallographic coordinates. Excited state chlorophyll *a* infrared absorption allows a real-space probe of exciton dynamics in the local, single pigment, basis. The dichroic amplitudes that result from photoselection are maintained on the 60 ps time-scale that corresponds to the dominant energy transfer process which provides compelling evidence for the transfer-to-the-trap limitation of the overall light-harvesting process. The alternative trap-limited model can be ruled out, since it predicts a more than one order faster decay of the dichroic amplitudes. On nanosecond time-scale non-crystallographic symmetry is developed for a structural time resolved measurement of the P_{680}^+/Pheo$^-$ charge separated state. The close correspondence of femtosecond infrared crystallography of exciton dynamics with structure-based theory of photosynthetic light harvesting demonstrates that excitation energy transfer in photosynthesis can be visualized in time and space by this new technique.

2. Femtosecond Infrared Crystallography

Experimental probes of ultrafast structural dynamics in proteins gain important structural information when conducted in crystalline form [Sage *et al.*, 2011], in combination with real-space analysis possible from recent high resolution X-ray structures. Using previously reported and characterized instrumentation that allows femtosecond time resolved pump-probe mid-infrared spectroscopy in diffraction-limited spots [Kaucikas *et al.*, 2013] we have performed polarization-resolved measurements in transmission mode of small oriented single crystals of PSII core complexes of *Synechococcus elongatus* with known index [Ferreira *et al.*, 2004; Loll *et al.*, 2005].

Crystal symmetry and the wave-vector determines whether double refraction will occur. For transparent (assuming weak extinction), non-conducting (neglecting current, *J*) anisotropic media the Maxwell's equations may be re-written to eliminate the magnetic field, resulting in Maxwell's generalised polarised wave equation of the electric field *E*.

$$\nabla \times \left(\nabla \times E \right) + \frac{1}{c^2} \frac{\partial^2 E}{\partial t^2} = -\frac{1}{c^2} \chi \frac{\partial^2 E}{\partial t^2} \tag{1}$$

Here c is the light velocity in the medium and χ is the susceptibility tensor.

There are generally two solutions for this equation in anisotropic media, which are given as two wave vector surfaces that represent the magnitude of two wave-vectors $k = n_{1,2,3}\,\omega/c$ for the allowed orthogonally polarised radiation modes, where $n_{1,2,3}$ are the three principal indices of refraction. The quadratic representation of these indices is known as the "Indicatrix", and allows finding the orthogonal polarisation directions for any propagation vector. With the exception of cubic symmetry, which are crystals that have point groups 432, $\overline{4}$3m, 23, m3 or m3m, all 27 other point group symmetries are doubly refracting. These are further separated in biaxial and uniaxial crystal classes that have three or two unique principal indices of refraction respectively. As a consequence, their Indicatrices are triaxial ellipsoids, or ellipsoids of revolution respectively. For weakly absorbing crystals, the two classes are trichroic and dichroic in nature. Point group symmetry determines whether the linear susceptibility tensor χ_1 includes off-diagonal elements, which is the case for triclinic and monoclinic symmetries that are both biaxial systems. The monoclinic crystals only include "zx" and "xz" off-diagonal elements whereas those involving the "y" coordinate are zero on account of the two-fold symmetry associated with the crystallographic b-axis. In contrast, orthorhombic, tetragonal, trigonal and hexagonal crystals all lack off-diagonal elements, which specifies that the principal susceptibility, axes of susceptibility that also determine the allowed polarisation directions, correspond with the crystallographic axes. Structural analysis on the basis of dichroic or trichroic extinction is known as pleochroism. It can be shown that in the case of uniaxial symmetries, the dichroic extinction depends only on the polar angle between the transition dipole moment and the high symmetry crystallographic axis which also corresponds to the optic axis. In contrast, in biaxial systems, the trichroic extinction depends on both azimuthal and polar angles and therefore provide more structural information than uniaxial systems. Orthorhombic crystals offer an additional advantage within the class of biaxial systems in that the three two-fold or mirror symmetries (point groups 222, mm2, mmm) force a wavelength independence. Because the wave-vector surfaces must survive the two-fold symmetry operations, the directions of the principal dielectric axes ($\varepsilon_x \neq \varepsilon_y \neq \varepsilon_z$) of orthorhombic crystals correspond to the crystallographic symmetry axes (a,b,c) and are wavelength independent [Born and Wolf, 1999]. By fortunate coincidence the available crystals of *S. elongatus* PSII core complexes are in orthorhombic space group $P2_12_12_1$. We have performed the first femtosecond time resolved infrared crystallography, or ultrafast infrared pleochroism, and presented the analysis recently [Kaucikas *et al.*, 2016].

The X-ray structure of *S. elongatus* PSII core complexes has initially been reported at 3.0 Å resolution, from which the coordinates of 74 chlorins in the dimer structure have been analysed [Loll *et al.*, 2005]. As summarised above, polarized infrared crystallography of orthorhombic space groups allows explicit analysis of the molecular transition dipole directions directly in the coordinate frame [Sage *et al.*, 2011]. We use synchrotron X-ray crystallographic indexing in order to determine the laboratory orientation of PSII crystals. Crystals were generally of cuboid morphology, with the thin dimension corresponding with the a-axis. Therefore the spectroscopically accessible surface was the {1 0 0} surface.

For oriented orthorhombic crystals with a *k*-vector in the direction of the crystallographic a-axis the two allowed polarized waves have the *E*-field in the directions that correspond to the crystallographic b- and c-axes [Born and Wolf, 1999]. For a molecule having a transition dipole moment (TDM) μ_1 the linear response for polarized waves with the *E* field in the e_1 and e_2 directions are then proportional to $(e_1.\mu_1)^2$ and $(e_2.\mu_1)^2$, respectively. Since the quantities $(e_1.\mu_1)^2$ and $(e_2.\mu_1)^2$ are invariant under each two-fold operation, all *n* copies of molecules present in the asymmetric unit are summed in order to find the linear response

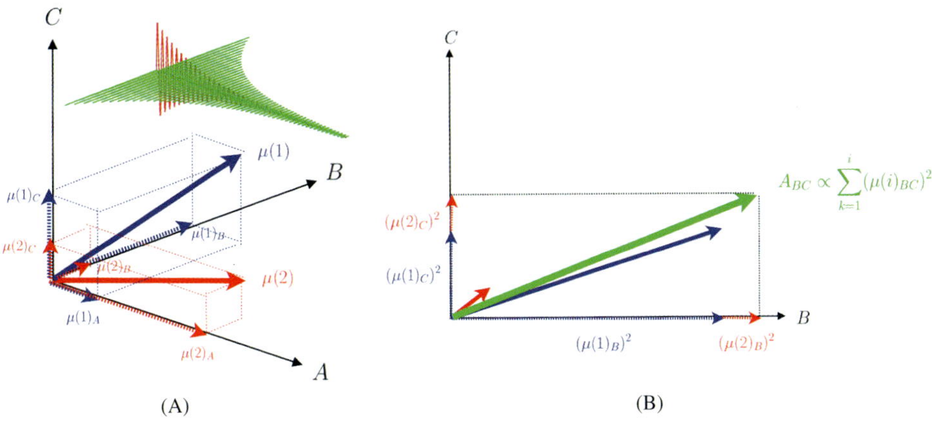

Figure 1. Vector presentation example for *two* dipoles μ(1) and μ(2) placed in the asymmetric unit of a crystal with orthorhombic symmetry and their projections on the A, B and C axes. For a ray with a k vector ‖ to the A axis, the allowed orthogonal polarisation directions E_1 and E_2 are wavelength independent, are associated with refractive indices n_1 and n_2 and experience attenuation coefficients proportional to $(\mu(1)_B)^2 + (\mu(2)_B)^2$ and $(\mu(1)_C)^2 + (\mu(2)_C)^2$.

for polarized waves for the unit cell in orthorhombic crystals. Although non-crystallographic symmetry is not perfect for *S. elongatus* PSII cores, *n* is taken to be 2 for the dimer structure in the asymmetric unit in the 2AXT pdb structure [Loll *et al.*, 2005].

3. Summary of Theory for Exciton Dynamics and Trapping by Electron Transfer for the Crystallographic Case

We show that femtosecond time resolved infrared crystallography provides a probe of electronic excitations in the local basis. This contrasts with a visible measurement, due to the excitonic nature of the excitations. Infrared measurements on the other hand can be analysed in the single pigment basis, if the direction of the measured infrared transition dipole is known relative to the nuclear coordinates. A requirement for this is that the vibrational delocation length is small. Experimental measurements of dichroic amplitudes of excited state chlorophyll *a* keto modes has provided clear evidence that for such high frequency modes this localised character is sufficiently accurate to allow for an analysis of excited state contributions in the single pigment basis. This evidence is taken from the correspondence of the predicted and experimental photoselection of dichroic amplitudes of excited state keto stretching modes. With multiple pigments contributing, it is seen that the dichroic amplitude maximises in the direction of the polarisation of the visible excitation. A fortuitous correspondence of the directions of transition dipole moments of the visible Qy transition, and the infrared keto C=O stretching mode dominates the photoselection that is seen. Both excitations are in the plane of the heme group and have a relatively small angle of ~10 degrees between them, as predicted by TD-DFT and dipole gradient from harmonic frequency calculations, respectively. Structure based calculations of exciton dynamics and charge separation for crystals of PSII are extended from methods and results reported previously [Renger and Müh, 2013; Raszewski and Renger 2008; Shibata *et al.*, 2013].

The first-principle calculations were done using non-Markovian density matrix theory for the optical excitation [Renger & Marcus, 2002], including Redfield theory for exciton relaxation within domains of strongly coupled pigments and generalized Förster theory for excitation energy transfer between these domains [Raszewski and Renger, 2008]. The site energies and excitonic couplings obtained in work reported previously [Shibata *et al.*, 2013] were used to parameterize the exciton Hamiltonian. In order to apply this theory to the crystallography case, the polarisation direction and resulting linear response is included for the spectroscopically accesible {1 0 0} face, for E-fields parallel to the crystallographic b- and c-axes.

The relative contributions $\Delta_i(t) = N_i(t)p_i$ of the pigments to the dichroic absorbance A_{bc} of the pump-probe signal depend on the polarization of the excitation laser and become time-dependent because of energy transfer between the pigments. These dependencies are included in the time-dependent populations $N_i(t)$ of electronically excited pigments. In order to calculate $N_i(t)$ we have to describe the absorption of an ultrashort optical pulse by the core complex that is oriented according to the crystal considered. Because of excitonic couplings between electronic transitions of different pigments the excited states of the core complex are partially delocalized. The delocalization is limited by differences in local optical transition energies and by static and dynamic disorder, caused by slow and fast protein dynamics, respectively. The differences in local optical transition energies are caused by the different protein environments in the different binding sites of the pigments, therefore these energies are often termed site energies. Protein dynamics that is slow compared to excited state lifetimes of the pigments (fs to ns) causes static disorder in site energies and is included by a Monte Carlo average that takes into account a Gaussian distribution function for the site energies. Any fast protein dynamics causes a homogeneous broadening of optical lines and also limits the extent of exciton delocalization. These dynamic localization effects are included in the present theory implicitly by introducing exciton domains of strongly coupled pigments and allowing exciton delocalization to occur only within these domains, as described in detail previously [Raszewski and Renger, 2008; Shibata *et al.*, 2013]. A delocalized exciton state $|M_d\rangle = \Sigma_{m_d} c_{m_d}^{(M_d)} |m_d\rangle$ is given as a linear combination of localized excited states $|m_d\rangle$ where pigment m_d is excited and all other pigments in this domain are in their electronic ground state. The initial populations $P_{M_d}^{(d)}(0)$ of the exciton state $|M_d\rangle$ in a given domain d are defined by the absorption of an ultrashort pump pulse that is assumed to have a Gaussian-shaped envelope of FWHM $2\tau_p\sqrt{2\ln 2}$, a carrier frequency Ω, and a polarization along the unit vector $\vec{e}_{pu}$, [Raszewski and Renger, 2008]:

$$P_{M_d}^{(d)}\left(0,\vec{e}_{pu}\right) \propto \left|\vec{e}_{pu}\cdot\vec{\mu}_{M_d}\right|^2 \int_{-\infty}^{\infty} d\tau\, e^{-i\left(\Omega-\omega_{M_d}\right)\tau}\, e^{G_{M_d}(t)-G_{M_d}(0)}\, e^{-\tau^2/\left(4\tau_p^2\right)}, \tag{2}$$

where $\vec{\mu}_{M_d} = \sum m_d c_{m_d}^{(M_d)}\vec{\mu}_{m_d}$ is the transition dipole moment for excitation of exciton state $|M_d\rangle$ from the electronic ground state, which is obtained as a linear combination of local transition dipole moments $\vec{\mu}_{m_d}$. The frequency ω_{M_d} denotes the transition frequency between the minima of the potential energy surfaces of the ground state and of exciton state $|M_d\rangle$. The function $G_{M_d}(t)$ describes excitation of the vibrational sideband of the exciton state $|M_d\rangle$ and is related to the mutual shift of the PES of this exciton state with respect to the PES of the ground state. It is determined by the spectral density $J(\omega)$ of the exciton vibrational coupling, the Bose–Einstein distribution function of vibrational quanta and the exciton coefficients $c_{m_d}^{(M_d)}$ [Raszewski and Renger, 2008]. $J(\omega)$ was extracted from experimental line narrowing data, as described before [Renger and Marcus, 2002]. Please note that in Raszewski and Renger, [2008] an average over random orientations of complexes in the sample has been carried out in the calculation of the initial populations $P_{M_d}^{(d)}(0)$. In the present case such an average is not needed since the complexes are oriented in a well-defined way in the crystal. The scalar product $\vec{e}_{pu}\cdot\vec{\mu}_{M_d}$ between the polarization vector of the pump-pulse and the exciton transition dipole moment determines the probability of excitation. In the experiments $\vec{e}_{pu}$ was either oriented along the crystallographic b- or c-axis.

In order to obtain the exciton energies and exciton coefficients an exciton matrix is diagonalized for every exciton domain d. The exciton matrix contains in the diagonal the site energies of the pigments that is the transition energies at which the pigments would absorb light if they were not coupled excitonically, and in the off-diagonal the excitonic couplings between pigments. The latter were obtained in Shibata *et al.* [2013] from the 1.9 Å resolution crystal structure [Umena *et al.*, 2011] by applying the Poisson-TrEsp method. We use the site energies obtained in Hall *et al.* [2016] and Shibata *et al.* [2013] from fitting optical spectra of the core complex and its subunits.

After creation of the initial populations of exciton states by the optical pulse, the excitation energy can relax within the exciton domains, it can be transferred between different domains, and it can be trapped by charge transfer in the reaction center domain. A system of master equations is solved to obtain the time-dependent exciton state populations $P_{M_d}^{(d)}(t)$, taking into account these different transfer processes.

$$\frac{d}{dt}P_{M_d}^{(d)}(t) = -\sum_{c,\,N_c}\left(k_{M_d\to N_c}P_{M_d}^{(d)}(t) - k_{N_c\to M_d}P_{N_c}^{(c)}(t)\right) -$$
$$\delta_{d,\mathrm{RC}}\,k_{M_{\mathrm{RC}}\to\mathrm{RP1}}P_{M_{\mathrm{RC}}}^{(\mathrm{RC})}(t). \tag{3}$$

In the above master equation the rate constant $k_{M_d\to N_c}$ for $d = c$ describes exciton relaxation between two delocalized states $|M_d\rangle$ and $|N_d\rangle$ in the same exciton domain and for $d \neq c$ excitation energy transfer between delocalized state $|M_d\rangle$ in one domain and delocalized state $|N_c\rangle$ in another domain. The former rate constant is taken from Redfield theory and the latter from generalized Förster theory, as described in detail in Raszewski and Renger [2008]. In the reaction center domain ($d = \mathrm{RC}$) the trapping of excitation energy by primary charge transfer is described by the rate constant

$$k_{M_{\mathrm{RC}}\to\mathrm{RP1}} = \left|c_{\mathrm{Chl}_{\mathrm{D1}}}^{(M_{\mathrm{RC}})}\right|^2 k_{\mathrm{intr}}, \tag{4}$$

where $\left|c_{\mathrm{Chl}_{\mathrm{D1}}}^{(M_{\mathrm{RC}})}\right|^2$ denotes the probability that the primary electron donor $\mathrm{Chl}_{\mathrm{D1}}$ is excited in exciton state $|M_{\mathrm{RC}}\rangle$ and k_{intr} is the intrinsic rate constant for primary charge transfer between the excited primary electron donor $\mathrm{Chl}_{\mathrm{D1}}^*$ and the primary electron acceptor $\mathrm{Pheo}_{\mathrm{D1}}$

$$\mathrm{Chl}_{\mathrm{D1}}^*\mathrm{Pheo}_{\mathrm{D1}} \xrightarrow{k_{\mathrm{intr}}} \mathrm{Chl}_{\mathrm{D1}}^+\mathrm{Pheo}_{\mathrm{D1}}^-.$$

In the calculations we take $k_{\mathrm{intr}} = (6\ \mathrm{ps})^{-1}$, as has been estimated from the fluorescence decay for closed reaction centers in Raszewski and Renger [2008]. Please note that because of the large free energy difference of the initial charge transfer reaction $\Delta G_{\mathrm{intr}}^{(0)} \geq 175\ \mathrm{meV}$, as has been estimated in Raszewski and Renger [2008], back electron transfer from the radical pair state has practically no influence on the decay of excited states in the time interval considered here (see Figure 3 in the supporting online information of Raszewski and Renger [2008]). Therefore, this back electron transfer is neglected and no further radical pair states need to be included in the present calculations. From the populations $P_{M_d}^{(d)}(t)$ of exciton state $|M_d\rangle$ in a given exciton domain d, we obtain the local population $N_i(t)$ of any pigment i in that domain by

$$N_i(t) = \left|c_i^{(M_d)}\right|^2 P_{M_d}^{(d)}(t). \tag{5}$$

The VIS/IR pump-probe signal in Figure 2 c/d is obtained from these local populations as

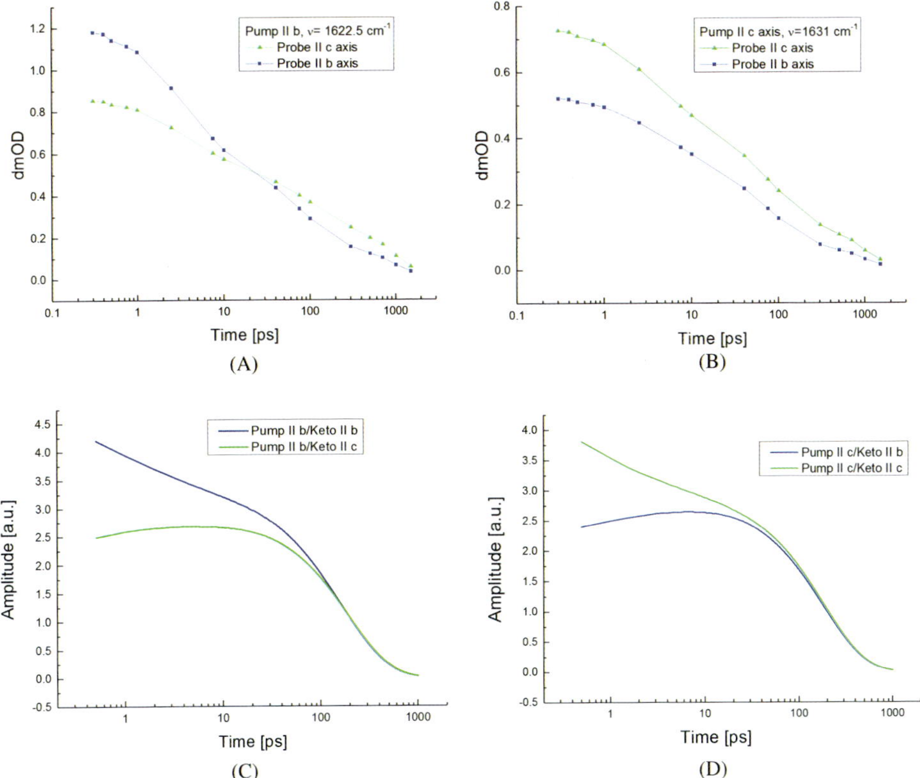

Figure 2. (A) Time traces for polarized signals at 1622.5 cm^{-1} in the b- and c-directions with optical excitation in the c-direction. (B) Time traces for polarized signals at 1631 cm^{-1} in the b- and c-directions with optical excitation in the b-direction. (C) Full simulation (Eq. 6) of infrared crystallographic signals in the b- and c-directions with optical excitation in the b-direction, summing all keto-modes and assuming local mode vibrational character and applying simulated populations from theory. (D) Full simulation of infrared crystallographic signals in the b- and c-directions with optical excitation in the c-direction, summing all keto-modes and assuming local mode vibrational character and applying simulated populations from theory.

$$\Delta OD \sim \sum_i N_i\left(t, \vec{e}_{\text{pu}}\right)\left|\vec{e}_i^{(\text{keto})} \cdot \vec{e}_{\text{pr}}\right|^2, \tag{6}$$

where the polarization vectors $\vec{e}_{\text{pu}}$ of the VIS pump pulse and $\vec{e}_{\text{pr}}$ of the IR probe pulse are taken at all possible combinations of the b- and c-crystallographic axes and $\vec{e}_i^{(\text{keto})}$ is the unit vector along the vibrational keto transition dipole moment of the pigments. Please note that $N_i\left(t, \vec{e}_{\text{pu}}\right)$ refers to the local excited state population of pigment i after optical excitation with a pump pulse that is polarized along $\vec{e}_{\text{pu}}$.

In order to see how the results change if the system would behave like suggested by the ERPE model, we have artificially increased the inter-domain exciton

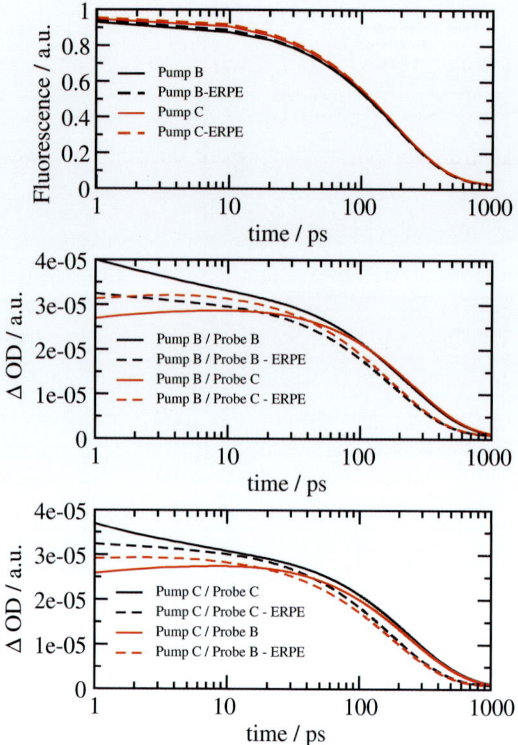

Figure 3. Upper part: Decay of isotropic fluorescence for different polarization directions of the excitation pulse (black lines: polarization along the b-axis, red lines: polarization along the c-axis) and in two different models (solid: original structure based model, dashed: fictitious ERPE type model). Middle and Bottom Parts: VIS Pump / IR probe signal calculated for different polarizations of the pump- and probe pulse (as given in the legends) and in two different models (as in upper part).

transfer rates by a factor of 50 and decreased the free energy difference of the initial charge transfer reaction to $\Delta G_{\mathrm{intr}}^{(0)} = 0.02\,\mathrm{eV}$, which makes this reaction reversible, that is we have

$$\mathrm{Chl}_{\mathrm{D1}}^{*}\mathrm{Pheo}_{\mathrm{D1}} \leftrightarrow \mathrm{Chl}_{\mathrm{D1}}^{+}\mathrm{Pheo}_{\mathrm{D1}}^{-},$$

where the ratio of forward and backward rate constant $k_{\mathrm{intr}}/k_{\mathrm{intr\text{-}back}}$ equals $\exp\left(\Delta G_{\mathrm{intr}}^{(0)}/kT\right)$ and we slightly decreased the intrinsic rate constant to $k_{\mathrm{intr}} = (8.8\,ps)^{-1}$.

In order to allow for an irreversible trapping of excitation energy the secondary electron transfer step

$$\mathrm{P}_{\mathrm{D1}}\mathrm{Chl}_{\mathrm{D1}}^{+}\mathrm{Pheo}_{\mathrm{D1}} \leftrightarrow \mathrm{P}_{\mathrm{D1}}^{+}\mathrm{Chl}_{\mathrm{D1}}\mathrm{Pheo}_{\mathrm{D1}}^{-}$$

was included with a larger free energy difference $\Delta G^{(0)}_{\text{RP1}\rightarrow\text{RP2}} = 0.138\,\text{eV}$ and a forward rate constant $k_{\text{RP1}\rightarrow\text{RP2}} = (6\ \text{ps})^{-1}$, where the back transfer rate is given as $k_{\text{RP2}\rightarrow\text{RP1}} = \exp\left(-\Delta G^{(0)}_{\text{RP1}\rightarrow\text{RP2}} / kT\right) k_{\text{RP1}\rightarrow\text{RP2}}$.

All these charge transfer parameters were chosen such that for this fictitious ERPE type system, the resulting fluorescence decays in exactly the same way as in our original structure-based model.

Hence, without structural information it would not be possible to distinguish between the two models solely from the fluorescence decay. In contrast, the calculation of VIS pump-IR probe spectra in the middle and bottom part of Figure 3 reveals significant differences of the two models. Due to the faster exciton transfer in the ERPE model, the system "forgets" the polarization of the initial excitation much faster than in the structure-based model. Consequently the anisotropy of the pump-probe signal is practically absent in the ERPE model but clearly visible in the structure-based model. By comparing the calculated with the measured pump probe signals it becomes clear that only the structure-based transfer-to-the-trap limited model can at least qualitatively explain the experimental data.

A word of caution should be added to the above calculations. A rescaling of the energy transfer times by a factor of 50 is, of course, not in agreement with the structure of PSII. Therefore, we have to consider the ERPE type dynamics as that of a fictitious system with a different structure. In the absence of any crystal structure information the decay of the fluorescence alone could not distinguish between the two models, but the present polarization resolved VIS/IR experiment on oriented single crystals could, as our calculations demonstrate.

A quantitative difference between calculation and experiment is that in the intermediate 5–20 ps time range the experimental signals decays faster than the calculated one.

Since for the present pump intensities on average 2.5 photons are absorbed per core complex, as has been estimated in the Kaucikas *et al.* [2016], the data are influenced also by exciton-exciton annihilation effects not included in the theoretical analysis so far. The Coulomb coupling between exciton domains that are both optically excited couples the transition between the one-exciton manifold and the two exciton manifold of one domain to the transition from the one exciton state to the ground state of an other domain. The population of the doubly excited exciton manifold afterwards non-radiatively decays back to the one exciton domain, caused by intra domain exciton-exciton annihilation and overall an excitation has been quenched. If initially the same exciton domain is doubly excited optically, it gets quenched directly. In the following we will take into account intra-domain exciton-exciton annihilation implicitly, by assuming that it is fast compared to inter-domain exciton relaxation and provide an explicit description of the latter. In addition, we will use the fact that intra-domain exciton transfer is fast compared to inter-domain exciton transfer, by assuming that inter-domain exciton transfer and inter-domain exciton annihilation starts from equilibrated intra-domain exciton states. In this case the inter-domain rate constant. $k_{d \to c}$ for transfer between exciton domains c and d follows as

$$k_{d \to c} = \sum_{M_c, N_d} f\left(M_c\right) k_{M_c \to N_d} \tag{7}$$

where $f(M_c)$ describes the equilibrium population of intra-domain exciton state $|M_c\rangle$ and $k_{M_c \to N_d}$ is the generalized Förster rate constant for inter-domain exciton transfer between the Mth exciton state of domain c and the Nth exciton state of domain d. The exciton-exciton annihilation process occurring between domains c and d, where an exciton in domain c is annihilated is approximated by the rate constant

$$k_{cd \to d} = \sum_{M_c, N_d, 2K_c} f\left(M_c\right) f\left(N_d\right) k_{M_c N_d \to 2K_c}, \tag{8}$$

where $k_{M_c N_d \to 2K_c}$ is the rate constant for exciton fusion between the Mth exciton state of domain c and the Nth exciton state of domain d giving rise to the creation of the Kth exciton state in the two-exciton manifold of domain c. This process is considered to be the rate limiting step of the exciton-exciton annihilation. In complete analogy to the Generalized Förster rate constant $k_{M_c \to N_d}$ for transfer between one-exciton states in different domains we obtain for the rate constant

$$k_{M_c N_d \to 2K_c} = \frac{2\pi}{\hbar^2} \left|V_{M_c \to 2K_c, N_d \to 0}\right|^2 O_{M_c \to 2K_c, N_d \to 0}, \tag{9}$$

where $O_{M_c \to 2K_c, N_d \to 0}$ is the spectral overlap between the lineshape functions of the $M_c \to 2K_c$ excitation in domain c and the $N_d \to 0$ transition in domain d, and $V_{M_c \to 2K_c, N_d \to 0}$ denotes the excitonic coupling between the involved electronic transitions in the two domains that can be expressed in terms of the excitonic couplings between individual pigments and coefficients of one-exciton states $|M_c\rangle = \sum_{m_c} c_{m_c}^{(M_c)} |m_c\rangle$ and two-exciton states $|2K_c\rangle = \sum_{m_c, k_c}^{k_c < m_c} c_{m_c k_c}^{(2K_c)} |m_c k_c\rangle + \sum_{m_c} c_{2m_c}^{(2K_c)} |2m_c\rangle$, where $|m_c\rangle$ denotes a singly excited state that is localized at the mth pigment in domain c, $|m_c k_c\rangle$ is a doubly excited state localized at the kth and the mth pigments, and in the state $|2m_c\rangle$ the mth pigment of domain c is doubly excited.

The two-exciton coefficients are obtained by diagonalizing the two-exciton matrix that contains in the diagonal the site energies $E_m^{(c)} + E_k^{(c)}$ of intermolecular two-exciton states $|m_c k_c\rangle$ and the site energies $E_{2m}^{(c)}$ of the intramolecular two exciton states $|2m_c\rangle$ which we approximate as $E_{2m}^{(c)} \approx 2E_m^{(c)}$.

The inter-domain exciton fusion coupling is given as

$$V_{M_c \to 2K_c, N_d \to 0} = \sum_{m_c, n_d} c_{m_c}^{(M_c)} c_{n_d}^{(N_d)} \left(\sum_{k_c}^{k_c < m_c} c_{m_c k_c}^{(2K_c)} V_{n_d, k_c} \right.$$
$$\left. + \sum_{k_c}^{k_c > m_c} c_{k_c m_c}^{(2K_c)} V_{n_d, k_c} + c_{m_c m_c}^{(2K_c)} V_{n_d, m_c \to 2m_c} \right). \tag{10}$$

The excitonic coupling V_{n_d, k_c} describes the coupling of the ground-to-first excited state transition densities of pigment n in domain d and pigment k in domain c and $V_{n_d, m_c \to 2m_c}$ that of the transition densities of the the ground-to-first excited state of the nth pigment in domain d and the first-to-higher excited state transition of pigment m in domain c. For simplicity, we assume $V_{n_d, m_c \to 2m_c} \approx V_{n_d, m_c}$ and we consider the upper limit of the rate constant $k_{M_c N_d \to 2K_c}$ for exciton-exciton annihilation in Eq. (9) by assuming perfect overlap of lineshape functions, that is, we set the overlap integral $O_{M_c \to 2K_c, N_d \to 0}$ to unity (1 fs).

In the limit of fast intra-domain exciton relaxation the populations of exciton states are Boltzmann equilibrated, that is, we have

$$P_{M_d}^{(d)}(t) = P_d(t) f(M_d), \tag{11}$$

for the singly excited states of the complex, containing the Boltzmann factors f, and

$$P_{M_c, N_d}^{(c,d)}(t) = P_{cd}(t) f(M_c) f(N_d), \tag{12}$$

for the doubly excited states that are localized in domains c and d. The singly-excited domain populations $p_d(t)$ fulfil the following master equation

$$\frac{d}{dt} P_d(t) = -\sum_c \left(k_{d \to c} P_d(t) - k_{c \to d} P_c(t) - k_{cd \to d} P_{cd}(t) \right), \tag{13}$$

where $k_{d \to c}$ denotes the inter-domain exciton transfer and includes now also primary charge transfer in the case that d equals RC and c equals RP1. As before, we neglect charge back transfer, that is, $k_{\text{RP1} \to \text{RC}} = 0$. As a new term, not included so far we have the population transfer from the doubly excited state localized at domains c and d to the singly excited state localized at domain d, described by the exciton annihilation rate constant $k_{cd \to d}$.

The population transfer in the doubly excited state manifold is obtained from the master equation

$$\frac{d}{dt} P_{cd}(t) = -\left(\sum_g \left(k_{c \to g} + k_{d \to g} \right) + k_{cd \to d} + k_{cd \to c} \right) \left(P_{cd}(t) \right.$$
$$\left. + \sum_g \left(k_{g \to c} P_{gd}(t) + k_{g \to d} P_{cg}(t) \right). \right. \tag{14}$$

As the initial condition we have $P_d(0) = \Sigma_{M_d} P_{M_d}^{(d)}(0)$ and $P_{cd}(0) = P_c(0)P_d(0)$ with the initial exciton populations $P^{(d)}_{M_d}(0)$ given in Eq. (2).

The annihilation rate constant $k_{cd \to d}^{M_d}$ in Eqs. (13) and (14) so far has been only defined for exciton domains c and d. Its known, however, that the chlorophyll cation (Chl^+) as it occurs in the radical pair states RP_i in the reaction centre is an efficient quencher of excitation energy [Schlodder *et al.*, 2005]. Assigning a microscopic rate constant for this process would require to assign transition dipole moments for the absorption of Chl^+. Here, we will implicitly take into account this quenching by assuming that it occurs after the excitation energy has been transferred to the RC. Since the bottleneck of this reaction is the transfer to the RC we can approximate the rate constant $k_{dRP_i \to RP_i}$ by the rate constant $k_{d \to RC}$ for excitation energy transfer from the antenna domain d to the respective reaction center, where the radical pair state is located.

We have checked first our assumption of fast intra-domain exciton relaxation by comparing the annihilation free dynamics with and without using this approximation. The resulting pump-probe curves reveal good quantitative agreement for delay times larger than 1 ps. Hence it is justified to use this approximation in the calculation of exciton-exciton annihilation.

The VIS/IR pump-probe spectra obtained for pump-polarization along the crystal b- and c-axes are compared in Figures. 4 and 5 to the annihilation free cases. The overall decay of the signals for the different probe polarizations is very similar with and without including annihilation. The annihilation process seems to accelerate the decay of the signals somewhat in the 2–20 ps time window. This time window corresponds to excitation energy transfer times between different domains in the same antenna subunit CP43 or CP47. After 20 ps the decay of the pump-probe signals is very similar, with and without annihilation, but the signal calculated with annihilation has a 20% smaller amplitude due to the earlier annihilation processes.

A qualitative estimation of the decrease in signal amplitude at late times due to exciton-exciton annihilation processes can be obtained in the following way. We first note that due to the large distance between the CP43/47 subunits and the RC at low light intensities the bottleneck for the decay of excited states is the transfer between these subunits and the RC [Raszewski and Renger, 2008]]. Since also the distance between the CP43, between the CP47 and between the CP43 and the CP47 subunits in PS2 dimers are large, exciton annihilation occurs predominantly between exciton domains located in the same antenna subunit. In Figure 6 the different realizations of multi-excitation states in the antenna subunits of PS 2 dimers are illustrated. For two excitations the latter might be located either in the same subunit or in two different ones. There are four microscopic realizations of the first type and six of the second one (as noted in brackets below the respective

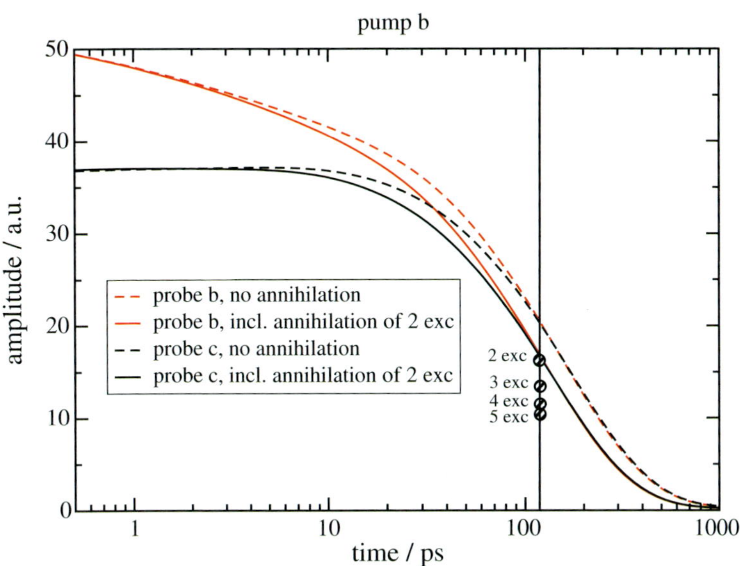

Figure 4. VIS Pump/IR probe signal calculated with (solid lines) and without (dashed lines) including exciton-exciton annihilation in the two excitation manifold for pump pulse polarization in B-direction and probe pulse polarized in b (red lines) and in c (black lines) directions. For comparison the annihilation free curves have been upscaled. The site energies of Shibata *et al.* [2013] were used in the calculations. The circles at 110 ps denote the amplitude of the signal expected from a simple statistical model, explained in the text, if exciton-exciton annihilation with the 2, 3, 4 and 5 exciton manifold are included.

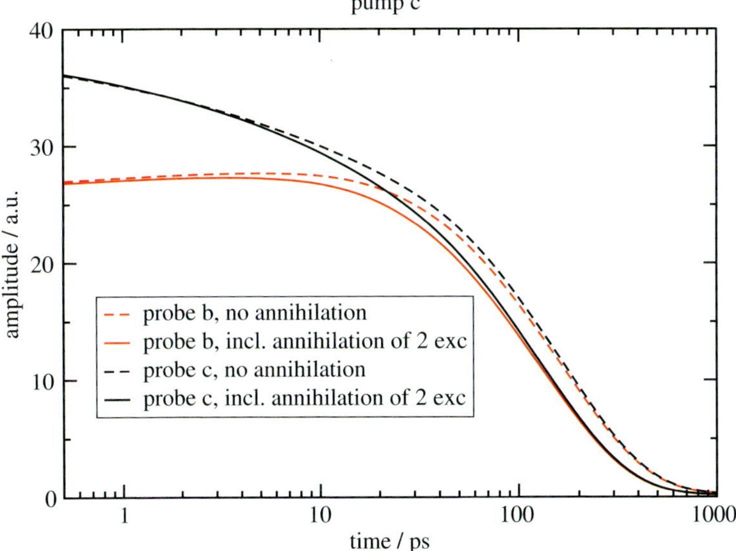

Figure 5. Same pump-probe signals as in Figure 4, but for pump-pulse polarized parallel to the crystal c-axis.

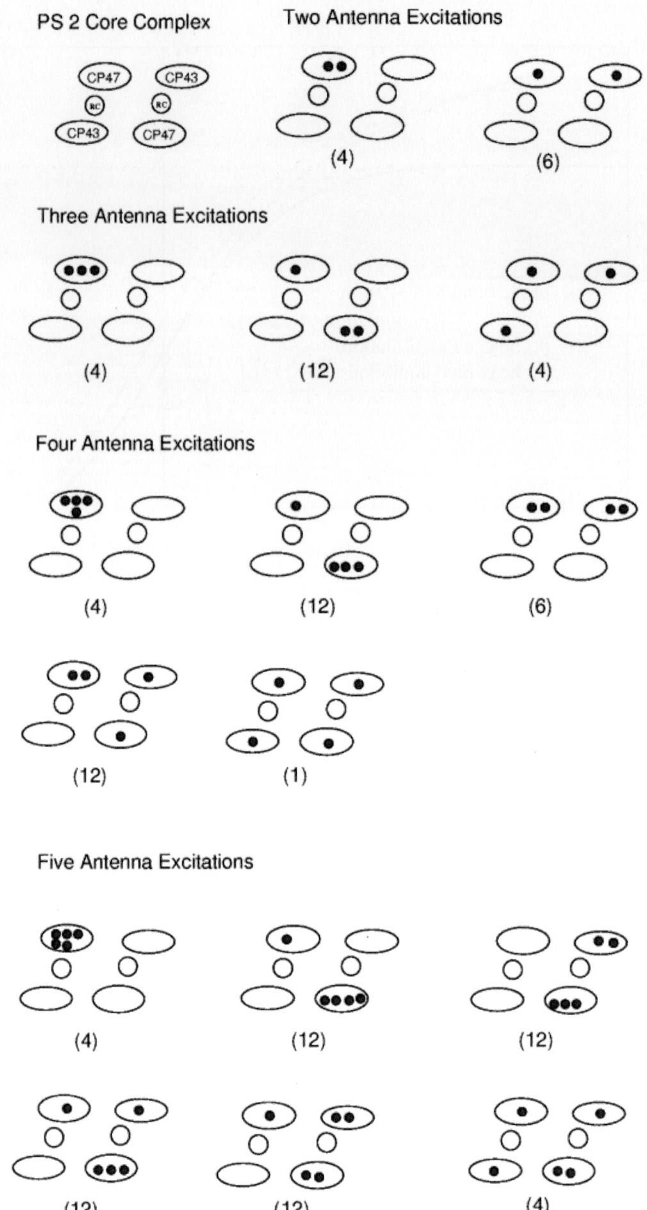

Figure 6. Illustration of different realizations of multiple antenna excitations states. The structure of the PS 2 core complex dimer (shown in the upper left corner) contains two reaction centers and one CP43 and CP47 antenna subunit per RC. A single red dot in a subunit represents a single excitation present in one of the exciton domains belonging to this subunit. Multiple red dots in the same subunit describe multiple excitations present in a subunit and are likely to experience exciton-exciton annihilation, as explained in the text. The numbers below the different states denote the statistical weight, which describes the number of microscopic realizations of that state. These statistical weights and the above states are used in the text to estimate the influence of exciton-exciton annihilation on the amplitude of the pump-probe signal at large delay times.

illustrations). Only the first type is influenced by exciton-exciton annihilation. The second type decays by transfer to the RC and electron transfer, since the inter-subunit distances are large. Since the inter-domain distances within a subunit (CP43 or CP47) are small compared to the distances between these subunits and the RC, two excitations present in the same subunit will react fast by annihilation and a single excitations is left that is afterwards quenched by transfer to the RC. Hence, in a subunit with multiple excitations only one will contribute to the amplitude of the slow 60 ps decay process in the experiment, reflecting excitation energy transfer from an antenna subunit to the RC. Hence, concerning the relative amplitude η of the slow decay process with and without annihilation for a given number of excitations, we may just count the subunits that have at least one excitation and divide by the number of excitations. Taking into account the different statistical weights of the different types of multi-excitation states in Figure 6, we obtain for the two antenna excitation states

$$\eta_{2exc} = \frac{4 \times 1 + 6 \times 2}{10 \times 2} = 0.8$$

in very good agreement with the relative amplitude of the signal at large delay times in Figure 4 obtained from our microscopic simulations with and without taking into account exciton-exciton annihilation of the two-excitation states. Since the inclusion of more than two excitations in these calculations increases the numerical costs dramatically, we have not taken into account these states so far. Nevertheless, our simple statistical model introduced above can be used to obtain an estimate for the change in relative signal amplitude at large times.

In the case of three antenna excitations we get as a new state that with three excitations in the same subunit. As in the case of the two excitation states we expect efficient quenching of all but one excitation before excitation energy transfer to the RC takes place. Therefore, we obtain

$$\eta_{3exc} = \frac{4 \times 1 + 12 \times 2 + 4 \times 3}{20 \times 3} = 0.67.$$

In complete analogy the relative amplitudes of the slow decay with and without annihilation due to contributions of the four and five excitation states are given as

$$\eta_{4exc} = \frac{4 \times 1 + 12 \times 2 + 6 \times 2 + 12 \times 3 + 1 \times 4}{35 \times 4} = 0.54,$$

and

$$\eta_{5exc} = \frac{4 \times 1 + 12 \times 2 + 12 \times 2 + 12 \times 3 + 12 \times 3 + 4 \times 4}{56 \times 5} = 0.50.$$

We have marked the expected signal due to the different number of excitations present in the annihilation process as circles at a large time (110 ps) in Figure 4. The relative weight for different number of excitations is determined by the intensity of the pump-pulse applied in the experiment. For the present experiment an average of 2.5 absorbed photons per PS 2 dimer has been estimated in Kaucikas *et al.* [2016]. Hence, it can be expected that at least the three excitation states will have a significant population. Since for larger number of excitations the average distance between excited domains within one subunit decreases, we expect, besides the smaller amplitude of the signal at large delay times between pump- and probe pulse, discussed above, also faster exciton-exciton annihilation. The larger weight of the shorter annihilation times will lead to an earlier decrease of the signal that will bring the latter closer to the experimental curves. Finally, we have investigated an alternative set of site energies of the CP47 pigments, suggested very recently by Hall *et al.* [2016] on the basis of fits of linear optical spectra including linear absorption, linear dichroism, circular dichroism and circularly polarized luminescence (CPL). The most striking new result from this study is that Chl 11 instead of Chl 29 of the earlier studies is suggested to be the lowest energy pigment in CP47. In case of CP43 the site energies of Shibata *et al.* [2013] were found to explain the linear spectra, including the newly measured CPL [Hall, 2016]. We have combined the Shibata *et al.* [2013] site energies for the reaction center pigments and the ones in CP43 with those suggested by Hall *et al.* [2016] for CP47 and recalculated the pump-probe decays with and without including exciton-exciton-annihilation for pump-polarization in b-direction as shown in Figure 7. The overall decay of the pump-probe spectra is somewhat slower than the one obtained for the site energies of Shibata *et al.* in Figure 4, but the overall behavior is qualitatively very similar in the case of the pump pulse polarization along the crystal b-axis. For pump pulse polarization along the c-axis the signals for the two probe polarizations cross at about 5 ps and the difference between the two decays on a similar time scale as for the first pump polarization. It is interesting to note that such a crossing of pump-probe signals also occurs in the experiment, but for pump polarization along the crystal b-axis (Figure 2a).

As before, including annihilation leads again to a faster decay for intermediate times (2–20 ps) that brings the signal somewhat closer to the experimental data. As before one may well expect that the agreement between theory and experiment will increase further by including larger number of antenna excitations in the calculations.

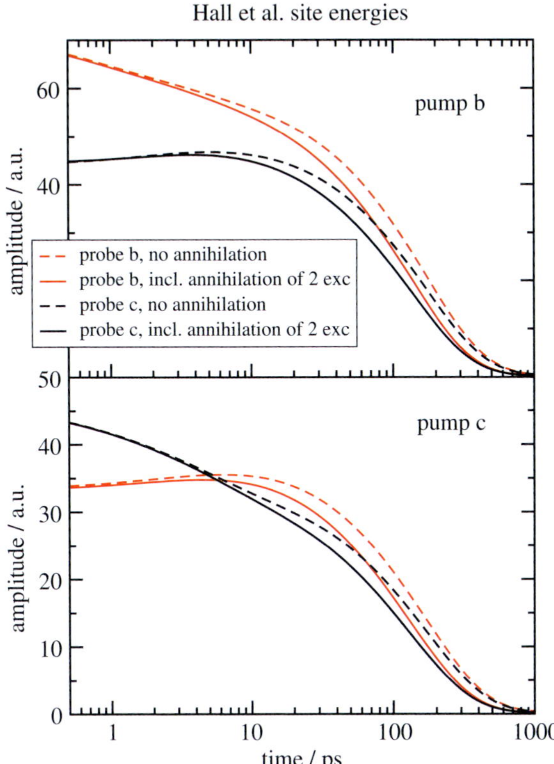

Figure 7. Same calculations as in Figures 4 and 5, except for a replacement of the site energies of CP47 by recent values of Hall *et al.*[2016]. These values (in corresponding wavelengths in nm) are: Chl 11: 687.5, Chl 12: 665, Chl 13: 672, Chl 14: 672, Chl 15: 662, Chl 16: 676, Chl 17: 676, Chl 21: 668, Chl 22: 669, Chl 23: 666.5, Chl 24: 683, Chl 25: 663, Chl26: 669, Chl 27: 679, Chl 28: 670, Chl 29: 689.5.

4. Femtosecond Infrared Crystallography of Exciton Dynamics

The femtosecond transient IR crystallographic measurements [Sage *et al.*, 2011] were performed with visible excitation (680 nm, 100 fs, 2 W/cm²) polarized along either of the two longest crystal edges, which corresponded to polarization directions parallel to the b- and c-axes with propagation along the crystallographic a-axis. The probing light (6 μm, 100 fs, 180 cm⁻¹ bandwidth) was much weaker and was polarized either parallel or orthogonal relative to the excitation.

Thus for each crystal, four time-resolved spectra were acquired at every delay point, and ultrafast pump-probe spectral data acquisition was conducted between 0 and 1500 ps.

Global analysis of the four datasets was performed to retrieve time constants. Generally, with soft limits, the measurements could be represented with three time constants with representative values of 3 ps, 61.5 ps and 832 ps (Figure 8). The 3 ps time constant describes the decay of the initial photoexcited population and represents intra-subunit exciton equilibration [Raszewski and Renger, 2008] and, because of the high light intensities applied, contains also exciton annihilation [Müller *et al.*, 1996]. The 61.5 ps time constant corresponds primarily to energy transfer processes between the subunits, in agreement with previously reported assignments [Raszewski and Renger, 2008; Shibata *et al.*, 2013] and the P_{680}^+/Pheo$^-$ charge separated state develops during the 832 ps phase [Pawlowicz, *et al.*, 2007].

Thus, the spectral and infrared crystallographic analysis of the dominant 61.5 ps time constant that describes the overall energy transfer process is central in describing the overall nature of energy transfer. Spectral analysis of transient infrared absorption in the 6 μm region for the early 3 ps and 61.5 ps time constants shows dominant contributions from excited state chrolophyll *a* response, including both ground state bleaches and induced excited state absorption. In agreement with previous TR-IR measurements reported for solutions of PSII cores, excited state absorption is very prominently seen, both as a result of frequency downshift relative to ground state bleaching, as well as a considerable gain in cross-section. Band-fitting procedures show several contributions to excited state signals, with a dominant difference signal at 1622(+)/1676(–) cm^{-1}. The 61.5 ps spectra were modelled using three pairs of negative and positive bands. The lowest energy bleach was paired with the lowest energy induced absorption and so on. Thus three different groups of Chl *a* molecules are assumed to be present in the sample. The position of each of the bleach and induced absorption was determined approximately and was allowed to vary by ± 7 cm^{-1} during the fitting. In addition to the Chl *a* 13-1-keto C=O bands several more bands were used in the modelling, mostly in the lower energy region (< 1610 cm^{-1}). The assignment of these bands is uncertain and they were not used in the following analysis. Furthermore three weak narrow bands were added to the model in the 1657–1672 range. These bands most likely correspond to the absorption of the emerging charge separated state as they become more prominent in the 832 ps spectrum. A further restriction was implemented to the amplitudes of the Chl *a* bands. The anisotropy of the paired negative and positive bands was kept equal.

The globally fitted spectra indicate a significant photoselection effect for excited state keto. Specifically, the b-polarised induced absorption of excited C=O keto modes near ~1625 cm^{-1} maximise with b-polarised 680 nm excitation in the 3 ps spectra, whereas the c-polarised induced absorption near ~1630–1640 cm^{-1}

maximises with c-polarised 680 nm excitation. While this photoselection is maximal at early time, remaining photoselection is seen also for later time, in the 61.5 ps spectra. As the induced absorptions are almost spectrally isolated, it is possible to consider and analyse single-wavelength kinetic traces, as indicative of polarised excited state response. Figure 2 shows that the initial dichroic amplitudes invert with polarisation direction of the 680 nm excitation, and the general decay profile, including remaining dichroic differences that are maintained at the ~60 ps time scale, that is indicative of the overall energy transfer process to the reaction center. These experimental traces can be compared with a full simulation. As described in the section above, structure-based simulations were done for the crystallographic case that were subsequently analysed in the local basis. Time-dependent concentration profiles of all 74 chlorin pigments in the dimer were multiplied with their corresponding real-space transition dipole orientations — squared in order to find the individual contributions of excited state keto mode absorptions. In the Figure 2, these are all summed in the b- and c-directions, thus neglecting frequency assignment of individual keto oscillators. The comparison with the experimental dichroic kinetics of dominant modes shows a qualitative correspondence between experiment and structure-based theory.

An analysis of the time dependence of the probability that individual chlorophyll pigments contribute to the excited state keto mode dichroic amplitudes has been presented [Kaucikas *et al.*, 2016]. Firstly, a probability can be expressed on the basis of experimental dichroic amplitudes only.

We define a unit vector parallel to TDM as μ_i (where $i = 1..N$, N — number of molecules) and its projections on b- and c-axes correspondingly as μ_{ib} and μ_{ic}. To evaluate the relative contribution of each molecule to the absorption, we introduce an effective TDM μ_E that represents the superposition of all individual absorption contributions. Then the experimentally measured dichroic ratio R between c-polarised probe beam absorption (A_c) and b-polarised probe beam absorption (A_b) is equal to

$$R = \frac{A_c}{A_b} = \frac{\sum_i \mu_{ic}^2}{\sum_i \mu_{ib}^2} = \frac{\mu_{Ec}^2}{\mu_{Eb}^2} \ . \tag{15}$$

Here μ_{Eb} and μ_{Ec} are the projections of μ_E on the b- and c-axes correspondingly. The squared magnitude of the effective vector μ_E is

$$\mu_E^2 = \mu_{Eb}^2 + \mu_{Ec}^2 = \sum_i \mu_{ib}^2 + \sum_i \mu_{ic}^2$$

Then, for measurements on the {1 0 0} plane the relative contribution of each individual molecule is proportional to the scalar product of the squared dipole projection in the b-c plane, μ_{ibc}, with the absorbance vector $A_{bc} = (A_b, A_c)$

$$p_i = \overrightarrow{\mu(i)^2_{bc}} \cdot \overrightarrow{A_{bc}} = \frac{\mu(i)^2_b \, A_b + \mu(i)^2_c \, A_c}{\sqrt{\mu(i)^4_a + \mu(i)^4_b + \mu(i)^4_c} \, \sqrt{A^2_b + A^2_c}}. \tag{16}$$

The relative contributions of each pigment C_i equals the ratio of p_i and the sum of all p_{-i},

$$C_i = \frac{p_i}{\sum\limits_{k=1}^{n} p_k}$$

but the equation above are normalised to the maximal possible vector projection for convenience. This projection furthermore incorporates the ambiguity for the two possible dipole directions ($2\pi - \phi$ and ϕ) for the azimuthal angle ϕ that results in the same values for R and p_i due to taking the square of the dipole component [Sage *et al.*, 2011].

The dichroic amplitudes can be analysed independently, and have additionally been combined with theoretical calculation as outlined in Section 3. A useful analysis evaluates as $\Delta_i(t) = N_i(t) * p_i$, which corresponds to a probability amplitude that includes both experimental keto-mode contribution p_i and calculated population N_i. The dominant ~ 61 ps time constant that represents the overall energy transfer process can be graphically represented to indicate loss and gain amplitude in the single pigment basis, for the two polarisation directions of the optical field. This requires that amplitudes are scaled according to the overall excited state population decay between 3 ps and 61 ps. The analysis of the single pigment loss and gain amplitudes in Figure 9 provides evidence for net energy transfer between CP43 and CP47 with a time constant of ~60 ps, as predicted for closed reaction centers (Raszewski and Renger, 2008).

5. Structural Measurement of the Charge Separated State $P_{680}^+/Pheo^-$

With a ~ 800 ps time constant, the formation of the charge separated state $P_{680}^+/Pheo^-$ is observed (Figure 8) Since the $P_{680}^+/Pheo^-$ difference spectrum only has contributions from two copies in the asymmetric unit, a specific structural measurement was possible on the basis of band-fitting which was seen to fit the X-ray coordinates well using published vibrational mode assignments for the P_{680} and pheophyin contributions. There are pronounced differences between absorption of the b-polarised and c-polarized probe in the 1690–1740 cm^{-1} region. The spectra no longer differ for two excitation polarizations, and the yield of charge separation is the same independent of the differing original excited state populations prepared with b- or c-polarised excitation. This observation also supports the notion that photosynthetic light harvesting and charge separation is highly efficient and

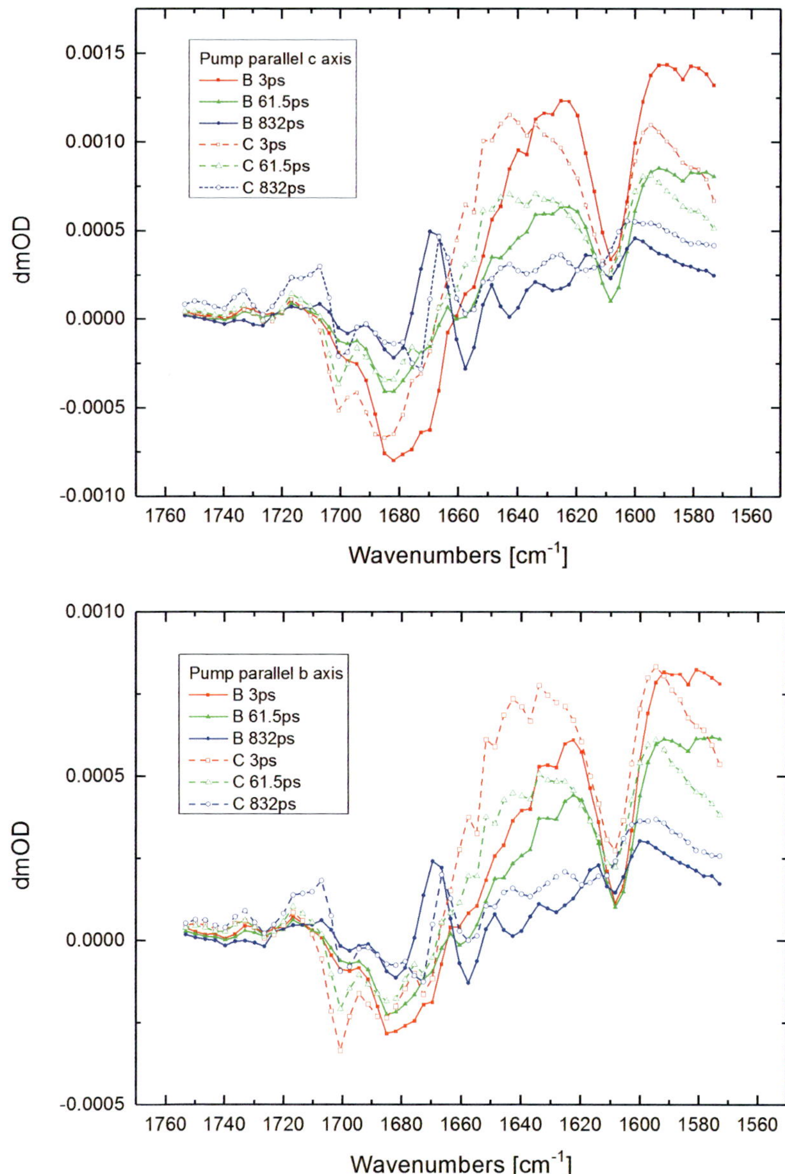

Figure 8. Global Analysis results of the PSII crystal TRIR. The results for the case when pump is parallel to c-axis are presented left, when pump is parallel b-axis — right. Solid line spectra correspond to probe parallel to b-axis, dashed — parallel to c-axis in both panels. For each case there are three spectra corresponding to 3 ps, 61.5 ps and 832 ps time constants.

104 M. Kaucikas et al.

adaptive under conditions of modified initial excitation. Furthermore, because at this stage both monomers in the crystallographic asymmetric unit are equally populated, there exists a fully developed non-crystallographic symmetry.

A band-fitting analysis was based on previously published FTIR difference spectra and assignments for P_{680}^+/P_{680} [Okubo *et al.*, 2007] and Phe$^-$/Pheo [Shibuya *et al.*, 2010]. A set of bands were simultaneously fitted to the P_{680}^+/P_{680} FTIR, Pheo$^-$/Pheo FTIR, the polarization resolved P_{680}^+/Pheo$^-$ crystal spectra as well as the P_{680}^+/Pheo$^-$ spectrum of isotropic sample acquired under identical conditions as crystallographic measurements. There is generally very good agreement between set of bands needed to fit TRIR and FTIR spectra. Both Chl *a* 13-1-keto C=O keto and Pheo 13-1-keto C=O as well as 13-2-ester C=O bands can be reliably fitted. The corresponding theoretically calculated TDM directions have been presented previously [Kaucikas *et al.*, 2016]. However, comparison of the experimental results with the theoretical calculations is complicated by the presence of

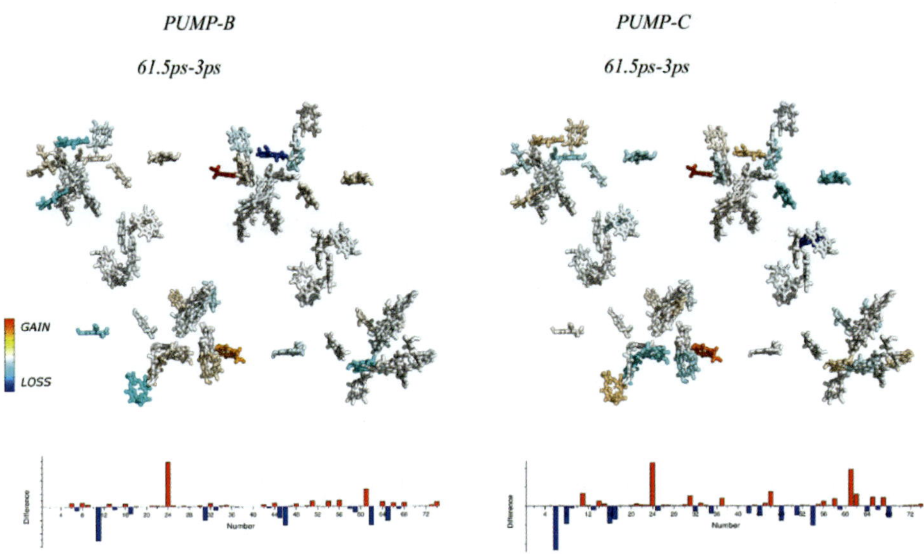

Figure 9. Probability amplitudes representing loss and gain of excited state population from decay-adjusted Δ_i (t) differences between 3 ps and 60 ps. The results can be summed for the following domains: Domain 1 (CP47-Chlz$_{D2}$) Pumping B: + 1.27 (gain), Pumping C: +1.05 (gain). Domain 2 (CP43-Chlz$_{D1}$) Pumping B: −1.77 (loss), Pumping C: +0.55 (gain). Domain 3 (D1D2 minus Chlz) Pumping B: +0.33 (gain), Pumping C: −1.32 (loss). Note that these values (pi times population) indicate net transfer from CP43 to CP47 but with c-excitation has high initial photoselection of pheophytin (but not in b-direction) which dominates the loss in domain 3. Note also that with pumping the c-direction the calculated population is significantly smaller, about 47% that of the population with pumping the b-direction.

non-crystallographic symmetry. As a result there are two copies of each TDM contributing to the every absorption band in the sets used to fit the spectra.

For example in the case of the 1706.5 cm^{-1} band that was assigned to P_{D1} 13-1-keto C=O cation vibration, there is a P_{D1} molecule in both monomers of the crystallographic asymmetric unit, 30 each with its distinct TDM directions. Thus the experimental dichroic ratio R for this band will reflect the direction of the vector sum of the two TDMs.

Due to this complication, further analysis had to rely on the numerical procedure that is described below. Briefly, for each vibrational band, the calculation looked for all possible orientations of TDM that satisfies two conditions. First, the angle between the two copies of the same TDM had to be equal to the equivalent angle from DFT calculations. The second condition required was that the R for the sum of the TDMs must be equal to the experimental value. The result of this analysis are assigned correspondingly to P_{D1} 13-1-keto C=O cation and Pheo 13-2-ester C=O anion vibrations. The results show that, even in the presence of non-crystallographic symmetry, an analysis of the three-dimensional orientations of vibrational TDM is possible.

The spectrum corresponding to decay constant 832 ps mainly represents the absorption of the charge separated state in the reaction centre. The state consists of $Pheo_{D1}^{-}$ anion and P_{680}^{+} cation [Pawlowicz *et al.*, 2007]. Pawlowicz and co-authors noted that it is not expected for Q_A to contribute to the charge separated IR spectrum, because due to the relatively high illumination levels, Q_A is in the reduced state [Di Donato *et al.*, 2008; Pawlowicz *et al.*, 2007]. Modelling of this spectrum is rather complicated, because of several overlapping positive and negative bands. To aid the modelling of the crystal spectra, FTIR spectra of Pheo$^-$/Pheo and P_{680}^{+}/P_{680} as well as the pump probe spectrum of the liquid PSII sample were used. As the vibrational spectra of different PSII preparations can vary, only the FTIR spectra from the *T. elongatus* with gene PsbA1 were used in further analysis. The Pheo$^-$/Pheo FTIR spectrum was provided by the authors of Shibuya *et al.* [2010] and the P_{680}^{+}/P_{680} spectrum by the authors of Okubo *et al.* [2007]. The liquid sample pump probe spectra were obtained with the current setup and in identical conditions as in the crystal spectra case. Furthermore, the focus of the following analysis was restricted to the 1690-1730 cm^{-1} region as it is well investigated using both FTIR and ultrafast spectroscopy methods.

5.1 *Vibrational mode assignments of the* P_{680}^{+}/P_{680} *and Pheo$^-$/Pheo spectra*

For mid-IR TRIR of PSII, the most investigated spectral region is the 1650–1750 cm^{-1} range. This is due to the fact that it contains absorption lines corresponding to the keto and ester C=O vibrations of Chl *a* and Pheo molecules. As these vibrations are usually localized on a single molecule and the centre frequency of each transition is indicative of the environment the molecule is in, there is a potential to track each excitation state of each separate molecule and monitor its change in time. However, protein amide I band can also contribute to the absorption spectrum in this region. Below a brief review of the vibrational band assignment in this region is presented.

One of the first analyses of possible band assignments of P_{680}^{+}/P_{680} spectrum in this region was carried out by Noguchi *et al.* [1998]. The assignment of the bands in FTIR difference spectrum of PSII RC sample from spinach was made by comparing it with resonant Raman (RR) spectra of the same sample. Based on this comparison, the authors identified the negative bands at 1679 and 1704 cm^{-1} in their spectrum as belonging to the keto C=O modes of neutral Chl or Pheo molecules. Furthermore, based on structural information, they proposed that the 1704 cm^{-1} band corresponds to both 13-1-keto C=O vibrations of P_{680}, but the authors also noted that this assignment was not conclusive. The bands at 1711 and 1729 cm^{-1} were proposed as candidates for 13-1-keto C=O modes of P_{680}^{+}. A strong negative band at 1654 cm^{-1} was assigned to amide I vibration because there was no corresponding band in the RR spectrum. Okubo *et al.* [2007] measured P_{680}^{+}/P_{680} FTIR difference spectra for several different PSII preparations, namely spinach PSII membranes, *T. elongatus* core complexes and spinach reaction centres. Negative peaks at 1701 and 1680 cm^{-1} were consistent between all three preparations with the 1701 cm^{-1} band showing a much bigger amplitude than the one at 1680 cm^{-1}. The authors assigned the stronger band to the 13-1-keto C=O mode of both neutral P_{D1} and P_{D2}. Their argument against assigning one of these modes to the 1680 cm^{-1} band is based on missing corresponding equally shifted positive band. Furthermore, the authors claim that peak location would indicate highly polar environment or a weak hydrogen bond and they did not observe this situation in the X-ray crystallographic structure. The positive peaks at 1711 and 1724 cm^{-1} were assigned to P_{680}^{+} keto modes. The authors note that the two peaks visible in membrane and core complex spectra merge into one in the reaction centre spectrum. They explain this fact by the charge redistribution between P_{D1} and P_{D2} in this sample compared to the other two. The negative band at 1736 is assigned to the 13-3-ester C=O vibration of neutral P_{680} and positive bands at 1743 and 1750 cm^{-1} to the cation P_{680}^{+}. Di Donato *et al.* [2008] also reported FTIR difference spectrum of PSII core complexes from a *Synechocystis* cyanobacterium. In contrast to previous findings, the negative bands at 1681 and 1699 cm^{-1} appeared to be of the same magnitude. This led the authors to suggest that the 1681 cm^{-1} band should be assigned to P_{D2}. From the X-ray crystallographic structure, they suggest that a particular hydrogen bond could be formed to lower the frequency of the absorption band. Furthermore, the same paper reported the FTIR spectrum of a mutant which was known to change charge distribution in P_{680}^+. This mutation seems to affect not only the 1709 and 1724 cm^{-1} bands usually assigned to the cation 13-1-keto C=O mode, but also both the 1699 and 1681 cm^{-1} bands. The authors suggested that this points to the fact that 1681 cm^{-1} also corresponded to P_{680} 13-1-keto C=O mode. The most thorough study aimed at determining the assignments of PSII vibrational bands was performed by Romero *et al.* [2011]. First a global N^{15} labelling of PSII reaction centre from spinach was performed. As protein modes were expected to downshift upon isotope exchange, this allowed identifying which FTIR bands belonged completely or were coupled to protein vibrations. The isotope exchange should not affect the modes corresponding to Chl *a* or Pheo vibrations as verified by DFT calculations performed by the authors. In the case of the P_{680}^+/P_{680} FTIR spectrum, the shift upon N^{15} labelling of the bands in the 1650–1750 cm^{-1} region is negligible. The only bands to shift by 1 cm^{-1} were those at 1720 cm^{-1} (positive) and 1702 cm^{-1} (negative). Thus the contribution of protein absorption in this region is negligible and the bands belong exclusively to P_{680} or P_{680}^+. While this confirms previous assignments for the bands in the 1670–1750 cm^{-1} region, the bands at 1667, 1652 and 1645 cm^{-1} were also attributed to P_{680}, contrary to the suggestions in earlier studies. However, the authors did not attempt to assign these bands to any specific vibration. Furthermore Romero [2011]. studied how the P_{680}^+/P_{680} difference spectrum changes for different spinach PSII preparations with varying antenna size, namely membrane, core complex and reaction centre. It was observed that the spectra in the 1650–1750 cm^{-1} region were identical for all three cases. This result contradicted the previous result of Okubo *et al.* [2007] that found the doublet at 1711 and 1724 cm^{-1} merge into single band in the case of reaction centre preparation. To check whether this inconsistency was due to the use of different redox mediator, Romero acquired the spectra using several combinations of the mediators, including the one used by Okubo *et al.* However, the resulting spectra still showed two peaks. The authors conclude that the charge distribution in the case of reaction centre preparations is the same as in more intact preparations. The same result was observed in core complex and reaction centre preparations of PSII from a *Synechocystis* cyanobacterium. However, Romero noted that in this case the negative band at 1681 cm^{-1} became much more prominent when compared to the corresponding spectra of spinach preparations. Based on previous assignment of this band to P_{D2} neutral 13-1-keto C=O mode, the authors suggested that the protein environment of the carbonyl groups of P_{D1} and P_{D2} is more asymmetric in this case. A summary of all the $P_{680}+/P_{680}$ vibration band assignments was presented previously [Kaucikas *et al.*, 2016].

In the case of Pheo$^-$/Pheo, Shibuya *et al.* [2010] measured the FTIR spectrum for different preparations derived from spinach and *T. elongatus* [10]. They proposed that the neutral state 13-1-keto C=O vibration corresponds to the band at around 1680 cm^{-1} for all preparations. Anion state absorption however is upshifted substantially and the shift depended on the hydrogen bonding strength. The observed band position varied from 1603 to 1587 cm^{-1}. The situation in the 1700–1740 cm^{-1} region is less clear. Shibuya *et al.* proposed that there are two sets of Pheo molecules in the reaction centre. The molecules with hydrogen bonding have 13-2-ester C=O neutral state band at 1720 cm^{-1} and the anion state band at 1699 cm^{-1}. On the other hand molecules that are hydrogen bond free have corresponding bands at 1741 and 1727 cm^{-1}. The authors did not determine what could be the cause of this heterogeneity. The previously mentioned study by Romero *et al.* [2011] also investigated Pheo$^-$/Pheo mode assignments. As in the case of P$_{680}$, global N^{15} labelling was performed and difference FTIR spectra were recorded. It was observed that none of the bands in the 1670–1750 cm^{-1} region shifted upon labelling, indicating that all the vibrational bands in this region belong to Pheo or Pheo$^-$. On the other hand bands in the 1640–1670 cm^{-1} region downshifted by 1 cm^{-1} and lead the authors to conclude that these bands belong to amide I modes, perturbed by Pheo$^-$ formation.

Furthermore, Romero investigated the effect of hydrogen bonding strength between 13-1-keto carbonyl of Pheo and glutamine residue in position D1-130 for cyanobacterium *Synechocystis* reaction centre. Three mutants were produced that had different residue in this position. The residues formed hydrogen bonds of different strength, varying from none in the case of the D1-Gln130Leu mutant to strong in the D1-Gln130Glu mutant.

As in previous study by Shibuya *et al.* [2010], the negative band at 1739 cm^{-1} was assigned to free 13-2-ester C=O vibration of Pheo. The authors note, that upon reduction this band downshifts with the size of the shift depending on the amino acid in position D1-130. The second negative band in this region — 1723 cm^{-1} was assigned to hydrogen bonded 13-2-ester C=O vibration of Pheo. The authors propose that Tyr126 could be the candidate for establishing the hydrogen bond with 13-2-ester C=O. However, they note that mutation at D1-130 also affected the position of this band and its relative intensity. Thus the authors suggest that the residue at D1-130 influenced the hydrogen bond network around 13-2-ester C=O of Pheo and maybe even formed a bond itself. The negative band at 1678 cm^{-1} was assigned to 13-1-keto C=O vibration of Pheo confirming a previous assignment by Shibuya *et al.* [2010]. Its position was constant for all mutants with varying hydrogen bond strength. However, the anion absorption band shift was very dependent on the bond strength. The positive band was at 1630 cm^{-1} for

D1-Gln130Leu that did not form a hydrogen bond, 1603 cm^{-1} for wild type and 1587 cm^{-1} for D1-Gln130Glu which corresponded to a strong hydrogen bond. Consequently the authors proposed that 13-1-keto C=O of Pheo is hydrogen bonded only in the reduced state. However, the presence of a weaker 1630 cm^{-1} band in all mutant spectra could indicate that there were two populations of Pheo molecules, thus confirming the interpretation of 13-2-ester C=O vibration assignment and consistent with suggestions of Shibuya *et al.*

5.2 *Fitting the P_{680}^{+}/P_{680} and Pheo$^-$/Pheo FTIR spectra*

The derivative method [Giese and French, 1955] was used to identify the band positions in the FTIR spectrum for both Pheo$^-$/Pheo and P_{680}^{+}/P_{680} spectra. Then a set of Gaussian bands at these positions was used to model the FTIR spectra by varying their widths and positions within several cm^{-1} range. The fit of the FTIR spectra with set of Gaussian bands is presented in the Figure 11. As expected in P_{680}^{+}/P_{680} spectrum the strongest bands belong to 13-1-keto C=O vibrations — negative at 1700 cm^{-1} and two positive at 1706.5 and 1723.5^{-1} A small band corresponding to the 13-3-ester C=O vibration was observed at 1742 cm^{-1}. The fit results of the Pheo$^-$/Pheo FTIR spectrum presents a more complicated situation. While the main 13-2-ester C=O bands at 1701 cm^{-1} (positive), 1722 cm^{-1} (negative), 1733 cm^{-1} (positive) and 1741 cm^{-1} (negative) are observed as expected, additional lines had to be added to produce acceptable fit results. The assignment of the added 1692 cm^{-1} and 1714 cm^{-1} bands is currently unknown.

5.3 *Fitting the 832 ps TR-IR crystallography spectra*

The resulting set of bands served as a starting point for the modelling of the 832 ps spectrum. An additional restriction was introduced concerning the relative amplitude of individual bands. The common amplitude multiplier for all bands coming from the same FTIR spectrum (either Pheo or P_{680}) could vary freely, but the multiplier of each individual band amplitude was restricted to the 0.5–2 range. This restriction helped avoiding physically unrealistic modelling results and maintained relative proportionality of the fitted bands. The result of the modelling of charge separated state spectra with the set of bands obtained above is presented in the Figure 10 . Some of the bands had to be moved relative to their positions in the FTIR spectrum, but the adjustment never exceeded 2 cm^{-1}. This refinement was needed because of different conditions for sample preparation in the crystal case and the better resolution of the crystal spectra. To confirm that the need for band adjustment was not due to crystalline nature of the current sample, a pump probe

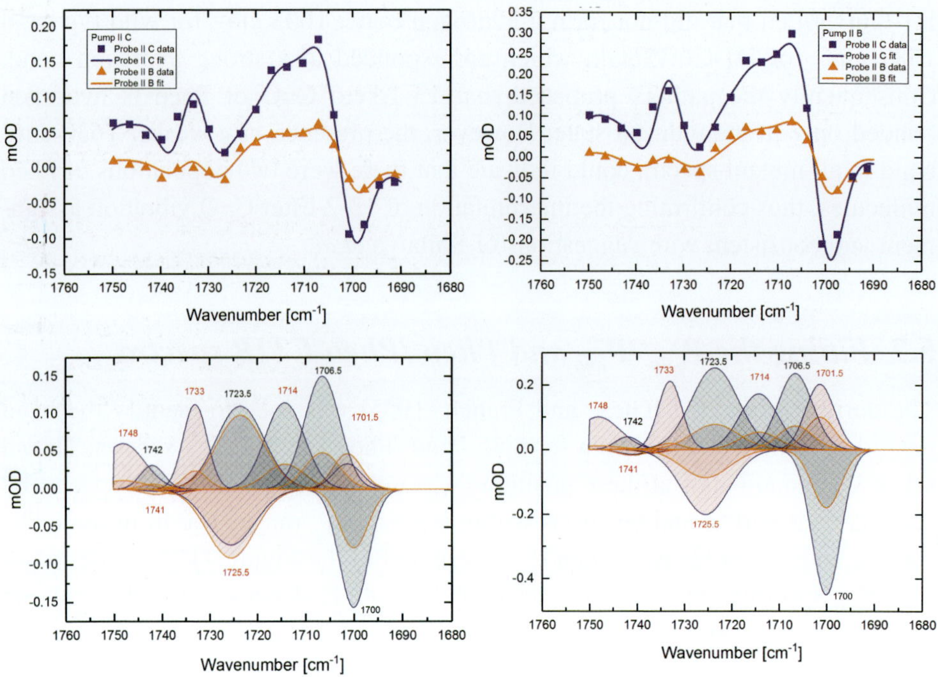

Figure 10. Results of band fitting the 832 ps spectra. In the left part the pump is parallel to the c-axis and in the right part — to the b-axis. In the upper panel the points represent Global Analysis 3 ps spectrum and solid line — band fitting result for the two probe polarizations: parallel to the c-axis (orange), parallel to b-axis (purple). The lower panel shows bands used to fit the spectrum. There are two versions of each band corresponding to the two spectra in the upper panel as indicated by the color of the band.

spectrum of the liquid PSII sample was modeled using exactly the same bands as in crystal case. The result clearly demonstrates that a very good fit could be achieved without the need for adjustment of the bands.

6. Implications for Light Harvesting Function of PSII Core Complexes

Since the 1980-ies physical models of light harvesting and charge separation have included kinetic schemes that place the overall bottleneck at the position of the initial charge separation. Known as the "exciton radical pair equilibrium model (ERPE)", initially developed by Holzwarth and co-workers [Schatz *et al.*, 1988], the assumption is that energy transfer is intrinsically very fast and the process of initial charge separation is slow, reversible and has small free energy difference — thus is shallow. In 2001 the first X-ray crystal structural details were reported for PSII, that indicated generally long distances between antenna

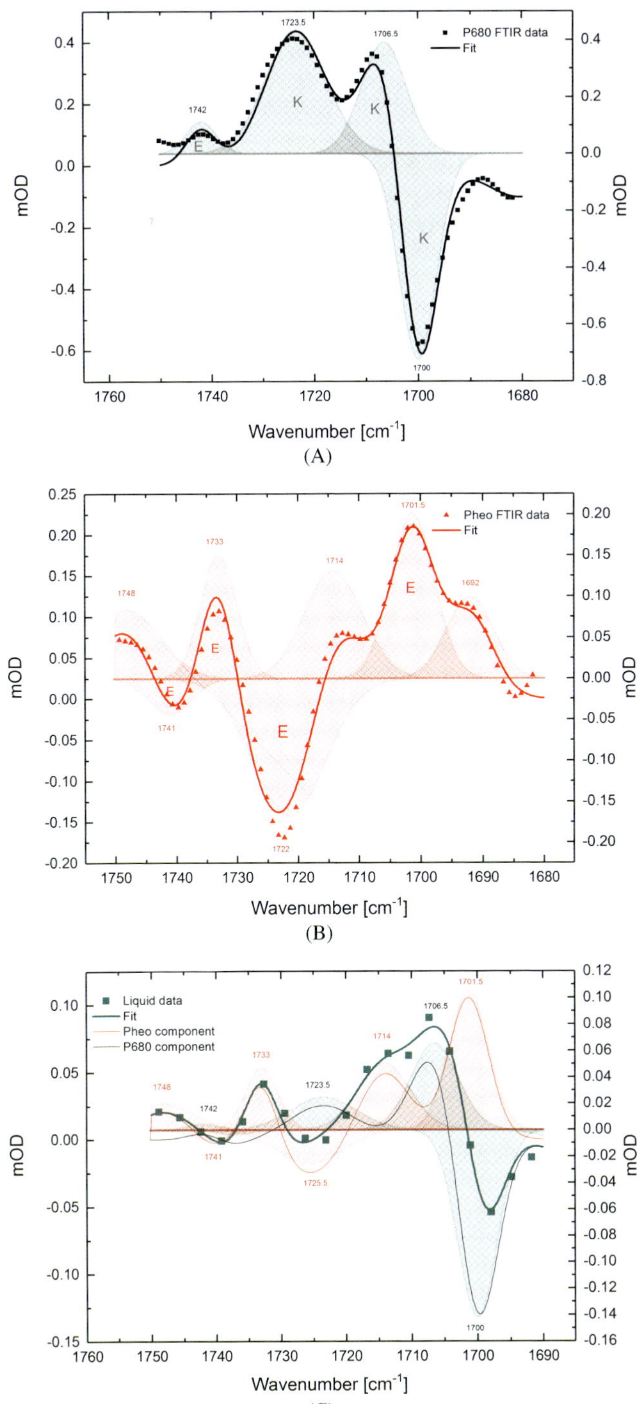

Figure 11. (A) Difference FTIR spectrum of P_{680}^+/P_{680} from (Ref.9) and its fit using set of Gaussian bands. (B) Difference FTIR spectrum of Pheo$^-$/Pheo from (Ref. 10) and its fit using set of Gaussian bands. (C) 1 ns TRIR difference spectrum of isotropic PSII sample and its fit using set of Gaussian bands.

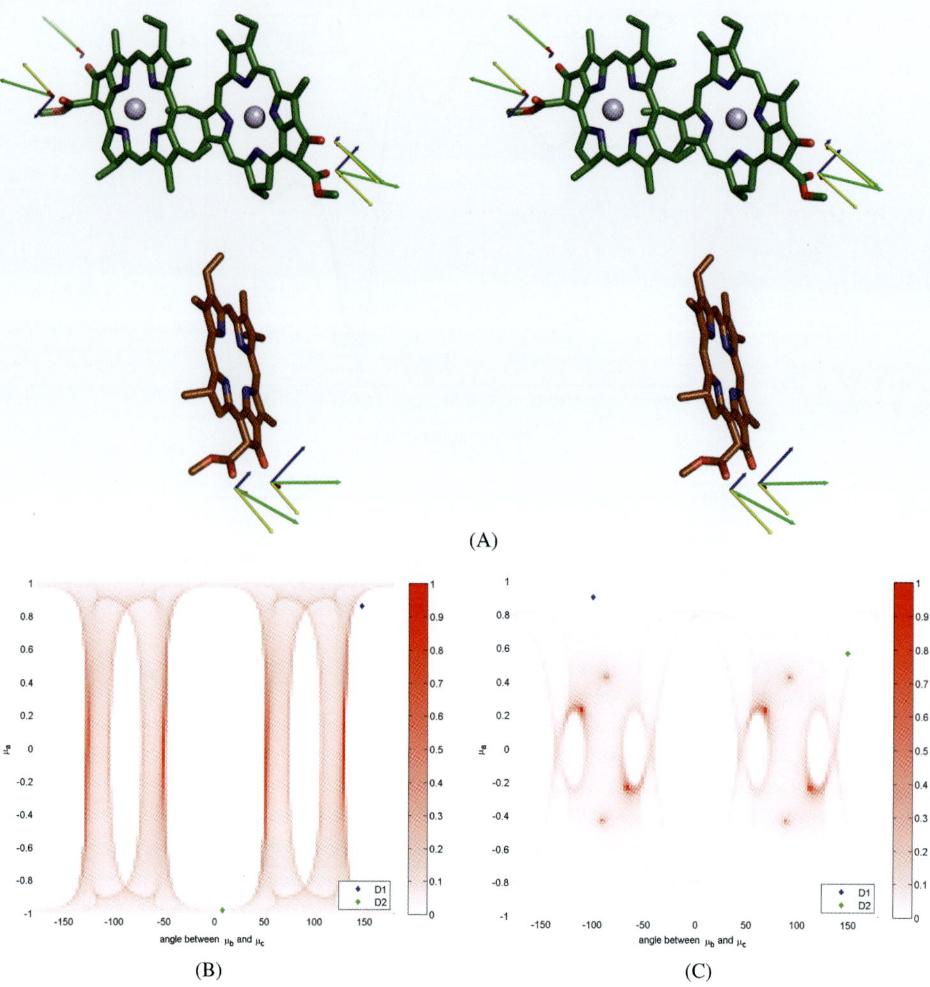

Figure 12. (A) Stereo figure of the 13-1 keto and ester TDM of the P_{680}^+/Pheo$^-$ C=O stretching modes. Shown are the non-crystallographic sums of squared TDM projections $\Sigma(\mu_a)^2$ (yellow), $\Sigma(\mu_a)^2$ (red) and $\Sigma(\mu_c)^2$ (blue) and the $\Sigma\mu$ (green),

pigments and the reaction centre [Zouni *et al.*, 2001]. From first principles it was proposed already in 2001 that the distances seen in modeled structures would not support the ultrafast rates of energy transfer that are implicit in the "ERPE" model [Vasil'ev *et al.*, 2001]. Since then, crystal structures of PSII cores have been reported with increasing resolution.

Coinciding with the increasingly accurate structural detail of complexes, support has grown for a physical model that places the overall bottleneck of light harvesting at the transfer step to the reaction center core, also known as the "transfer-to-the-trap" limited model. A definitive experimental test is provided from ultrafast infrared crystallography for the first time, where previously only structure-based theory supported the case for the "transfer-to-the-trap" limited model [Raszewski and Renger, 2008].

The IR measurements of the exciton annihilation and energy transfer stages of Photosynthesis show that b-polarized and c-polarized excitation results in very different excited state populations and the time dependent analysis of the dichroic amplitudes are possible from the discovery of the local mode character of the vibrational TDMs of the keto C=O stretching mode of Chl *a*. The measurements also demonstrate the flexibility of the light harvesting and charge separation processes that enable electron transfer with very high quantum efficiency with initial populations prepared with either b- or c-polarized excitation. Without structural information it would not be possible to distinguish between the two light-harvesting/charge transfer models solely from the fluorescence decay. In contrast, the calculation of VIS pump- IR probe spectra (Figure 3) reveals significant differences of the two models. Due to the faster exciton transfer in the ERPE model, the system "forgets" the polarization of the initial excitation much faster than in the structure-based transfer-to-the-trap limited model. The dichroic amplitude difference of the pump-probe signal on oriented single crystals is practically absent in a model that simulates the "ERPE" case, but is clearly present in the structure-based model. The key observation that the experimental dichroic amplitudes differ significantly at time-zero, and that differences are maintained for the duration that includes the dominant ~ 60 ps energy transfer reactions, shows that only the structure-based transfer-to-the-trap limited model can at least qualitatively explains the experimental data. The present experiments provide the closest approach to visualize exciton relaxation in time and space so far. Convincing evidence is obtained for 50–100 ps exciton equilibration between CP43 and CP47 subunits across the intervening reaction center in PSII core complexes for closed reaction centers. In the latter case excitation energy is proposed to accumulate at Chl 29 in CP47 [Raszewski and Renger, 2008; Shibata *et al.*, 2013] or at Chl 11 [Hall *et al.*, 2016] which are situated at the periphery of the complex and might represent photoprotective sites, where the excitation energy is quenched. In the case of open reaction centers, excitons do not equilibrate between CP43 and CP47 but are trapped by electron transfer in the reaction center [Raszewski and Renger, 2008]. The bottleneck, however, remains the same, namely transfer between the core antennae and the reaction center. As noted before, this slow transfer allows the reaction center to trap every exciton arriving in the reaction center by ultrafast electron transfer. The switch of the core complex into the photo-protective mode for closed reaction centers is triggered by the slowing down of primary charge transfer and promoted by the large entropy of excited antenna states that gives rise to a factor three larger rate constant of energy transfer from the RC to the antenna subunits than vice versa [Raszewski and Renger, 2008].

Furthermore, we consider that our results for energy transfer in PSII are also in agreement with a recent structure-based theoretical study [Bennett *et al.*, 2013] on light-harvesting in PSII supercomplexes. In this study an effective time constant of 100 ps has been inferred for transfer between the core antennae and the reaction center and an effective time constant of 110 ps for exciton diffusion in the whole super complex, containing peripheral light-harvesting complexes besides the core antennae [Bennett *et al.*, 2013]. Similar to the arguments developed in our study on PSII core complexes, in the latter study it was emphasized that a fit of time-resolved fluorescence data by kinetic models, without taking into account structural details, can lead to ambiguous conclusions about mechanistic details of the light-harvesting reaction. The present experiment removes this ambiguity for PSII core complexes.

References

Bennett, D.I.G., Amarnath, K. and Fleming, G.R. (2013). A structure-based model of energy transfer reveals the principles of light harvesting in photosystem II supercomplexes, *J. Am. Chem. Soc.*, 135, 9164–9173.

Born, M. and Wolf, E. (1999). *Principles of Optics: Electromagnetic Theory of Propagation, Interference and Diffraction of Light*, 7th ed. (Cambridge University Press, Cambridge).

Di Donato, M., Cohen, R.O., Diner, B.A, Breton, J., van Grondelle, R. and Groot, M.L. (2008). Primary charge separation in the photosystem II core from synechocystis: a comparison of femtosecond visible/midinfrared pump-probe spectra of wild-type and two P680 mutants, *Biophys. J.*, 94, 4783–4795.

Ferreira, K.N., Iverson, T.M., Maghlaoui, K., Barber, J. and Iwata, S. (2004). Architecture of the photosynthetic oxygen-evolving center, *Science*, 303, 1831–1838.

Giese, A.T. and French, C.S. (1955). The analysis of overlapping spectral absorption bands by derivative spectrophotometry, *Appl. Spectrosc.*, 9, 78–96.

Hall, J., Renger, T., Müh, F., Picorel, R. and Krausz, E. (2016). The lowest-energy chlorophyll of photosystem II is adjacent to the peripheral antenna: emitting states of CP47 assigned *via* circularly polarized luminescence, *Biochim. Biophys. Acta*, 1857, 1580–1593.

Hall, J., Renger, T., Picorel, R. and Krausz, E. (2016). Circularly polarized luminescence spectroscopy reveals low-energy excited states and dynamic localization of vibronic transitions in CP43, *Biochim. Biophy. Acta*, 1857, 115–128.

Holzwarth, A.R., Müller, M.G., Reus, M., Nowaczyk, M., Sander, J. and Rögner, M. (2006). Kinetics and mechanism of electron transfer in intact photosystem II and in the isolated reaction center: pheophytin is the primary electron acceptor, *Proc. Natl. Acad. Sci. USA*, 103, 6895–6900.

Kaucikas, M., Barber, J. and Van Thor, J.J. (2013). Polarization sensitive ultrafast mid-IR pump probe micro-spectrometer with diffraction limited spatial resolution, *Opt. Express*, 21, 8357–8370.

Kaucikas, M., Maghlaoui, K., Barber, J., Renger, T. and van Thor, J.J. (2016). Ultrafast infrared observation of exciton equilibration from oriented single crystals of photosystem II, *Nat. Commun.*, 7, 13977.

Loll, B., Kern, J., Saenger, W., Zouni, A. and Biesiadka, J. (2005). Towards complete cofactor arrangement in the 3.0 A resolution structure of photosystem II, *Nature*, 438, 1040–1044.

Müller, M.G., Hucke, M., Reus, M. and Holzwarth, A.R. (1996). Annihilation processes in the isolated D1-D2-cyt-b559 reaction center complex of photosystem II. An intensity-dependence study of femtosecond transient absorption, *J. Phys. Chem.*, 100, 9537–9544.

Noguchi, T., Tomo, T. and Inoue, Y. (1998). Fourier transform infrared study of the cation radical of P680 in the photosystem II reaction center: evidence for charge delocalization on the chlorophyll dimer, *Biochemistry*, 37, 13614–13625.

Okubo, T., Tomo, T., Sugiura, M. and Noguchi, T. (2007). Perturbation of the structure of P680 and the charge distribution on its radical cation in isolated reaction center complexes of photosystem II as revealed by fourier transform infrared spectroscopy, *Biochemistry*, 46, 4390–4397.

Pawlowicz, N.P., Groot, M.-L., van Stokkum, I.H.M., Breton, J. and van Grondelle, R. (2007). Charge separation and energy transfer in the photosystem II core complex studied by femtosecond midinfrared spectroscopy, *Biophys. J.*, 93, 2732–2742.

Raszewski, G. and Renger, T. (2008). Light harvesting in photosystem II core complexes is limited by the transfer to the trap: can the core complex turn into a photoprotective mode? *J. Am. Chem. Soc.*, 130, 4431–4446.

Renger, T. and Marcus, R.A. (2002). On the relation of protein dynamics and exciton relaxation in pigment–protein complexes: an estimation of the spectral density and a theory for the calculation of optical spectra, *J. Chem. Phys.*, 116, 9997–10019.

Renger, T. and Müh, F. (2013). Understanding photosynthetic light-harvesting: a bottom up theoretical approach, *Phys. Chem. Chem. Phys.*, 15, 3348–3371.

Romero, E. (2011). *The Electronic Structure of Photosystem II: Charge Separation Dynamics*. PhD thesis, VU University Amsterdam.

Sage, J.T., Zhang, Y., McGeehan, J., Ravelli, R., Weik, M. and van Thor, J.J. (2011). Infrared protein crystallography, *Biochim. Biophys. Acta*, 1814, 1–18.

Schatz, G.H., Brock, H. and Holzwarth, A.R. (1988). Kinetic and energetic model for the primary processes in photosystem II, *Biophys. J.*, 54, 397–405.

Schlodder, E., Çetin, M., Byrdin, M., Terekhova, I.V. and Karapetyan, N.V. (2005). P700$^+$- and ^{3}P700-induced quenching of the fluorescence at 760 nm in trimeric photosystem I complexes from the cyanobacterium *Arthrospira platensis*, *Biochim. Biophys. Acta*, 1706, 53–67.

Shibata, Y., Nishi, S., Kawakami, K., Shen, J.-R. and Renger, T. (2013). Photosystem II does not possess a simple excitation energy funnel: time-resolved fluorescence spectroscopy meets theory, *J. Am. Chem. Soc.*, 135, 6903–6914.

Shibuya, Y., Takahashi, R., Okubo, T., Suzuki, H., Sugiura, M. and Noguchi, T. (2010). Hydrogen bond interactions of the pheophytin electron acceptor and its radical anion in photosystem II as revealed by Fourier transform infrared difference spectroscopy, *Biochemistry*, 49, 493–501.

Umena, Y., Kawakami, K., Shen, J.-R. and Kamiya, N. (2011). Crystal structure of oxygen-evolving photosystem II at a resolution of 1.9 Å, *Nature*, 473, 55–60.

Vasil'ev, S., Orth, P., Zouni, A., Owens, T.G. and Bruce, D. (2001). Excited-state dynamics in photosystem II: insights from the x-ray crystal structure, *Proc. Natl. Acad. Sci. USA*, 98, 8602–8607.

Zouni, A., Witt, H.T., Kern, J., Fromme, P., Krauss, N., Saenger, W. and Orth, P. (2001). Crystal structure of photosystem II from *Synechococcus elongatus* at 3.8 A resolution, *Nature*, 409, 739–743.

Chapter 6

Bioenergetics, Water Splitting and Artificial Photosynthesis

James Barber

*Imperial College London, South Kensington Campus
Sir Ernst Chain Bld., SW7 2AZ, UK
j.barber@imperial.ac.uk*

About three billion years ago an enzyme emerged which would dramatically change the chemical composition of our planet and set in motion an unprecedented explosion in biological activity. This enzyme used solar energy to power the thermodynamically and chemically demanding reaction of water splitting. In so doing, it provided biology with an unlimited supply of reducing equivalents needed to convert carbon dioxide into the organic molecules of life while at the same time produced oxygen to transform our planetary atmosphere from an anaerobic to an aerobic state. The enzyme that facilitates this reaction and therefore underpins virtually all life on our planet, is known as photosystem II (PSII). It is a pigment-binding, multisubunit protein complex embedded in the lipid environment of the thylakoid membranes of plants, algae and cyanobacteria. Today we have a detailed understanding of the structure and functioning of this key and unique enzyme. The journey to this level of knowledge can be traced back to the discovery of oxygen itself in the 18th century. Since then there has been a sequence of mile stone discoveries, which makes a fascinating story, stretching over 200 years. However, it is the last few years that have provided the level of detail necessary to reveal the chemistry of water oxidation and O–O bond formation. In particular, the crystal structure of the isolated PSII enzyme has been reported with ever-increasing improvement in resolution. Thus the organisational and structural details of the water splitting site were revealed as a cluster of four Mn ions and a Ca ion surrounded by amino acid side chains, of which seven provide direct ligands to the metals. The metal cluster is organised as a cubane structure composed of three Mn ions (Mn1, Mn2 and Mn3) and a Ca^{2+} linked by oxo-bonds with the fourth Mn ion (Mn4) connected to the cubane by two oxos (O4 and O5). This structure seems to be very robust and thus its catalytic activity is proposed to take place on its surface and involve a nucleophilic attack of a hydroxyl (OH) ligated to the Ca^{2+} on to a highly electrophilic terminal oxo of Mn4 in a high valence state (possibly Mn^V). Thus, the overall structure of the catalytic site has given a framework on which to build a mechanistic scheme for photosynthetic dioxygen generation and at the same time provide a blue-print and incentive to develop catalysts for artificial photo-electrochemical systems to split water and generate renewable solar fuels.

1. Introduction

It is often not appreciated, or realized, that our planet is powered by solar energy resulting from the nuclear fusion reaction occurring in the Sun about 150 million km away. Unlike other planets in the solar system, visible light is absorbed by Earth and its energy made available to do work and create order. For us the energy of the Sun is a source of "negative-entropy". Through the process of photosynthesis, solar energy is captured and stored in chemical bonds of organic molecules. These molecules are the building blocks of all living organisms and are the origin of our fossil fuels. When these molecules are "burnt", either by respiration or combustion, energy is released which originated from our Sun. The Sun and solar system formed about 4.6 billion years ago. A billion years later our planet had cooled allowing life to appear as evidenced by the formation of stromatolites shown in Figure 1, recently photographed by Leslie Dutton.

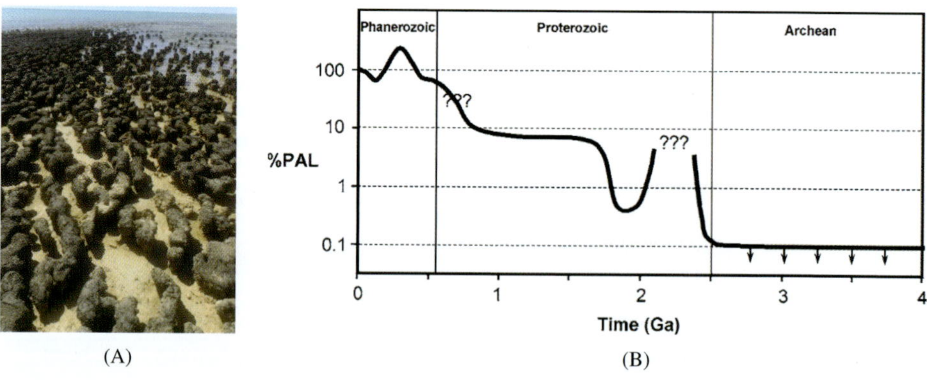

(A)   (B)

Figure 1.   (A) Stromatolites at Hamelin pool — in the Shark Bay area, W. Australia. Photographed by Leslie Dutton, September 2016. (B) Rise in atmospheric O_2 beginning 2.5 billion (Ga) years to present day (taken from Canfield [2005]). The question marks indicate periods when oxygen concentrations are particularly uncertain. This reconstruction is far less certain than the relatively few question marks would seem to indicate.

It took at least another half a billion years for organisms to evolve the capacity to derive reducing equivalents from splitting water using solar energy and release oxygen into the atmosphere. Prior to this, biological organisms had been dependent on hydrogen/electron donors such as H_2S, NH_3, organic acids and Fe^{2+}, that were in limited supply compared with the "oceans" of liquid water with which planet Earth is blessed. The water-splitting enzyme of photosynthesis is called photosystem II (PSII) [Barber, 2003, 2016a]. Its appearance and linkage, *via* an electron transfer chain, to the second light reaction, photosystem I (PSI), heralded the Big Bang of Evolution. The release of oxygen as a by-product over eons of time (Figure 1B) had dramatic consequences for the development of life, since it created an aerobic atmosphere and at the same time allowed the ozone layer to form. With oxygen available, the efficiency of cellular metabolism increased substantially, since for a given amount of substrate, aerobic respiration provides in the region of 20 times more energy than anaerobic respiration. It was probably this improved efficiency of cellular bioenergetics, which paved the way for the subsequent evolution of eukaryotic cells and multicellular organisms. The establishment of the ozone layer provided a shield against harmful UV radiation allowing organisms to exploit new ecological niches including the terrestrial environment. Consequently, life was able to prosper and diversify on an enormous scale as testified by the fossil records and by the extent and variety of living organisms on our planet today. Therefore, the energy flow of the planet became almost cyclical. Photosynthesis stored solar energy in the chemical bonds of organic molecules while the oxidation of these molecules, by either respiration or combustion, led to the recombination of the "stored hydrogen" with oxygen to form water with the consequential release of energy. However, as depicted in Figure 2, this wonderful cycle was biased to the storage of energy, resulting in the laying down of fossil fuels and depletion of carbon dioxide from the atmosphere with a concomitant rise in atmospheric oxygen.

Equation of photosynthesis (Eq. 1) below shows that for every molecule of O_2 produced one molecule of CO_2 is converted to carbohydrate (CH_2O).

$$2H_2O + CO_2 \xrightarrow{8h\nu} O_2 + (CH_2O) + H_2O \tag{1}$$

Today, there is 21% O_2 in the atmosphere and assuming all of it is derived from photosynthesis, then the amount of reduced carbon on our planet is enormous since the atmospheric CO_2 level is only 0.04%. This means that there is essentially an almost unlimited supply of fossil fuels although presumably a great deal of the reduced carbon will not be readily obtainable. If it were, and if it was all burnt, then our planet would return to its original anaerobic state with a very high level of CO_2 in the atmosphere. I will touch on this issue again later in this chapter in the context of artificial photosynthesis, but firstly I will summarize our

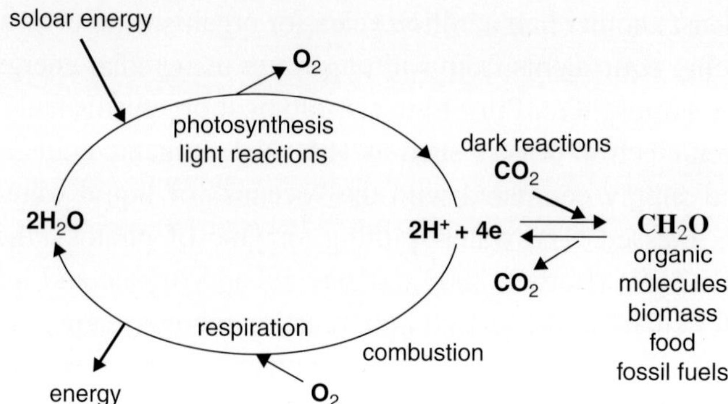

Figure 2. A diagrammatic representation of energy flow in biology. The light reactions of photosynthesis (light absorption, charge separation, water splitting, electron/proton transfer) provides the reducing equivalents in the form of protons (H^+) and energized electrons (e) to convert carbon dioxide (CO_2) to sugars and other organic molecules which make up living organisms (biomass) including those that provide humankind with food. The same photosynthetic reactions gave rise to the fossil fuels formed millions of years ago. The burning of these organic molecules either by respiration (controlled oxidation within our bodies) or by combustion of biomass and fossil fuels to power our technologies, is the reverse of photosynthesis, releasing CO_2 and combining the stored "hydrogen" back with oxygen to form water. In so doing energy is released, energy which originated from sunlight. Over the eons of time, there was a net accumulation of carbon (thick arrow) until modern times when the use of fossil fuels as an energy source has shifted the balance to the net release of CO_2.

current state of knowledge on how PSII catalyses the light driven splitting of water and the formation of the O–O bond. The importance of this reaction cannot be over stated since it is the entry point into the energy cycle of our planet. Moreover, the splitting of water into its elemental constituents is thermodynamically and chemically demanding especially when achieved in a delicate biological environment. Because of this, PSII has to repair itself from time to time with a half time of about 30 minutes in bright illumination [Barber and Andersson, 1992].

2. Photosystem II (PSII)

PSII is a pigment-binding, multi-subunit protein complex embedded in the lipid environment of the thylakoid membranes of plants, algae and cyanobacteria. Today we have detailed understanding of the structure and functioning of this unique enzyme [Barber, 2016a]. The journey to this level of knowledge can be traced back to the discovery that plants produced oxygen by Joseph Priestley in the 18th century [1772]. Since then there has been a sequence of milestone discoveries which makes a fascinating story. However, it is the last few years that have provided the level of detail necessary to reveal the chemistry of water splitting and O–O bond formation. In particular, the crystal structure of the isolated PSII enzyme has been reported with ever-increasing improvement in resolution. Thus, the organizational and structural details of its many subunits and cofactors are now well-understood including the catalytic centre where water is split.

2.1. *The catalytic centre*

It has long been known that the catalytic water splitting site of PSII contains four Mn ions [Debus, 1992], and understanding their organisation and interaction during catalysis of water oxidation has been a "holy grail" of bioinorganic chemistry. It is accomplished by the relatively low energy content of four photons of visible light centred around 680 nm and absorbed by chlorophyll (Eq. 2).

$$2H_2O \xrightarrow{\ 4h\nu\ } O_2 + 4e + 4H^+ \tag{2}$$

The four high energy electrons, together with $4H^+$, are used to reduce plastoquinone (PQ), the terminal electron acceptor of PSII, to plastoquinol (PQH_2) (Eq. 3):

$$4e + 4H^+ + PQ \xrightarrow{\ 4h\nu\ } 2PQH_2 \tag{3}$$

PQH_2 passes its reducing equivalents to an electron transfer chain which feeds into PSI and gains the additional reducing potential from a second light reaction necessary to drive CO_2 reduction [Barber, 2009; 2016a].

For many years, it was agreed that there were four Mn ions in the catalytic centre and that they function to accumulate the four oxidising equivalents necessary to split two water molecules and thus generate a molecule of dioxygen. That each photon absorbed drives each oxidation step at one catalytic centre, was clearly shown by monitoring oxygen release in a series of single turnover flashes of light where the maximum yield of oxygen followed a period of four [Joliot *et al.*,1969]. This finding gave rise to a scheme known as the S-state cycle (S_0 to S_4) (see Figure 3) [Kok *et al.*, 1970]. As shown in Figure 3, as the catalytic cycle proceeds from S_0 to S_4, protons are released at each S-state step except for the S_1 to S_2 transition where the metal cluster accumulates one positive charge. The redox active Y_Z is a neutral tyrosyl radical where its phenolic proton is donated to a nearby base, D1His190, to which it is H-bonded. Therefore, the neutral Y_Z radical is an ideal candidate for facilitating a proton coupled electron transfer (PCET) from the catalytic centre, as originally suggested by Hoganson and Babcock [1997]. An important study [Haumann *et al.*, 2005], indicated the coupling is sequential, with a strictly alternating removal of electrons and protons rather than hydrogen atom transfer as suggested by Tommos and Babcock [1998]. The absence of proton release during the S_1 to S_2 transition has been explained by the formation of an additional intermediate by deprotonation prior to electron transfer

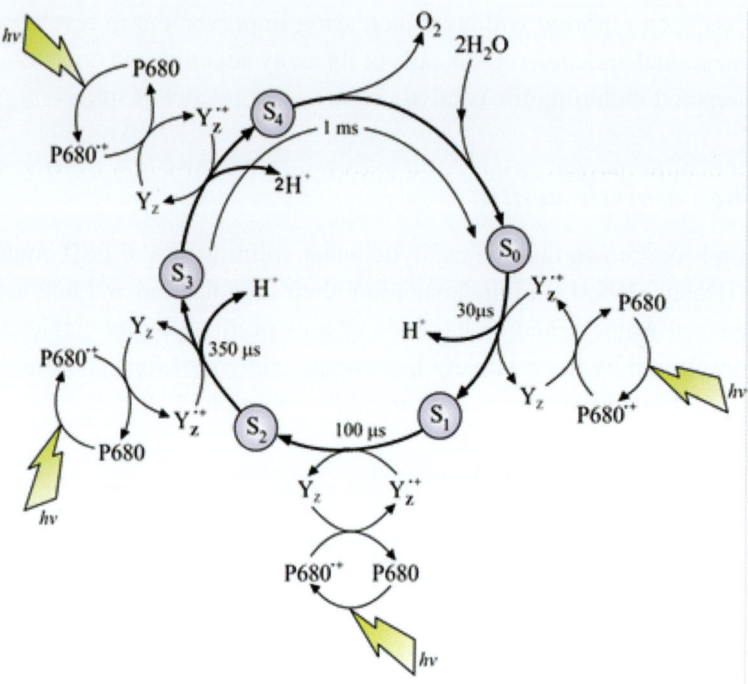

Figure 3. The S-state cycle showing how the absorption of four photons (4 hν) of light by P680 drives the splitting of two water molecules and formation of O_2 through a consecutive series of five intermediates (S_0, S_1, S_2, S_3 and S_4). Protons (H^+) are released during this cycle except for the S_1 to S_2 transition. Electron donation from the catalytic centre (Mn_4Ca^{2+}) to P680$^{\cdot+}$ is aided by the redox active tyrosine Y_Z. Each step involves a single oxidation of a Mn in the cluster, starting at S_0 with 3 x Mn^{III} plus Mn^{IV} advancing to S_3 with 4 x Mn^{IV}. The exact oxidation state of S_4 is unknown but could be 3 x Mn^{IV} plus Mn^{V} or 3 x Mn^{IV} plus Mn^{IV}-oxyl radical (see below). Also shown are half-times for the various steps of the cycle.

to Y_Z resulting in the storage of a positive charge. A similar event may occur during the S_4 to S_0 transition. The nature of these two possible intermediates is unknown [Dau and Haumann, 2007].

A whole range of techniques, particularly magnetic resonance spectroscopy, have shown that each step in this cycle involves an increase in the oxidation of a Mn ion, starting at S_0 with 3 $\times$ Mn^{III} plus Mn^{IV} advancing to S_3 with 4 $\times$ Mn^{IV} [Cox *et al.*, 2014]. The precise details of the final oxidation state, are not clear because O–O bond formation has to be very fast to avoid side reactions. This cycle is powered by the initial oxidation of a chlorophyll known as P680 generated by light driven primary charge separation in the reaction centre of PSII which is coupled to a redox active tyrosine (Y_Z) serving as an intermediate electron carrier between P680$^+$ and the Mn_4Ca^{2+}-cluster.

Not surprisingly, it has been one of the greatest challenges of photosynthesis research to determine the structure of the Mn_4-containing cluster of the PSII oxygen evolving complex (OEC) and thus reveal the molecular mechanism of the water splitting reaction. Over the years there have been many postulates of its structure mainly derived from X-ray absorption fine-structure (XAFS) spectroscopy [Yachandra, 2002] and electron paramagnetic resonance (EPR) studies [Britt *et al.*, 2004; Haddy, 2007]. However, in 2004 a complete fully refined structure of PSII was determined at 3.5 Å resolution using X-ray crystallography by my colleagues and I at Imperial College London and published in *Science* [Ferreira *et al.*, 2004].

We were able to assign, for the first time, over 5000 amino acid side chains of this huge dimeric membrane complex (700 kD) with each monomer consisting of 19 different protein subunits. In so doing, we answered many outstanding questions as well as revealing a wide range of important details hitherto unknown. The native electron density map, together with anomalous diffraction data collected at wavelengths of 1.89 Å and 2.25 Å, provided electron density profiles for Mn and Ca^{2+} respectively. This information was then used to build a model of the metalcluster. The anomalous electron density attributed to the four Mn ions was "pearshaped" indicative of the 3 + 1 organisation and thus one Mn was assigned to the small domain and three in the large globular domain, whereas the 2.25 Å wavelength map covered one metal ion in the large domain of the native density. From this data the three Mn ions (Mn1, Mn2 and Mn3) and the Ca^{2+}, located in the large domain, were modelled with a cubane geometry having bridging oxygens, an organisation which was compatible with the native electron density. The fourth Mn ion (Mn4), located in the small domain was modelled so that it was linked to the cubane by one of its bridging oxygens (dangling Mn4) (see Figure 4A). Based on values determined by XAFS measurements and comparison with other Mn-containing proteins, the three Mn-di-μ-oxy-Mn bonds of the cubane were spaced at 2.7 Å while the three Mn-di-μ-oxy-Ca^{2+} bonds were assumed to be 3.4 Å. The dangler Mn4 was positioned 3.3 Å from the closest Mn ion of the cubane and about 4 Å from the Ca^{2+}.

At that time, there was no similar chemical structure known in biochemistry or inorganic chemistry and therefore the model was greeted with some uncertainty, especially since it had been conceived from relatively low resolution data. However, the main opposition to this model was the argument that the metal cluster, and surrounding amino acid ligands, would be seriously disrupted by radiation damage during data collection [Yano *et al.*, 2005]. Indeed the cubane structure did not seem to be consistent with polarised XAFS measurements made at much lower radiation levels [Yano *et al.*, 2006]. Unfortunately, the claim that the Ferreira *et al.'s* cubane model was invalid caused a substantial level of confusion for several years, both for

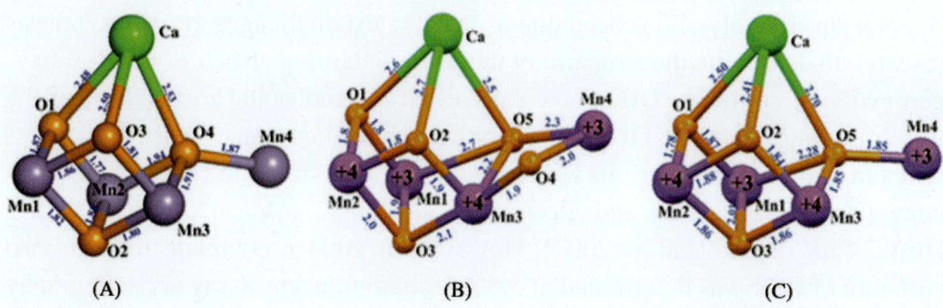

Figure 4. (A) Comparison of the $Mn_4Ca^{2+}O_4$ cubane model from coordinates (PDB 1S5L) deposited by Ferreira *et al.* in 2004 with (B) the 1.9 Å structure of the $Mn_4Ca^{2+}O_5$ cluster from Umena *et al.* [2011] (PDP 3WU2) and (C) the chemically synthesised $Mn_4Ca^{2+}O_4$ cubane by Zhang *et al.* [2015]. Note that unfortunately, the numbering of the Mn and O are different in (B) and (C) to that originally given in (A).

specialists and non-specialists. However, it has now become clear that the 2004 cubane model of Ferreira *et al.* was valid within the limitation of its resolution and therefore the extensive radiation damage argument and associated alternative postulated XAFS structures for the Mn-cluster [Yano *et al.*, 2006] were totally wrong. Indeed, doubts in the interpretation of the polarised XAFS data reported by Yano *et al.* [2006] were explored by Sproviero *et al.* [2008] who concluded that the polarized XAFS spectra were consistent with a cubane model which they went on to refine using QM/MM calculations and which closely matched the Ferreira *et al.* model.

In 2011, seven years after the coordinates of the cubane model by Ferreira *et al.* were deposited, the uncertainty of the cubane model was finally lifted with the report of 1.9 Å PSII crystal structure from Umena *et al.* [2011]. At this resolution the electron densities of individual metal ions could be resolved and bridging oxygens inferred. As shown in Figure 4B, the resulting model was overall similar to that of Ferreira *et al.* [2004] except an additional oxo-bridge was proposed to link the Mn outside the Mn_3Ca^{2+}cubane (Mn4) to a Mn ion of the cubane. Importantly this additional oxo-bridge had already been predicted by Siegbahn [2008] and by Dau *et al.* [2008] based on modification of the original 2004 Ferreira *et al.* cubane model using Density Function Theory (DFT) and other quantum mechanical methods. Nevertheless, the 1.9 Å diffraction data of Umena *et al.* [2011], together with high resolution diffraction data obtained by femtosecond X-ray free-electron laser (XFEL) pulses [Suga *et al.*, 2015, 2017; Young *et al.*, 2016] has provided reliable atomic structures of the OEC at cryo- and at room-temperatures, particularly in its dark-stable S_1 state, which has given validity to the detailed quantum mechanical calculations for the OEC and its functioning [Luber *et al.*, 2011; Yamanaka *et al.*, 2011; Cox *et al.*, 2014; Lohmiller *et al.*, 2014] and more rigorous interpretation of experimental results [Ames *et al.*, 2011; Rapatski *et al.*, 2012; McConnell *et al.*, 2012; Bovi *et al.*, 2013].

2.2. *Synthetic cubane mimics*

Of considerable importance was the report that the PSII cubane cluster can be synthesised *in situ* as a $Mn_4Ca^{2+}O_4$ molecule [Zhang *et al.*, 2015] whose structure is similar to that of the OEC (compare Figure 4C with 4B) and remarkably like the original model of Ferriera *et al.* (compare Figure 4C with 4A). As with the Ferreira *et al.* structure, the synthetic $Mn_4Ca^{2+}O_4$ molecule was also missing the additional oxo-bridge to Mn4. The chemical synthesis of the $Mn_4Ca^{2+}O_4$ structure in organic solvent shows that this cluster does not require protein to assemble and as long as there is an appropriate source of ligands, the structure is stable in an oxidised state. Moreover, cyclic voltammetry conducted in 1,2-dichloroethane showed that it can advance through the S-state cycle to S_3 [Zhang *et al.*, 2015]. However, no oxygen evolution was detected when a small amount of water was present. Nevertheless, this extraordinary finding could have implications for the evolutionary origin of the OEC and provides some explanation for the robustness of the *in vivo* cluster to site directed mutations (see below) and X-ray radiation.

Before the synthesis of the $Mn_4Ca^{2+}O_4$ cluster by Zhang *et al.* [2015], Christou and colleagues [Mukherjee *et al.*, 2012] reported the synthesis of a $Mn_3Ca_2O_4$ molecule which is structurally similar to that of $Mn_4Ca^{2+}O_4$ (see Figure 5C). Another important study was the synthesis of a $Mn_3Ca^{2+}O_4$ cubane by Agapie and colleagues [Kanady *et al.*, 2011] (see Figure 5D). However neither showed redox properties like that of the $Mn_4Ca^{2+}O_4$ synthesised by Zhang *et al.* [2015].

That a Mn_4CaO_4 structure like the one proposed by Ferreira *et al.* can also exist in the lattice of a macromolecular Mn-Ca molecule was shown by Misra

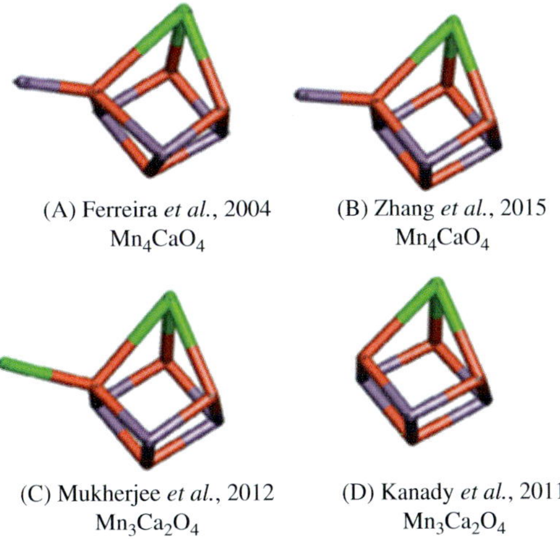

(A) Ferreira *et al.*, 2004
Mn_4CaO_4

(B) Zhang *et al.*, 2015
Mn_4CaO_4

(C) Mukherjee *et al.*, 2012
$Mn_3Ca_2O_4$

(D) Kanady *et al.*, 2011
$Mn_3Ca_2O_4$

Figure 5.   Diagrammatic comparison of synthesised cubane structures (B), (C) and (D) with the Ferreira *et al.* X-ray model (A). Mn purple, Ca^{2+} green and red are oxo-bonds. Taken from Barber [2016b].

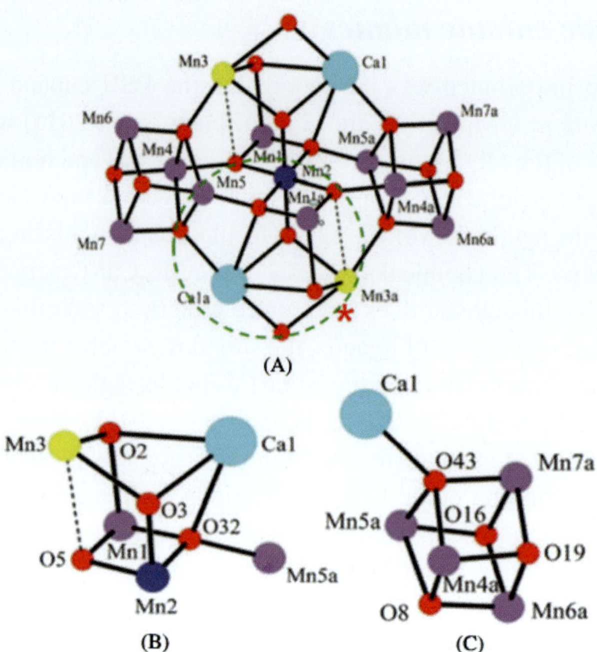

Figure 6. (A) [$Mn_{13}Ca_2O_{10}(OH)_2(OMe)_2(O_2CPh)_{18}(H_2O)_4$] macromolecule synthesised by Misra *et al.* [2005] composed of two building blocks shown in (B) and (C). Ca is turquoise, Mn purple and yellow, and oxygen in red.

et al. [2005]. They synthesised a [$Mn_{13}Ca_2O_{10}(OH)_2(OMe)_2(O_2CPh)_{18}(H_2O)_4$] molecule in a highly oxidised state (Figure 6A) which consisted of two Mn_4Ca building blocks. As can be seen in Figure 6, one form is the well-known Mn_4O_4 cube [Ruttinger *et al.*, 1999] with a Ca attached to the cube by one of its bridging oxos (Figure 6C) while the other building block (Figure 6B) has a similar cubane structure to that of Mn_4CaO_4 of PSII. This study was the first to show that Mn and Ca can form discrete molecular complexes and also gave credibility to the Mn_4CaO_4 cubane model of Ferreira *et al.* [2004] well before the publication of the 1.9 Å structure of Umena *et al.* [2011].

3. Mechanism of Water Splitting

3.1. *Base-catalysed nucleophilic attack mechanism*

Prior to the 2004 crystal structure, there had been many postulates of how PSII splits water to form dioxygen. One idea suggested by Messinger *et al.* [1995] and Pecorara *et al.* [1998] was that O–O bond formation in PSII results from a based catalysed nucleophilic attack of a hydroxyl group onto an electrophilic oxo derived from the deprotonation of the second substrate water molecule. In particular, Pecorara *et al.* [1998] conducted studies on synthesised Mn complexes and despite there being no crystal structure of the Mn_4Ca^{2+} cluster at that time they proposed that *a "nucleophilic attack by a calcium-ligated hydroxide on an electrophilic oxo group ligated to a high-valent manganese to achieve the critical O–O bond formation"*. The high oxidation state was taken to be Mn^V in line with the suggestion of Limburg *et al.* [1997] but could also be a Mn^{IV}-oxyl radical [Brudvig, 2008]. The Ca^{2+} would act as a weak Lewis acid and align a specific H_2O molecule for an efficient nucleophilic attack of an OH on the electrophylic oxo. The cubane structure of the $Mn_4Ca^{2+}O_4$ cluster of PSII derived from X-ray crystallography in 2014 by my colleagues and I was consistent with this proposed mechanism and discussed in our papers at that time [Ferreira *et al.*, 2004; Barber *et al.*, 2004]. Importantly, the dangler Mn4 was found to be immediately adjacent to the Ca^{2+} and close to side chains of several key amino acids, including the redox active tyrosine, Y_Z, and therefore we suggested that Mn4 and Ca^{2+} form a "catalytic" surface outside the cubane for binding the substrate water molecules and their subsequent oxidation. The progression through the S-state cycle would lead to deprotonation of these water molecules and the formation of a highly electrophilic Mn^V-oxo or Mn^{IV}-oxyl radical, and to a nucleophilic hydroxyl on the Ca^{2+}.

The formation of the O–O bond of dioxygen would result from the based catalysed nucleophilic attack of hydroxyl onto the electophilic oxo of Mn4 with a possible peroxide intermediate. The electrophilicity of Mn^V-oxo would be enhanced by the high oxidation level of the cubane ($3 \times Mn^{IV}$) plus a net positive charge (formed during the S_1 to S_2 transition), combining to form an "oxidising battery" for the terminal oxo on Mn4 (see Figure 10). The deprotonation of the substrate waters would be aided by nearby bases such as CP43Arg357 and by the weak Lewis acidity of Ca^{2+}. Indeed, there is an extensive H-bonding network leading from Yz to the lumenal side of OEC [Ferreira *et al.*, 2004; Murray and Barber, 2007; Umena *et al.*, 2011]. However, as stated above, the dangler Mn4 and Ca^{2+} each bind two water molecules according to the high resolution structure of PSII [Umena *et al.*, 2011] (see Figure 7). This may indicate a "carousel mechanism" providing efficient delivery of substrate water molecules to the active sites and allow maximum turnover rates for the S-state cycle rather than being restricted by the rate of water diffusion to the catalytic site just prior to the formation of the S_0-state. This may explain the action of ammonia, often considered to be an analogue of water. It has been shown that NH_3 binds strongly to Mn4 [Oyala *et al.*, 2015] but it does not inhibit production of O_2 but rather slows the turnover of the S-state cycle [Boussac *et al.*, 1990; Navarro *et al.*, 2013]. This would be consistent with it inhibiting the carousel process but not the chemistry of O–O bond formation. The idea of a carousel mechanism was first proposed

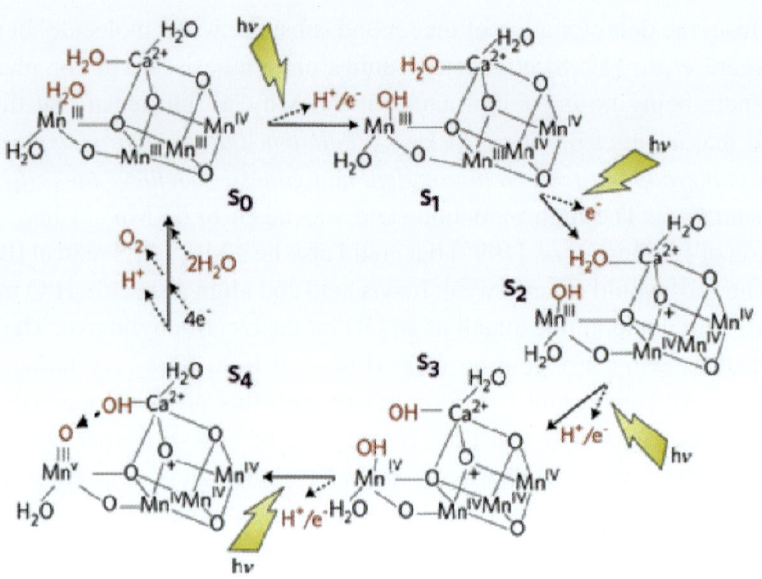

Figure 7. Diagrammatic representation of a mechanistic scheme for water splitting and dioxygen formation in PSII reproduced from Barber [2017]. The substrate water molecules and products of the oxidation reactions for each S-state are shown in red. Intermediates that may exist between the S-state transitions [Dau and Haumann, 2007] are not depicted nor is the possible peroxide intermediate just prior to O–O bond formation. Although the electrophilic oxo in the S_4-state is shown attached to Mn^V it could equally be incorporated in a terminal Mn^{IV}-oxyl.

by Batista and colleagues but for a different mechanism for the S_3 to S_4 transition [Askerka *et al.*, 2016].

Figure 7 is a mechanistic scheme for water splitting and dioxygen formation in PSII based on the acid-base, hydroxyl/water nucleophilic attack mechanism for O–O bond formation discussed above and taken from Barber [2017].

3.2. *Indirect support for the nucleophilic mechanism for O–O bond formation in PSII*

Although the nucleophilic attack mechanism for O–O bond formation in PSII makes good chemical sense [Vinyard *et al.*, 2015] and fits well with the geometry of the $Mn_4Ca^{2+}O_{4/5}$ cluster and its positioning within the complex [Barber *et al.*, 2004], there is no direct experimental evidence for it and there are a number of alternative mechanisms advocated recently with focus on the O5, the oxygen linking Mn4 to the cubane (see Figure 4B) [Umena *et al.*, 2011; Suga *et al.*, 2015, 2017; Young *et al.*, 2016; Askerka *et al.*, 2016; Cox *et al.*, 2014; Pérez-Navarro *et al.*, 2016; Siegbahn, 2013]. However, below I present some convincing indirect evidence which support the nucleophilic mechanism for water splitting by PSII as detailed in Figure 7.

3.3. *Comparison with Fe–Ni carbon monoxide dehydrogenase (CODH)*

There are anaerobic prokaryotic organisms, which use the oxidation of carbon monoxide as an energy source to split water. The enzyme that catalyses this reaction is carbon monoxide dehydrogenase (CODH). There are some interesting similarities between PSII and the Fe–Ni CODH, which support the nucleophilic attack mechanism for photosynthetic O–O bond formation [Barber, 2017]. As shown in Eq. 4, products of CODH are carbon dioxide and two reducing equivalents.

$$CO + 2H_2O \rightarrow CO_2 + 2e + 2H^+ \tag{4}$$

This chemistry is the well-known "water-gas shift reaction" discovered in 1780 by the Italian physicist, Felice Fontana [1780], and today adopted as a large scale industrial process to make pure hydrogen gas from carbon monoxide. This commercial process requires shifts in reaction temperature from high (approximately 400°C with an iron-chromium oxide catalyst) to low (approximately 225°C with copper-based catalysts [Smith *et al.*, 2010].

In contrast, in the case of Fe–Ni CODH, the generated reducing equivalents are produced at ambient temperatures and are used to drive the reductive chemistry of the organism rather than producing hydrogen gas, usually by electron transport involving nearby Fe_4–S_4 centres [Svetlitchnyi *et al.*, 2004; Gong *et al.*, 2008]. Remarkably, the catalytic centre of Fe–Ni CODH has a similar geometry to that of Mn_4CaO_5 of PSII (Figure 8). X-ray crystallography at high resolutions [Dobbek

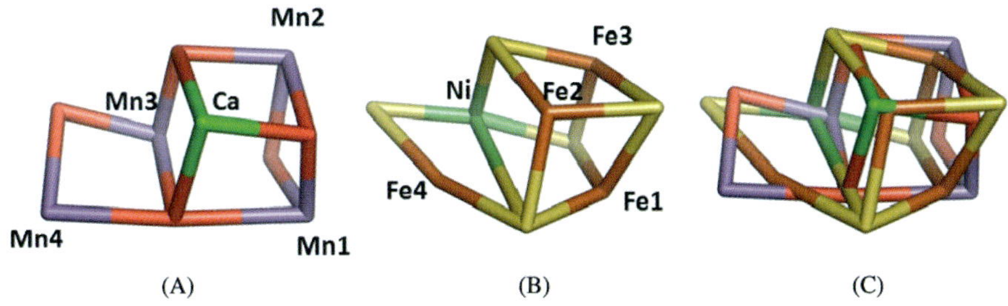

Figure 8. Comparison of (A) the $Mn_4Ca^{2+}O_5$ cluster of PSII at 1.9 Å using PDB 3WU2 [Umena *et al.*, 2011] with (B) $Fe_4Ni^{2+}S_5$ cluster of CO dehydrogenase at 1.03 Å using PDB 4UDX [Fesseler *et al.*, 2015]. (C) Overlay of the two structures with a Root Mean Square Deviation (RMSD) of 0.65 between the metal atoms. Oxo-bonds in red, sulphur bonds in yellow, manganese ions in purple, calcium and nickel ions in green as labelled and iron ions in orange.

et al., 2001; Jeoung and Dobbek, 2007] has shown that Fe–Ni CODH consists of a Fe_3NiS_4 cubane with a fourth "dangler" Fe attached to the cubane *via* a bridging S of the cubane and an additional S bridge to make a Fe_4NiS_5 cluster as shown in Figure 8B. Despite the two catalytic centres being composed of different elements, the overlay of their structures, shown in Figure 8C, is remarkably good with a RMSD of 0.65 between the metal atoms. In fact, they are the only known examples of metallo-catalytic centres having heterocubanes with an additional metal in exo.

It is known that in CODH, the CO is activated by binding to the Ni^{2+} of the cubane, which causes a shift in the coordination of the Ni atom from square-planar to a square-pyramidal geometry [Jeoung and Dobbek, 2007] and increases the electrophilicity of CO. Similarly, water is activated by forming a complex with the dangler Fe4^{2+}. This induces a movement of the cysteine Ni ligand leading to the possible formation of a carboxyl bridge between the Ni^{2+} and Fe4. Thus, the mechanism is a base-catalysed nucleophilic attack by the Fe4 bound water/hydroxyl group onto CO (see Figure 9). A decarboxylation of the intermediate leads to CO_2 release and the reduction of the metal cluster [Jeoung and Dobbek, 2007]. The oxidation back to the active state involves electron transfer to nearby Fe_4-S_4 centres including ferredoxin or to a closely bound acetyl-CoA synthase depending on species [Gong *et al.*, 2008].

Given the striking geometrical similarities between the catalytic centres of CODH and PSII, together with the fact they both catalyse the removal of reducing equivalents from water, it would be reasonable to conclude that there may be common features in their catalytic mechanisms. Of course, PSII water splitting

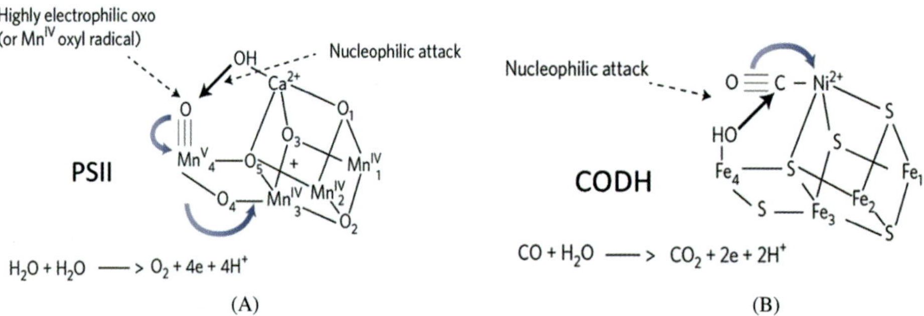

Figure 9. A structurally-based diagrammatic comparison of a base nucleophilic attack of a hydroxyl onto an electrophile leading to oxygen atom transfer for (A) O–O formation in S_4 state of PSII and (B) CO conversion to CO_2 in CODH. Note that the numbering of Fe ions in (B) are different to those in PDB 4UDX [Fesseler *et al.*, 2015] so as to compare with numbering of the Mn ions in (A) taken from PDB 3WU2 [Umena *et al.*, 2011]. Unfortunately, the numbering of the $Mn_4Ca^{2+}O_5$ in PDB 3WU2 [Umena *et al.*, 2011] differs from that of the earlier $Mn_4Ca^{2+}O_4$ in PDB 1S5L [Ferreira *et al.*, 2004]. Note that in the S_4-state there is the storage of a net positive charge generated during the S_1 to S_2 transition (see Figure 8).

involves high valent oxidative chemistry while CODH is low valent organometallic chemistry. Moreover, the water splitting reaction of PSII is non-reversible while that of CODH is. For these reasons, I wish to emphasise that the comparison of PSII with Fe–Ni CODH does not imply any direct evolutionary link but a similarity in their function to obtain reductive power from water.

3.4. *Ligands to the* $Mn_4Ca^{2+}O_5$ *cluster*

The Ferreira *et al.,* structure [2004] identified seven amino acids as ligands to the Mn_4Ca^{2+} cluster, six from the D1 reaction centre protein: Asp170, Glu189, His332, Ala344, Glu333 and Asp342, and Glu354 of the inner antenna chlorophyll-binding protein, CP43. That these seven amino acids are ligated to the Mn_4Ca^{2+} cluster has been confirmed in the 1.9 Å structure of Umena *et al.* [2011] and the precise details of the ligation pattern revealed (see Figure 10). The coordination properties of the three Mn ions of the cubane are totally satisfied by amino acid ligands. However, Umena *et al.* [2011] assigned two water ligands for the non-cubane dangler Mn4 and also for the Ca ion as shown in Figure 10. Ferreira *et al.* [2004] had already suggested that water molecules could act as ligands to Mn4 and Ca^{2+} and that this may be the site for dioxygen formation.

The amino acids which serve as ligands to the metal cluster are highly conserved in PSII over the whole cyanobacterial, plant and algal kingdoms [Murray, 2012]. Yet there are a large number of reports that their replacement by other residues using site directed mutagenesis, does not inhibit oxygen evolution completely and sometimes not at all [Debus, 2008]. One example is the replacement of the negatively charged D1-Glu189 by positively charged Lys, which made no difference to the rate of oxygen evolution [Clausen *et al.*, 2001]. This was also true for other D1-Glu189 mutants leading to the logical conclusion that it cannot be a ligand to the MnCa-cluster. Another example is the replacement of acidic CP43-Glu354 by neutral glutamine where O_2 evolution persisted at a reduced rate of

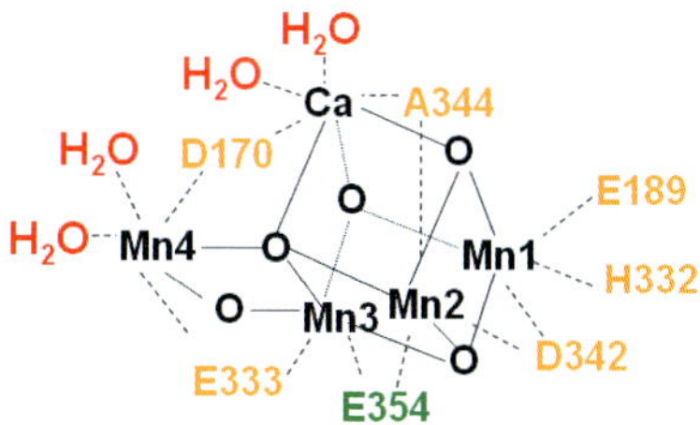

Figure 10. Diagrammatic representation of the amino acid side chains acting as ligands to the Mn_4CaO_5 cluster of the OEC together with four water molecules providing ligands to Ca and Mn4 (constructed from Umena *et al.* crystal structure [2011]). D1 residues shown in orange and green depict the CP43 residue Glu354.

10–18% compared to the wild type [Service *et al.*, 2010]. It is also possible to totally remove CP43 and yet assemble the OEC [Büchel *et al.*, 1999]. Thus compared to many other metallo-enzymes, the assembled Mn_4CaO_5 cluster is rather robust to changes in its first coordination sphere. There are exceptions since, for example D1-Asp170 also acts as a high affinity site for Mn^{2+} binding [Diner *et al.*, 1991] so that the photo-oxidation of P680 to $P680^+$ can facilitate its conversion to Mn^{3+}. In this way, as long as $CaCl_2$ is present, the Mn_4CaO_5 is assembled step-by-step to form the OEC, a process known as photoactivation [Ananyev and Dismukes, 1996].

4. Synthesized Chemical Model Systems

Ruthenium model chemistry was the first to achieve homogeneous water splitting and has subsequently become the best characterized. More recently, functional manganese model complexes have been synthesised and have the added advantage of being more biologically relevant to the OEC.

4.1. *Mechanism of O_2 production from organo-Ru complexes*

The blue dimer, cis,cis-$[(bpy)_2(H_2O)Ru^{III}ORu^{III}(H_2O)(bpy)_2]^{4+}$, was the first designed, well-defined molecule known to function as a catalyst for water oxidation [Gersten *et al.*, 1982]. As in the case of PSII, it met the stoichiometric requirements for water oxidation by utilizing PCET reactions in which both electrons and protons are transferred. This avoided charge buildup, allowing for the accumulation of multiple oxidative equivalents at the Ru–O–Ru core. Since then, a wide range of both dinuclear-Ru and mononuclear-Ru organo-complexes have been synthesised which show high rates of O_2 evolution using Ce^{4+} as an oxidant. Although radical coupling is a dominant path to dioxygen formation there is a significant number of them, particularly the mononuclear Ru molecules, that generate the O–O bond by nucleophilic attack of an electrophilic oxo of $Ru^V{=}O$ [Murakami *et al.*, 2011; Tong *et al.*, 2011; Sala *et al.*, 2013; Duan *et al.*, 2012, 2015]. This mechanism and the rate of turnover are dependent on the design of the ligands (Concepcion *et al.*, 2009). One example is $[Ru^{II}(bpc)(bpy)(OH_2)]^+$, (where bpc = 2,2′-bipyridine-6-carboxylate and bpy = 2,2′-bipyridine). A proposed mechanism for these types of mononuclear Ru catalysts at low pH is shown in Fig. 10. The $[Ru^{II}{-}OH_2]^+$ is first oxidized to $[Ru^{III}{-}H_2]^{2+}$, followed by a 2H$^+$/1e PCET oxidation step generating $[Ru^{IV}{=}O]^+$. $[Ru^{IV}{=}O]^+$ is not reactive toward water and therefore has to be further oxidized to generate the active species $[Ru^V{=}O]^{2+}$. Thereafter, solvent water nucleophilic attack on the Ru^V-oxo occurs, forming the

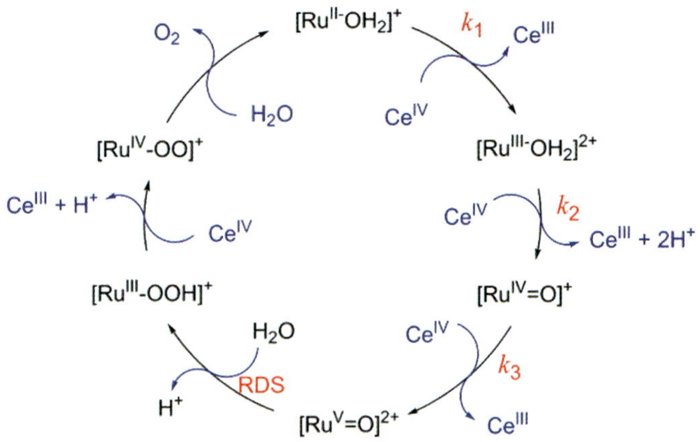

Figure 11. (A) Mechanistic scheme for the generation of dioxygen by treating [Ru^{II}(bpc)(bpy)(OH$_2$)]$^+$ (where bpy = 2,2-bipyridine and bpc = 2,2-bipyridine-6-carboxylate) with Ce^{IV}. A water nucleophilic attack on to Ru^V=O is proposed for the rate determining step (RDS). Taken from Duan *et al.* [2015].

hydroperoxo intermediate [Ru^{III}–OOH]$^+$ with oxygen evolving from [Ru^{IV}–OO]$^+$. The carbonate ligand (bpc) enhances electron transfer steps without obviously influencing the PCET [Duan *et al.*, 2015].

This type of rational design of Ru-complexes with a general bipyridine-dicarboxylate ligand motif has produced many molecular catalysts having exceptionally high turnover frequencies (TOF). For example, the mononuclear ruthenium complex [Ru(bda)(isoq)^2] (where bda = 2,2′-bipyridine-6,6-dicarboxylic acid and isoq = isoquinoline) has a high catalytic activity for water oxidation when using Ce^{IV} as the oxidant under acidic conditions [Duan *et al.*, 2015]. Indeed with some constructs, high rates of O$_2$ formation are achieved which are much faster than PSII (1000 s^{-1} based on rate constants for the S-state cycle and not restrained by the reductive chemistry), with turnover rates of 8000 s^{-1} at pH 7.0 and 50,000 s^{-1} at pH 10 and have a high chemical stability [Matheu *et al.*, 2015]. These very impressive turnover rates are obtained with Ru complexes having the general formula [Ru^n(tda)(py)$_1$]$^{m+}$ with tda^{2-} being, [2,2:6′,2″-terpyridine]-6,6″-dicarboxylate including complex [Ru^{IV}(OH)(tda–K–N^{3}O)2(py)$_2$]$^+$. Again, the mechanism is thought to involve a nucleophilic attack by water on to the oxo of Ru^V as in Figure 11. The high rates are attributed to (a) easy access to high oxidation states, provided by tda^{2-} ligand through the anionic nature of the carboxylates and the capability of stabilizing a seven coordination to Ru in high oxidation states and (b) the ability of the dangling carboxylate to be a proton acceptor in an intramolecular fashion, while incoming substrate water molecule undergoes the nucleophilic attack.

4.2. *Mechanism of O_2 production from organo-Mn complexes*

In context of the water/hydroxide nucleophilic attack mechanism for PSII O–O bond formation, a number of efforts have been made to prepare Mn^V-oxo complexes which relate to OEC of PSII. The groups of both Crabtree and Brudvig [Cady *et al.*, 2008; Limburg *et al.*, 1997, 1999, 2001; Vrettos *et al.*, 2001] and Pecoraro *et al.* [1994, 1998] have provided evidence for Mn^V-oxo intermediates related to the OEC that were very short-lived. One catalyst, $[Mn_2^{III/IV} (\mu\text{-}O)_2(terpy)_2(OH_2)_2](NO_3)_3$ (where terpy = 2,2′:6′2″-terpyridine) has provided a model system for the oxygen-evolving complex of PS II [Limburg *et al.*, 1999, 2001]. These studies have used a range of added oxidants and often involve radical coupling of two adjacent oxos.

However, both porphyrins and corroles and some related complexes have been shown to produce dioxygen by a nucleophilic mechanism by generating a Mn^V-oxo as the electrophilic species [Gao *et al.*, 2007, 2009] (see Figure 12). A hydroxyl nucleophilic attack of water on the Mn^V-oxo yielded dioxygen (step 2 to 3 in Figure 12) as deduced from DFT calculation, spectroscopy and isotopic O^{18} exchange experiments.

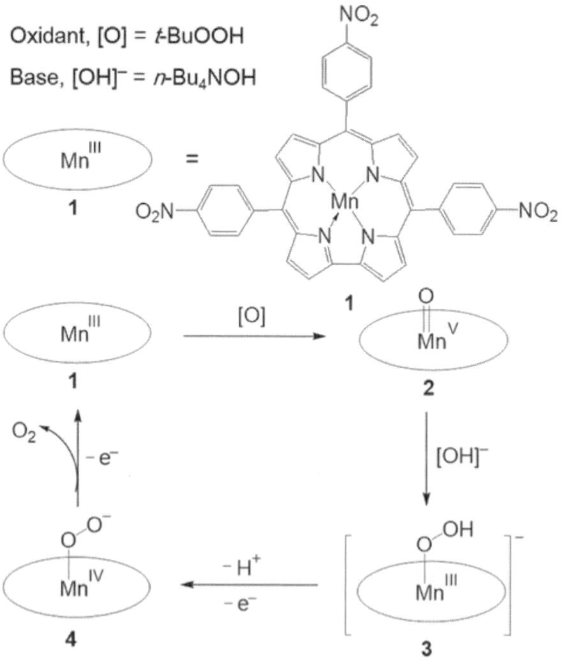

Figure 12. Mechanism for a base induced hydroxyl nucleophilic attack on to Mn^V=O of nitrophenylcorrole (structure shown and depicted as Mn^{III} in scheme) using *t*-BuOOH as the oxidant and *n*-Bu₄NOH as the nucleophilic base. Taken from Gao *et al.* [2009].

In another study, it was shown that both mononuclear and dinuclear "Pacman" Mn^{IV} corrole complexes could act as a catalyst for water splitting [Gao *et al.*, 2007]. In principle, the dinuclear Mn complexes could react *via* oxygen–oxygen radical coupling, but in these systems this route was found to be energetically less favorable than the nucleophilic attack mechanism.

5. Alternative Mechanisms are Less Compelling

An alternative to the base-catalysed nucleophilic mechanism for O–O bond formation in PSII, is radical pair coupling which was postulated by Babcock and colleagues [Hoganson and Babcock,1997; Tommas and Babcock, 1998] prior to the elucidation of the OEC structure. DFT calculations and X-ray diffraction data has proved to be remarkably good for studying the OEC as emphasised by Siegbahn who predicted, using the Ferreira *et al.'s* cubane model as the starting point, the existence and structure of the $Mn_4Ca^{2+}O_5$ cluster three years before the 1.9 Å crystal structure of Umena *et al.* was published [Siegbahn, 2008]. This powerful DFT approach has also been used extensively by Siegbahn [2006, 2009] to explore the mechanism of water oxidation and particularly the nature of S_4 in the S-state cycle.

Initially Siegbahn favored a surface radical coupling mechanism involving a terminal Mn^{IV}-oxyl radical (Siegbahn and Crabtree, 1999) but according to his calculation this mechanism has a higher energy barrier than an oxyl-oxo coupling mechanism. He therefore concluded that the most likely mechanism is the formation of a terminal oxyl radical within the cubane at S_4 which attacks a nearby bridging oxo, O5 to form the O-O bond (see Figure 13). This proposed mechanism requires significant conformational changes due to the insertion of a substrate water molecule into the cubane during the S_2 to S_3 transition and reconstruction of the cubane during the S_4 to S_0 transition. However, a recent X-ray Free Electron Laser (XFEL) study (Young *et al.*, 2016) obtained room temperature diffraction data which did not provide any evidence of structural changes in the S_3 state as

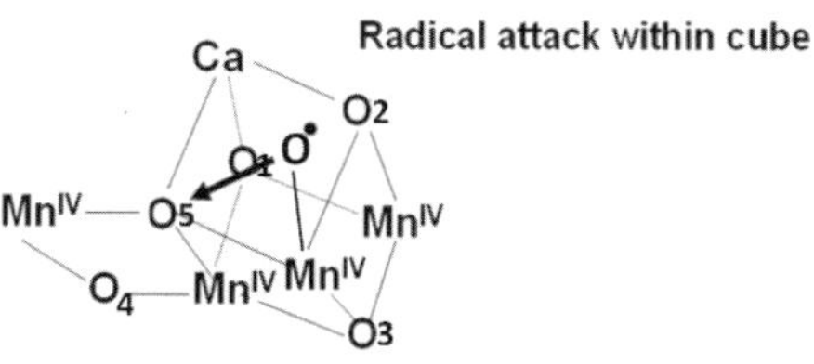

Figure 13. A mechanism for O–O bond formation during the S_4 to S_0 transition postulated by Siegbahn [2006, 2008, 2009, 2013] whereby formation of a terminal oxyl radical originating from a substrate water incorporated into the cubane during the S_2 to S_3 transition leads to a radical attack of an adjacent oxo ligand (O5) incorporated within the Mn_3CaO_4-cubane from the first substrate water during the reassembly of the cubane during the S_4 to S_0 transition.

required by the Siegbahn oxyl-oxo coupling mechanism. Thus this alternative mechanism is unlikely although very recently a XFEL study did detect some structural changes close to O5, which were suggested to be involved in O–O bond formation [Suga *et al.*, 2017]. Nevertheless, as discussed above the asymmetric cubane with a metal ion in exo seems to be a very stable and robust structure. To my mind, this favours a surface reaction rather than a mechanism involving disruption of the cubane bridging oxo-bonds during the S-state cycle, especially given the low over potential of the PSII water splitting reaction.

6. Artificial Photosynthesis

As outlined above, some progress has been made in mimicking photosynthesis in artificial systems, particularly involving ruthenium complexes, but as yet there is no system identified which is both efficient and robust for incorporation into a working technology for capturing and storing solar energy in chemical bonds on a large scale as does natural photosynthesis.

To date the main focus of research has been to design and synthesise molecular catalysts which can be linked to a light-driven charge separation system [Tran *et al.*, 2012]. Dyes have been used for the latter but inorganic semiconductors offer a more realistic and robust approach for providing the redox reducing potentials necessary to split water and power reductive chemistry.

Insights gleaned from the recent structural determination of PSII have initiated considerable efforts to identify artificial catalysts for water oxidation and hydrogen production using solar energy [Eisenberg and Gray, 2008; Hunter *et al.*, 2016]. The hydrogen produced could be used directly as a source of energy but could also be used, as it is in natural photosynthesis, to reduce CO_2 to other types of fuels such as methane, methanol, *etc.* The challenge is to have a molecular arrangement so that the artificial catalysts efficiently utilize light energy to split water and concomitantly provide reducing potential for hydrogen gas production or CO_2 reduction. It has been demonstrated that catalysts based on Mn or on Mn doped with Ca^{2+} are capable of water splitting leading to the generation of dioxygen [Limberg *et al.*, 1999; Najafpour, 2011; Najafpour *et al.*, 2010, 2012; Tagore *et al.*, 2008; Zaharieva *et al.*, 2011]. Frei's group reported that nanostructured manganese oxide clusters supported on mesoporous silica efficiently evolved oxygen in aqueous solution under mild conditions [Jiao and Frei, 2010]. The recently synthesised $Mn_4Ca^{2+}O_4$-cluster [Zhang *et al.*, 2015] shows charge storage under strong oxidation conditions similar to those observed in the natural PSII-cluster progressing to S_3 but has yet to be shown that it can catalyse the generation of oxygen from water.

Complementing the work on Mn has been the earlier discovery mentioned above, that ruthenium based catalysts, such as the "blue dimer" can photooxidise water to dioxygen [Gersten *et al.*, 1982; Liu *et al.*, 2008] and the recent spectacular work of Duan *et al.* [2012] and Maheu *et al.* [2015] have led to rationally designed Ru-based catalysts with unprecedented turnover rates mentioned above. However, perhaps the most practical catalysts for water splitting are based on cobalt, which is a relatively abundant element. Kanan and Nocera [2008] have described a self-assembling catalyst composed of Co and phosphate ions, which can produce molecular oxygen from water at neutral pH with a low over potential akin to that which operates in the OEC of PSII. Dau and his colleagues [Risch *et al.*, 2009] have revealed important structural details and properties of this Co-based catalyst and found it to have a molecular organisation remarkably similar to the Mn_3Ca-cubane of PSII. Du and Eisenberg [2012] have reviewed recent progress and future challenges of using Co as a water splitting catalyst compared with other transition metals.

Hematite (α-Fe_2O_3) is a semiconductor which is also capable of photoelectochemically splitting water to molecular oxygen. It has a favorable optical band gap ($Eg = 2.2$ eV), excellent chemical stability in aqueous environments, natural abundance and low cost [Bassi *et al.*, 2014; Sivula *et al.*, 2011]. Indeed, hematite has been theoretically predicted to achieve a water oxidation efficiency of 16.8% [Murphy *et al.*, 2006]. However, the reported efficiencies of hematite are lower than this predicted value, mainly due to the very short lifetime of photo-generated charge carriers, short hole diffusion length (2–4 nm), slow oxygen evolution reaction kinetics, low flat band potential and significant reduction in the absorption cross-section at wavelengths approaching the band gap value. Another fundamental limitation of the hematite system is the need for externally applied bias because the conduction band of hematite is lower than the potential required to reduce protons to hydrogen. Nevertheless, it is a system which is receiving considerable attention at the present time and the recent construction of a tandem cell incorporating perovskite and Co-doped hematite nanorods and requiring no external electrical bias for photo-driven water splitting [Gurudayal *et al.*, 2015], is an example of the potential for this type of photoanode for constructing an "artificial leaf" for solar fuel production (see Figure 14).

The next step will be to couple these oxygen producing systems to non-platinum catalysts which will use the protons and high-energy electrons derived from the water splitting reaction to produce hydrogen gas or reduce CO_2. In the case of the former, considerable progress is being made [Tran and Barber, 2012; Wang *et al.*, 2012], in part by mimicking the natural hydrogenase enzymes found in a wide variety of microorganisms [Tran *et al.*, 2010]. Also a number of inorganic catalysts have been identified with activities which are almost as good as platinum [Tran and Barber, 2012]. One such class of catalysts are based on sulphides of Mo and W [Merki and Hu, 2011; Tran *et al.*, 2013, 2016; Zong *et al.*, 2011] and

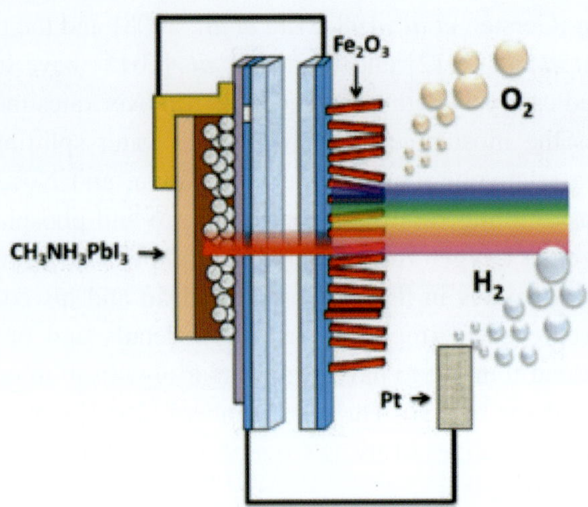

Figure 14. Diagrammatic representation of a tandem photoelectrochemical cell incorporating perovskite ($CH_3NH_3PbI_3$) and hematite (Fe_2O_3) nanorods and requiring no external electrical bias for photo-driven water splitting [Gurudayal *et al.*, 2015].

another is an alloy composed of nickel, molybdenum and zinc [Reece *et al.*, 2011]. Identifying catalysts for the reduction of CO_2, however, is more difficult because multielectron reactions are required and the emergence of catalysts for generating useful carbon fuels will require considerable effort [Arakawa *et al.*, 2001; Fujita, 2000].

The most successful coupling of catalysts using a semiconductor for light capture and charge separation was reported by Nocera and colleagues [Reece *et al.*, 2011]. They used a triple junction amorphous Si wafer as the semiconductor, the CoPi catalyst for water splitting and the NiMoZn alloy for the cathodic hydrogen producing catalyst as shown in Figure 15.

This demonstration is a significant step towards the development of an efficient, robust, low-cost and scalable photocatalytic device for water splitting at neutral pH to generate molecular hydrogen using only sunlight as the energy source.

7. Conclusions

Several lines of evidence suggest that in its oxidized state, the $Mn_4Ca^{2+}O_4$ unit of the water splitting catalytic centre of PSII is a stable chemical and robust entity able to assemble in the absence of protein. The additional oxo (O4) bridging between dangler Mn4 and the cubane is not present in the synthesised cluster suggesting it is more labile and for this reason may be involved in the chemistry of

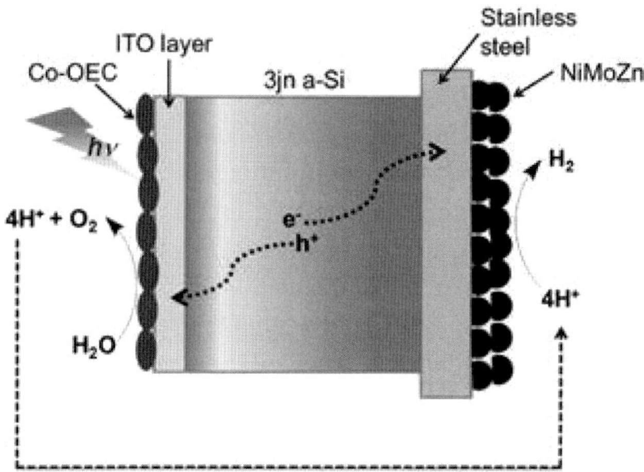

Figure 15. Diagrammatic representation of Nocera's photocatalytic "artificial leaf" device for water splitting consisting of a triple n-junction amorphous silicon wafer, Co-based oxygen evolving catalyst and NiMoZn H_2-evolving catalyst as reported in Reece *et al.* [2011].

O–O bond formation. Nevertheless, the robustness of the $Mn_4Ca^{2+}O_4$ unit suggests that it maintains its structure throughout the S-state cycle. If this is the case, then it would rule out the Siegbahn mechanism which involves the insertion of substrate water molecules into the cubane structure during the S_2 to S_3 transition and during the re-formation of cubane geometry associated with the S_4 to S_0 step of the S-state cycle. Moreover, these large conformational changes in the catalytic centre will be energy demanding and thus likely to be incompatible with the low over potential of water splitting reaction of PSII. On the other hand, the nucleophilic mechanism presented in Figure 7 occurs on the surface of the $Mn_4Ca^{2+}O_5$ cluster and therefore involves no large energy demanding conformational changes of the cubane. Moreover this pathway for O–O bond formation seems well established in synthesized complexes of Ru and Mn activated by the addition of chemical oxidants. The similarity of PSII with Fe–Ni CODH to split water to obtain reducing equivalents is also persuasive of a nucleophilic surface mechanism. Note that the bridging O4 could also be a target for a nucleophilic attack without disrupting the cubane structure but its acidity would be lower than that of a terminal Mn^V-oxo and therefore a weaker electrophile [Winkler and Gray, 2011]. However identifying the precise mechanism of O–O bond formation in PSII will be a significant challenge since the reactants; H_2O, OH and O differ only by one or two protons. Although recent studies directed at this challenge using femtosecond XFEL diffraction are technically impressive, the resolutions of the data obtained (approximately 2 to 2.5 Å), is far too low to distinguish between these reactants, which will need to be at least 1 Å or less.

References

Ames, W., Pantazis, D.A., Krewald, V., Cox, N., Messinger, J., Lubitz, W. and Neese, F. (2011). Theoretical evaluation of structural models of the S_2 state in the oxygen evolving complex of photosystem II: protonation states and magnetic interactions, *J. Am. Chem. Soc.,* 133, 19743–19757.

Ananyev, G.M. and Dismukes, G.C. (1996). Assembly of the tetra-Mn site of photosynthetic water oxidation by photoactivation: Mn stoichiometry and detection of a new intermediate, *Biochemistry*, 35, 4102–4109.

Arakawa, H., Aresta, M., Armor, J.N., Bartaux, M.A., Beckman, E.J., Bell, A.T., … Tumas, W. (2001). Catalysis research of relevance to carbon management: progress, challenges, and opportunities, *Chem. Rev.,* 101, 953–996.

Askerka, M., Wang, J., Vinyard, D.J., Brudvig, G.W. and Batista, V.S. (2016). S_3 state of the O_2-evolving complex of photosystem II: insights from QM/MM, EXAFS, and femtosecond X-ray diffraction, *Biochemistry*, 55, 981–984.

Barber, J. (2003). Photosystem II: the engine of life, *Q. Rev. Biophys.,* 36, 71–89.

Barber, J. (2009). Photosynthetic energy conversion: natural and artificial, *Chem. Soc. Rev.,* 38, 185–196.

Barber, J. (2016a). Photosystem II: the water splitting enzyme of photosynthesis and the origin of oxygen in our atmosphere, *Q. Rev. Biophys.,* 49, 1–21.

Barber, J. (2016b). Mn_4Ca cluster of photosynthetic oxygen-evolving center: structure, function and evolution, *Biochemistry,* 55, 5901–5906.

Barber, J. (2017). A mechanism for water splitting and oxygen production in photosynthesis, *Nat. Plants*, 3, 17041–17046.

Barber, J. and Andersson, B. (1992). Too much of a good thing: light can be bad for photosynthesis, *Trends Biochem. Sci.,* 17, 61–66.

Barber, J., Ferreira, K., Maghlaoui, K. and Iwata, S. (2004). Structural model of the oxygen-evolving centre of photosystem II with mechanistic implications, *Phys. Chem. Chem. Phys.,* 6, 4737–4742.

Bassi, P.S., Wong, L.H. and Barber, J. (2014). Iron based photoanodes for solar fuel production, *Phys. Chem. Chem. Phys.,* 16, 11834–11842.

Boussac, A., Rutherford, A.W. and Styring, S. (1990). Interaction of ammonia with the water splitting enzyme of photosystem II, *Biochemistry,* 29, 24–32.

Bovi, D., Narzi, D. and Guidoni, L. (2013). The S_2 state of the oxygen evolving complex of photosystem II explored by QM/MM dynamics: spin surfaces and metastable states suggest a reaction path towards the S_3 state, *Angew. Chem. Int. Ed. Engl.,* 52, 11744–11749.

Britt, R.D., Campbell, K.A., Peloquin, J.M., Gilchrist, M.L., Aznar, C.P., Dicus, M.M., Robblee, J. and Messinger, J. (2004). Recent pulsed EPR studies of the photosystem II oxygen-evolving complex: implications as to water oxidation mechanisms, *Biochim. Biophys. Acta,* 1655, 158–171.

Brudvig, G.W. (2008). Water oxidation chemistry of photosystem II, *Phil. Trans. R. Soc. Lond. B Biol. Sci.,* 363, 1211–1218.

Büchel, C., Barber, J., Ananyev, G., Eshaghi, S., Watt, R. and Dismukes, C. (1999). Photoassembly of the manganese cluster and oxygen evolution from monomeric and dimeric CP47-reaction centre photosystem II complexes, *Proc. Natl. Acad. Sci. USA,* 96, 14288–14293.

Cady, C.W., Crabtree, R.H. and Brudvig, G.W. (2008). Functional models for the oxygen-evolving complex of photosystem II, *Coord. Chem. Rev.,* 252, 444–455.

Canfield, D.E. (2005). The early history of atmospheric oxygen: homage to Robert M. Garrels, *Annu. Rev. Earth Planet Sci.,* 33, 1–36.

Clausen, J., Winkler, S., Hays, A.M.A., Hundelt, M., Debus, R.J. and Junge, W. (2001). Photosynthetic water oxidation in *Synechocystis* sp. PCC6803: mutations D1-E189K, R and Q are without influence on electron transfer at the donor side of photosystem II, *Biochim. Biophys. Acta,* 1506, 224–235.

Concepcion, J.J., Jurss, J.W., Brennaman, M.K., Hoertz, P.G., Patrocinio, A.O.T., Murakami Iha, N.Y., Templeton, J.L. and Meyer, T.J. (2009). Making oxygen with ruthenium complexes, *Acc. Chem. Res.,* 42, 1954–1965.

Cox, N., Retegan, M., Neese, F., Pantais, D.A., Boussac, A. and Lubitz, W. (2014). Electronic structure of the oxygen-evolving complex in photosystem II prior to O-O bond formation, *Science,* 345, 804–808.

Dau, H. and Haumann, M. (2007). Eight steps preceding O-O bond formation in oxygenic photosynthesis—a basic reaction cycle of the photosystem II manganese complex, *Biochim. Biophys. Acta,* 1767, 472–483.

Dau, H., Grundmeier, A., Loja, P. and Haumann, M. (2008). On the structure of the manganese complex of photosystem II: extended-range EXAFS data and specific atomic-resolution models for four S-states, *Philos. Trans. R. Soc. Lond. B Biol. Sci.,* 363, 1237–1243.

Debus, R.J. (1992). The manganese and calcium-ions of photosynthetic oxygen evolution, *Biochim. Biophys. Acta,* 1102, 269–352.

Debus, R.J. (2008). Protein ligation of the photosynthetic oxygen-evolving center, *Coord. Chem. Rev.* 252, 244–258.

Diner, B.A., Nixon, P.J. and Farchaus, J.W. (1991). Site-directed mutagenesis of photosynthetic reaction centers, *Curr. Opin. Struct. Biol.,* 1, 546–554.

Dobbek, H., Svetlitchnyi, V., Gremer, L., Huber, R. and Meyer, O. (2001). Crystal structure of a carbon monoxide dehydrogenase reveals a [Ni-4Fe-5S] cluster, *Science,* 293, 1281–1285.

Du, P. and Eisenberg, R. (2012). Catalysts made of earth-abundant elements (Co, Ni, Fe) for water splitting: recent progress and future challenges, *Energy Environ. Sci.,* 5, 6012–6021.

Duan, L., Bozoglian, F., Mandal, S., Stewart, B., Privalov, T., Llobet, A. and Sun, L. (2012). A molecular ruthenium catalyst with water-oxidation activity comparable to that of photosystem II, *Nat. Chem.,* 4, 418–423.

Duan, L., Wang, L., Li, F., Li, F. and Sun, L. (2015). Highly efficient bioinspired molecular Ru water oxidation catalysts with negatively charged backbone ligands, *Acc. Chem. Res.,* 48, 2084–2096.

Eisenberg, R. and Gray, H.B. (2008). Preface on making oxygen, *Inorg. Chem.,* 47, 1697–1699.

Ferreira, K.N., Iverson, T.M., Maghlaoui, K., Barber, J. and Iwata, S. (2004). Architecture of the photosynthetic oxygen-evolving center, *Science,* 303, 1831–1838.

Fesseler, J., Jeoung, J.-H. and Dubbed, H. (2015). How the $[NiFe_4S_4]$ cluster of CO dehydrogenase activates CO_2 and NCO(-), *Angew. Chem. Int. Ed. Engl.,* 54, 8560–8564.

Fontana, F. (1780). Accounts of air extracted from different kinds of water, *Philos. Trans. R. Soc. Lond. B Biol. Sci.,* 69, 432–453.

Fujita, E. (2000). Carbon dioxide reduction. In *McGraw-Hill Yearbook of Science & Technology,* Licker, M.D., ed. (New York: McGraw-Hill Book), pp. 71–74.

Gao, Y., Åkermark, T., Liu, J., Sun, L. and Åkermark, B. (2009). Nucleophilic attack of hydroxide on a MnV-oxo complex: a model of the O-O bond formation in the oxygen evolving complex of photosystem II, *J. Am. Chem. Soc.,* 131, 8726–8727.

Gao, Y., Liu, J., Wang, M., Na, Y., Åkermark, B. and Sun, L. (2007). Synthesis and characterization of manganese and copper corrole xanthene complexes as catalysts for water oxidation, *Tetrahedron,* 63, 1987–1994.

Gersten, S.W., Samuels, G.J. and Meyer, T.J. (1982). Catalytic oxidation of water by an oxo-bridged ruthenium dimer, *J. Am. Chem. Soc.,* 104, 4029–4030.

Gong, W., Hao, B., Wei, Z., Ferguson, D.J., Jr., Tallant, T., Krzycki, J.A. and Chan, M.K. (2008). Structure of the $\alpha 2\varepsilon 2$ Ni-dependent CO dehydrogenase component of the *Methanosarcina barkeri* acetyl-CoA decarbonylase/synthase complex, *Proc. Natl. Acad. Sci. USA,* 105, 9558–9563.

Gurdayel, Sabba, D., Kumar, M.H., Wong, L.H., Barber, J., Grätzel, M. and Mathews, N. (2015). Perovskite-hematite tandem cells for efficient overall solar driven water splitting, *Nano Lett.,* 15, 3833–3839.

Haddy, A. (2007). EPR spectroscopy of the manganese cluster of photosystem II, *Photosynth. Res.,* 92, 357–368.

Haumann, M., Liebisch, P., Müller, C., Barra, M., Grabolle, M. and Dau, H. (2005). Photosynthetic O_2 formation tracked by time-resolved X-ray experiments, *Science,* 310, 1019–1021.

Hoganson, C.W. and Babcock, G.T. (1997). A metalloradical mechanism for the generation of oxygen from water in photosynthesis, *Science,* 277, 1953–1956.

Hunter, B.M., Gray, H.B. and Luller, A.M. (2016). Earth-abundant heterogeneous water oxidation catalysts, *Chem. Rev.* 116, 14120–14136.

Jeoung, J.H. and Dobbek, H. (2007). Carbon dioxide activation at the Ni, Fe-cluster of anaerobic carbon monoxide dehydrogenase, *Science,* 318, 1461–1464.

Jiao, F. and Frei, H. (2010). Nanostructured manganese oxide clusters supported on mesoporous silica as efficient oxygen-evolving catalysts, *Chem. Comm.,* 46, 2920–2922.

Joliot, P., Barbieri, G. and Chabaud, R. (1969). Un nouveau modele des centres photochimiques du systeme II, *Photochem. Photobiol.,* 10, 309–329.

Kanady, S., Tsui, E., Day, M. and Agapie, T. (2011). A synthetic model of the Mn_3Ca subsite of the oxygen-evolving complex in photosystem II, *Science* 333, 733–736.

Kanan, M.W. and Nocera, D.G. (2008). *In situ* formation of an oxygen-evolving catalyst in neutral water containing phosphate and Co^{2+}, *Science*, 321, 1072–1075.

Kok, B., Forbush, B. and McGloin, M. (1970). Cooperation of charges in photosynthetic O_2 evolution. 1. A linear four-step mechanism, *Photochem. Photobiol.* 11, 467–475.

Li, X. and Siegbahn, P.E. (2015). Alternative mechanisms for O_2 release and O-O bond formation in the oxygen evolving complex of photosystem II, *Phys. Chem. Chem. Phys.*, 17, 12168–12174.

Limburg, J., Brudvig, G.W. and Crabtree, R.H. (1997). O_2 evolution and permanganate formation from high-valent manganese complexes, *J. Am. Chem. Soc.*, 119, 2761–2762.

Limburg, J., Vrettos, J.S., Chen, H., de Paula, J.C., Crabtree, R.H. and Brudvig, G.W. (2001). Characterization of the O_2-evolving reaction catalyzed by [(terpy)(H_2O) Mn^{III} (O) $2Mn^{IV}$ (OH_2)(terpy)]$(NO_3)^3$ (terpy = 2,2′:6,2″-terpyridine), *J. Am. Chem. Soc.*, 123, 423–430.

Limburg, J., Vrettos, J.S., Liable-Sands, L.M., Rheingold, A.L., Crabtree, R.H. and Brudvig, G.W. (1999). A functional model for O-O bond formation by the O_2-evolving complex in photosystem II, *Science*, 283, 1524–1527.

Lohmiller, T., Krewald, V., Navarro, M.P., Retegan, M., Rapatskiy, L., Nowaczyk, M.M., Boussac, A., Neese, F., Lubitz, W., Pantazis, D.A. and Cox, N. (2014). Structure, ligands and substrate coordination of the oxygen-evolving complex of photosystem II in the S_2 state: a combined EPR and DFT study, *Phys. Chem. Chem. Phys.*, 16, 11877–11892.

Luber, S., Rivalta, I., Umena, Y., Kawakami, K., Shen, J.R., Kamiya, N., Brudvig, G.W. and Batista, V.S. (2011). S_1-state model of the O_2-evolving complex of photosystem II, *Biochemistry,* 50, 6308–6311.

Liu, F., Concepcion, J.J., Jurss, J.W., Cardolaccia, T., Templeton, J.L. and Meyer, T.J. (2008) Mechanisms of water oxidation from the blue dimer to photosystem II, *Inorg. Chem.,* 47, 1727–1752.

Matheu, R., Ertem, M.Z., Benet-Buchholz, J., Coronado, E., Batista, V.S., Sala, X. and Llobet, A. (2015). Intramolecular proton transfer boosts water oxidation catalyzed by a Ru complex, *J. Am. Chem. Soc.,* 137, 10786–10795.

McConnell, I.L., Grigoryants, V.M., Scholes, C.P., Myers, W.K., Chen, P.Y., Whittaker, J.W. and Brudvig, G.W. (2012). EPR-ENDOR characterization of (^{17}O, ^1H, ^2H) water in manganese catalase and its relevance to the oxygen-evolving complex of photosystem II, *J. Am. Chem. Soc.,* 134, 1504–1512.

Merki, D. and Hu, X. (2011). Recent developments of molybdenum and tungsten as hydrogen evolution catalysts, *Energy Environ. Sci.,* 4, 3878–3888.

Messinger, J., Badger, M. and Wydrzynski, T. (1995). Detection of one slowly exchanging substrate water molecule in the S_3 state of photosystem II, *Proc. Natl. Acad. Sci. USA,* 92, 3209–3213.

Misra, A., Wernsdorfer, W., Abboud, K.A. and Christou, G. (2005). The first high oxidation state manganese-calcium cluster: Relevance to the water oxidizing complex of photosynthesis, *Chem. Commun.,* 1, 54–56.

Mukherjee, S., Stull, J.A., Yano, J., Stamatatos, T., Pringouri, K., Stich, T.A., Abbroud, K.A., Britt, R.D., Yachandra, V.K. and Christou, G. (2012). Synthetic model of the asymmetric [Mn$_3$CaO$_4$] cubane core of the oxygen-evolving complex of photosystem II, *Proc. Natl. Acad. USA,* 109, 2257–2262.

Murakami, M., Hong, D., Suenobu, T., Yamaguchi, S., Ogura, T. and Fukuzumi, S. (2011). Catalytic mechanism of water oxidation with single-site ruthenium-heteropolytungstate complexes, *J. Am. Chem. Soc.,* 133, 11605–11613.

Murphy, A.B., Barnes, P.R.F., Randeniya, L.K., Plumb, I.C., Grey, I.E., Horne, M.D. and Glasscock, J.A. (2006). Efficiency of solar water splitting using semiconductor electrodes, *Int. J. Hydrogen Energy,* 31, 1999–2017.

Murray, J.W. (2012). Sequence variation at the oxygen-evolving centre of photosystem II: a new class of 'rogue' cyanobacterial D1 proteins, *Photosynth. Res.,* 110, 177–184.

Murray, J.W. and Barber, J. (2007). Structural characteristics of channels and pathways in Photosystem II including the identification of an oxygen channel, *J. Struct. Biol.* 159, 228–237.

Najafpour, M.M. (2011). Hollandite as a functional and structural model for the biological water oxidizing complex: manganese-calcium oxide minerals as a possible evolutionary origin for the CaMn$_4$ cluster of the biological water oxidizing complex, *Geomicrobiol. J.,* 28, 714–718.

Najafpour, M.M., Ehrenberg, T., Wiechen, M. and Kurz, P. (2010). Calcium manganese (III) oxides (CaMn$_2$O$_4$ × H$_2$O) as biomimetic oxygen-evolving catalysts, *Angew. Chem. Int. Ed. Engl.,* 49, 2233–2237.

Najafpour, M.M., Rahimi, F., Aro, E.M., Lee, C.H. and Allakhverdiev, S.I. (2012). Nano-sized manganese oxides as biomimetic catalysts for water oxidation in artificial photosynthesis: a review, *J. R. Soc. Interface,* 9, 2383–2395.

Navarro, M.P., Ames, W.M., Nilsson, H., Lohmiller, T., Pantazis, D.A., Rapatskiy, L., Nowaczyk, M.M., Neese, F., Boussac, A., Messinger, J. and Lubitz, W. (2013). Ammonia binding to the oxygen-evolving complex of photosystem II identifies the solvent-exchangeable oxygen bridge (μ-oxo) of the manganese tetramer, *Proc. Natl. Acad. Sci. USA,* 110, 15561–15566.

Oyala, P.H., Stich, T.A., Debus, R.J. and Britt, R.D. (2015). Ammonia binds to the dangler manganese of the photosystem II oxygen-evolving complex. *J. Am. Chem. Soc.,* 137, 8829–8837.

Pecoraro, V.L., Baldwin, M.J., Caudle, M.T., Hsieh, W.Y. and Law, N.A. (1998). A proposal for water oxidation in photosystem II, *Pure Appl. Chem.,* 70, 925–929.

Pecoraro, V.L., Baldwin, M.J. and Gelasco, A. (1994). Interaction of manganese with dioxygen and its reduced derivatives, *Chem. Rev.,* 94, 807–826.

Pérez-Navarro, M., Neese, F., Lubitz, W., Pantazis, D.A. and Cox, N. (2016). Recent developments in biological water oxidation, *Curr. Opin. Chem. Biol.,* 31, 113–119.

Preistley, J. (1772). Observations on different kinds of air, *Phil. Trans. R. Soc. Lond.,* 62, 147–264.

Rapatskiy, L., Cox, N., Savitsky, A., Ames, W.M., Sander, J., Nowaczyk, M.M., Rögner, M., Boussac, A., Neese, F., Messinger, J. and Lubitz, W. (2012). Detection of the water-binding sites of the oxygen-evolving complex of photosystem II using W-band ^{17}O electron-electron double resonance-detected NMR spectroscopy, *J. Am. Chem. Soc.*, 134, 16619–16634.

Reece, S.Y., Hamel, J.A., Sung, K., Jarvi, T.D., Esswein, A.J., Pijpers, J.J. and Nocera, D.G. (2011). Wireless solar water splitting using silicon-based semiconductors and earth-abundant catalysts, *Science*, 334, 645–648.

Risch, M., Khare, V., Zaharieva, I., Gerencser, L., Chernev, P. and Dau, H. (2009). Cobalt-oxo core of a water-oxidizing catalyst film, *J. Am. Chem. Soc.*, 131, 6936–6937.

Ruettinger, W.F., Ho, D.M. and Dismukes, G.C. (1999). Protonation and dehydration reactions of the $Mn_4O_4L_6$ cubane and synthesis and crystal structure of the oxidized cubane $[Mn_4O_4L_6]^+$: A model for the photosynthetic water oxidizing complex, *Inorg. Chem.*, 38, 1036–1037.

Sala, X., Maji, S., Bofill, R., García-Antón, J., Escriche, L. and Llobet, A. (2013). Molecular water oxidation mechanisms followed by transition metals: state of the art, *Acc. Chem. Res.*, 47, 504–516.

Service, R.J., Yano, J., McConnell, I., Hwang, H.J., Niks, D., Hille, R., Wydrzynski, T., Burnap, R.L., Hillier, W. and Debus, R.J. (2010). Participation of glutamate-354 of the CP43 polypeptide in the ligation of manganese and the binding of substrate water in photosystem II, *Biochemistry*, 50, 63–81.

Siegbahn, P.E. (2006). O–O bond formation in the S_4-state of the oxygen evolving complex in photosystem II, *Chem. Eur. J.*, 12, 9217–9237.

Siegbahn, P.E. (2008). A structure-consistent mechanism for dioxygen formation in photosystem II, *Chemistry,* 14, 8290–8302.

Siegbahn, P.E. (2009). Structures and energetics for O_2 formation in photosystem II, *Acc. Chem. Res.*, 42, 1871–1880.

Siegbahn, P.E. (2013). Water oxidation mechanism in photosystem II, including oxidations, proton release pathways, O–O bond formation and O_2 release, *Biochim. Biophys. Acta*, 1827, 1003–1019.

Siegbahn, P.E. and Crabtree, R.H. (1999). Manganese oxyl radical intermediates and OO bond formation in photosynthetic oxygen evolution and a proposed role for the calcium cofactor in photosystem II, *J. Am. Chem. Soc.*, 121, 117–127.

Sivula, K., Le Formal, F. and Grätzel, M. (2011). Solar water splitting: progress using hematite (aFe_2O_3) photoelectrodes, *ChemSusChem,* 4, 432–449.

Smith, R.J., Loganathan, M. and Shantha, M.S. (2010). A review of the water gas shift reaction kinetics, *Int. J. Chem. Reactor Eng.*, 8. doi:10.2202/1542-6580.2238.

Sproviero, E.M., Gascón, J.A., McEvoy, J.P., Brudvig, G.W. and Batista,V. (2008). Quantum mechanics/molecular mechanics study of the catalytic cycle of water splitting in photosystem II, *J. Am. Chem. Soc.*, 130, 3428–3442.

Suga, M., Akita, F., Hirata, K., Ueno, G., Murakami, H., Nakajima,Y., Shimizu, T.,Yamashita, K., Yamamoto, M., Ago, H. and Shen, J.R. (2015). Native structure of photosystem II at 1.95 A resolution viewed by femtosecond X-ray pulses, *Nature*, 517, 99–103.

Suga *et al.* (2017). Light-induced structural changes and the site of O–O bond formation in PSII caught by XFEL, *Nature*, 543, 131–135.

Svetlitchnyi, V., Dobbek, H., Meyer-Klaucke, W., Meins, T., Thiele, B., Römer, P., Huber, R. and Meyer, O. (2004). A functional Ni–Ni–[4Fe–4S] cluster in the monomeric acetyl-CoA synthase from *Carboxydothermus hydrogenoformans*, *Proc. Natl. Acad. Sci. USA*, 101, 446–451.

Tagore, R., Crabtree, R.H. and Brudvig, G.W. (2008). Oxygen evolution catalysis by a dimanganese complex and its relation to photosynthetic water oxidation, *Inorg. Chem.*, 47, 1815–1823.

Tommos, C. and Babcock, G.T. (1998). Oxygen production in nature: a light-driven metalloradical enzyme process, *Acc. Chem. Res.*, 31, 18–25.

Tong, L., Duan, L., Xu, Y., Privalov, T. and Sun, L. (2011). Structural modifications of mononuclear ruthenium complexes: a combined experimental and theoretical study on the kinetics of ruthenium catalyzed water oxidation, *Angew. Chem. Int. Ed. Engl.*, 50, 445–449.

Tran, P.D., Artero, V. and Fontecave, M. (2010). Water electrolysis and photoelectrolysis on electrodes engineered using biological and bio-inspired molecular systems, *Energy Environ. Sci.*, 3, 727–747.

Tran, P.D. and Barber, J. (2012). Proton reduction to hydrogen in biological and chemical systems, *Phys. Chem. Chem. Phys.*, 14, 13772–13784.

Tran, P.D., Chiam, S.Y., Boix, P.P., Ren, Y., Pramana, S.S., Fize, J., Artero, V. and Barber, J. (2013). Novel cobalt/nickel-tungsten-sulfide catalysts for electrocatalytic hydrogen generation from water, *Energy Environ. Sci.*, 6, 2452–2459.

Tran, P.D., Tran, T.V., Orio, M., Torelli, S., Truong, Q.D., Nayuki, K., Sasaki, Y., Chiam, S.Y., Yi, R., Honma, I. Barber, J. and Artero, V. (2016). Coordination polymer structure and revisited hydrogen evolution catalytic mechanism for amorphous molybdenum sulfide, *Nat. Mater.*, 15, 640–646.

Tran, P.D., Wong, L.H., Barber, J. and Loo, J.S. (2012). Recent advances in hybrid photocatalysts for solar fuel production, *Energy Environ. Sci.*, 5, 5902–5918.

Umena, Y., Kawakami, K., Shen, J.R. and Kamiya, N. (2011). Crystal structure of oxygen-evolving photosystem II at a resolution of 1.9 Å, *Nature*, 473, 55–60.

Vrettos, J.S., Limburg, J. and Brudvig, G.W. (2001). Mechanism of photosynthetic water oxidation: combining biophysical studies of photosystem II with inorganic model chemistry, *Biochim. Biophys. Acta*, 1503, 229–245.

Vinyard, D.J., Khan, S. and Brudvig, G.W. (2015). Photosynthetic water oxidation: binding and activation of substrate waters for O–O bond formation, *Faraday Discuss.*, 185, 37–50.

Wang, M., Chen, L. and Sun, L. (2012). Recent progress in electrochemical hydrogen production with earth-abundant metal complexes as catalysts, *Energy Environ. Sci.*, 5, 6763–6778.

Winkler, J.R. and Gray, H.B. (2011). *Molecular Electronic Structures of Transition Metal Complexes I* (Springer Berlin, Heidelberg), pp. 17–28.

Yachandra, V.K. (2002). Structure of the Mn complex in Photosystem II: Insights from X-ray spectroscopy, *Philos. Trans. R. Soc. Lond. B Biol. Sci.*, 357, 1347–1358.

Yamanaka, S., Isobe, H., Kanda, K., Saito, T., Umena, Y., Kawakami, K., Shen, J.R., Kamiya, N., Okumura, M., Nakamura, H. and Yamaguchi, K. (2011). Possible mechanisms for the O-O bond formation in oxygen evolution reaction at the $CaMn_4O_5(H_2O)_4$ cluster of PSII refined to 1.9 Å X-ray resolution, *Chem. Phys. Lett.*, 511, 138–145.

Yano, J., Kern, J., Irrgang, K.D., Latimer, M.J., Bergmann, U., Glatzel, P., Pushkar, Y., Biesiadka, J., Loll, B., Sauer, K. and Messinger, J. (2005). X-ray damage to the Mn_4Ca complex in single crystals of photosystem II: a case study for metalloprotein crystallography, *Proc. Natl. Acad. Sci. USA*, 102, 12047–12052.

Yano, J., Kern, J., Sauer, K., Latimer, M.J., Pushkar, Y., Biesiadka, J., Loll, B., Saenger, W., Messinger, J., Zouni, A. and Yachandra, V.K. (2006). Where water is oxidized to dioxygen: structure of the photosynthetic Mn_4Ca cluster, *Science*, 314, 821–825.

Young *et al.* (2016). Structure of photosystem II and substrate binding at room temperature, *Nature*, 540, 453–457.

Zaharieva, I., Najafpour, M.M., Wiechert, M., Haumann, M., Kurz, P. and Dau, H. (2011). Synthetic manganese-calcium oxides mimic the water-oxidizing complex of photosynthesis functionally and structurally, *Energy Environ. Sci.*, 4, 2400–2408.

Zhang, C., Chen, C., Dong, H., Shen, J.R., Dau, H. and Zhao, J. (2015). A synthetic Mn_4Ca-cluster mimicking the oxygen-evolving center of photosynthesis, *Science*, 348, 690–693.

Zong, X., Han, J., Ma, G., Yan, H., Wu, G. and Li, C. (2011). Photocatalytic H_2 evolution on CdS loaded with WS_2 as cocatalyst under visible light irradiation, *J. Phys. Chem. C*, 115, 12202–12208.

Chapter 7

A Quest for the Atomic Resolution of Plant Photosystem I

Nathan Nelson

*Department of Biochemistry and Molecular Biology,
The George S. Wise Faculty of Life Sciences,
Tel Aviv University, Tel Aviv, 69978, Israel*
nelson@post.tau.ac.il

Plant photosystem I (PSI) is one of the most intricate membrane complexes in nature. It comprises two distinct building blocks; the reaction center and the light-harvesting complexes (LHC). They contain 16 subunits and over 200 prosthetic groups, including chlorophylls, carotenoids, quinines, phospholipids, glycolipids, and three iron-sulfur clusters. Our group was the first to determine the crystal structure of plant PSI. That first structure was solved to 4.4 Å resolution. Over the last 14 years, we have constantly improved the purification procedure, crystallization conditions, and phasing, by combining numerous datasets. Currently, we have approached 2 Å resolution, which well defines the presence of chlorophylls *a* and *b* and the various carotenoids. The crystal structure of plant PSI was solved from two distinct crystal forms. The first was crystallized at pH 6.5 and exhibited P21 symmetry. The second was crystallized at pH 8.5 and exhibited P212121 symmetry. The surfaces involved in binding plastocyanin and ferredoxin were identical in both crystal forms.

1. Introduction

Currently, we know enough about the basic rules underlying atoms and their interactions to rule out miracles, telekinesis, and life after death, among other phenomena. Disappointingly, life consists of only three essential particles: electrons, protons, and neutrons. These particles interact *via* three forces: nuclear force, electromagnetism, and gravity. However, it is difficult to explain bioenergetic processes with simple physical and chemical principals. Half a century ago, biochemical and biophysical studies revealed that membrane-bound protein complexes catalyze the most fundamental processes in nature, respiration and photosynthesis. The three great scientists, Leslie Dutton, John Walker, and Jan Anderson that we honor in this book, came from different scientific camps, and each has their own unique flavor. All three were the subject of my envy for their command of the language. My long-lasting quest to understand the structure and function of chloroplast membrane complexes was initiated by reading Jan's paper [Anderson and Boardman, 1966]. Jan was able to isolate photochemically active particles from spinach chloroplasts. Later, I met Leslie Dutton at the 1971 Bioenergetics Gordon Conference. He and I were young members of the two "friendly", but opposing camps of Britton Chance and Efraim Racker, respectively. Eventually, Les gave us the physical "ruler" for measuring electron tunneling in biological processes [Page *et al.*, 1999]. John struck like thunder, first with his work in molecular biology, and then, with his demonstration of the marvels of structural biology [Walker *et al.*, 1982, 1984; Abrahams *et al.*, 1994; van Raaij *et al.*, 1996].

The photochemical functions of oxygenic photosynthesis are performed by two photosystems, known as photosystems I (PSI) and II (PSII) [Barber, 2004; Nelson and Ben-Shem, 2004]. PSII mediates the transfer of electrons from water, the initial electron donor, to the plastoquinone pool. PSI mediates electron transfer from plastocyanin (PC) to ferredoxin (Fd), thereby generating the reducing power needed for CO_2 fixation in the form of NADPH. The architecture of oxygenic photosynthesis in cyanobacteria has been largely determined, and the structures of both photosystems were solved at high resolution [Jordan *et al.*, 2001; Kurisu *et al.*, 2003; Ferreira *et al.*, 2004; Loll *et al.*, 2005; Umena *et al.*, 2011; Mazor *et al.*, 2014, 2017]. Although PSII is unique in its ability to extract electrons from water, PSI is arguably the most efficient photoelectric apparatus in nature; it exhibits nearly 100% quantum efficiency in its utilization of light for electron transport [Nelson, 2009; Nelson and Junge, 2015]. The ability of PSI to convert sunlight to energy depends on the precise spatial arrangement of the protein subunits and the relative positions of the cofactors. Consequently, understanding these mechanisms at a molecular level requires detailed knowledge of the three-dimensional arrangement of these complexes.

2. The Structure of Plant PSI

The solution of the crystal structure of the trimeric PSI from thermophilic cyanobacterium, *Thermosynechococcus elongatus* [Jordan *et al.*, 2001], provided an unprecedented breakthrough in understanding the light-induced excitation transfer in PSI. The structure revealed that, in the PSI trimer, each monomer is composed of 12 protein subunits and 96 chlorophylls. That finding led to detailed insights into the molecular architecture of this complex [Jordan *et al.*, 2001]. Recently, we solved the structure of the PSI in the mesophilic cyanobacterium, *Synechocystis* sp. PCC6803 [Mazor *et al.*, 2014; Malavath *et al.*, 2017]. Several differences were detected between the thermophilic and mesophilic PSI structures.

The crystal structure of the plant PSI was solved with two distinct crystal forms. The first form was crystallized at pH 6.5, and it displayed P21 symmetry [Ben-Shem *et al.*, 2003; Amunts *et al.*, 2007; Qin *et al.*, 2015]; the second form was crystallized at pH 8.5, and it displayed P212121 symmetry [Mazor *et al.*, 2015, 2017]. The composition and oligomeric state of PSI changed as the eukaryotic lineage evolved. New subunits, such as PsaH and PsaG appeared, and other subunits, such as PsaM, were lost [Mazor *et al.*, 2012]. Ultimately, this process resulted in an exclusively monomeric complex in eukaryotes. The newly exposed surfaces presented by the monomeric complex resulted in the eventual appearance of a super-complex, which includes the higher-plant PSI core, stably associated with an antenna complex, known as the light harvesting complex I (LHCI) [Ben-Shem *et al.*, 2003; Amunts *et al.*, 2007, 2010]. The road to solving the plant PSI was paved by an enormous biochemical work. These data, when combined with the crystal structure, revealed that the LHCI is composed of four nuclear gene products (Lhca1 to Lhca4), which assemble into two dimers, arranged in a series, thus creating a half-moon shaped belt, that docks to the core on the PsaF side. The LHCI belt contributes a mass of 180 kDa to the total 650 kDa of the holo-complex [Nelson and Yocum, 2006; Croce and van Amerongen, 2013]. Taken together, these structures represent a cellular stepping-stone, from which other, even larger cellular complexes are assembled. These complexes are formed as part of the transition between state I and state II in eukaryotes, and they may include additional complexes dedicated to cyclic electron flow around PSI [Lunde *et al.*, 2000; Scheller *et al.*, 2001; Standfuss *et al.*, 2005; Bellafiore *et al.*, 2005; Kouril *et al.*, 2005; Kargul *et al.*, 2005; Rochaix, 2011; Rochaix *et al.*, 2012].

The road to refining the resolution from 4.4 Å to nearly 2 Å was paved by good ideas, hard work, serendipity, and luck. The first crystal structure of the plant PSI was published in 2003, at a resolution of 4.4 Å [Ben-Shem *et al.*, 2003]. To my knowledge, it was the first plant membrane protein solved with X-ray crystallography. The structural features of the complex included two bound complexes: the core complex, which was highly homologous to the cyanobacterial PSI [Jordan *et al.*, 2001], and the light harvesting complex (LHc), which was

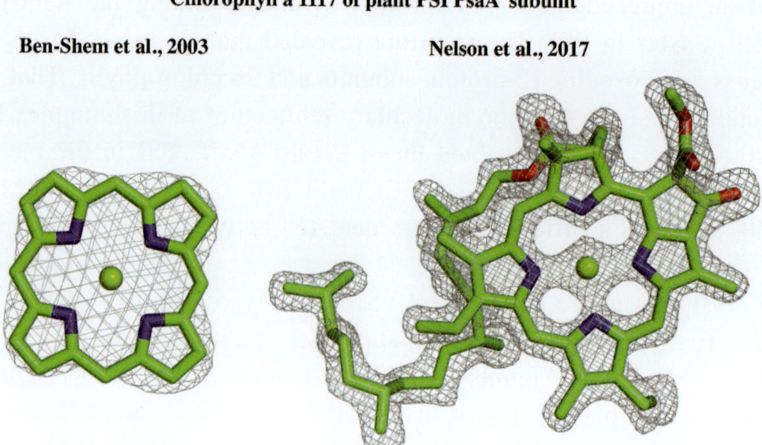

Figure 1.   Electron density maps of chlorophyll *a*: then and now. Electron density maps and model of chlorophyll *a* 1117 of PsaA then (Ben-Shem *et al.*, 2003) and now (Nelson *et al.*, unpublished).

composed of four distinct chlorophyll-protein subunits bound together in a unique way [Ben-Shem *et al.*, 2003]. Moreover, the identity and arrangement these subunits within the complex were determined. Two subunits connected to the core complex, PsaG and PsaH, were identified in the structure, and their evolutionary appearance and function was suggested [Nelson and Ben-Shem, 2004; Nelson and Yocum, 2006; Amunts *et al.*, 2007; Amunts and Nelson, 2009; Nelson, 2013; Nelson and Junge, 2015]. The resolution of the plant PSI structure improved over the years, and currently, we have approached a resolution of 2 Å [Amunts *et al.*, 2007, 2010; Mazor *et al.*, 2015, 2017; Nelson *et al.*, unpublished]. Figure 1 illustrates the electron density maps of one of the chlorophylls in 2003 and today. Figure 2 shows the structure of the plant PSI at a 2.6 Å resolution [Mazor *et al.*, 2017].

Two factors were pivotal in achieving high quality electron density maps, such as the one depicted in Figure 1, for the chlorophyll *a* molecule. The first factor was that enormous amounts of complete datasets had been collected. A cluster analysis performed on unit-cell-based data sets produced a merged dataset that comprised over 150 complete data sets. The second factor that contributed to high quality electron density maps was that proper phasing had been achieved by utilizing anomalous diffraction spots related to the sulfates and phosphates in the complex [Mazor *et al.*, 2015, 2017]. Serendipity played major role in the improved phasing of the plant PSI structure. We aimed to improve the phases with single-wavelength anomalous diffraction phasing of the PSI.

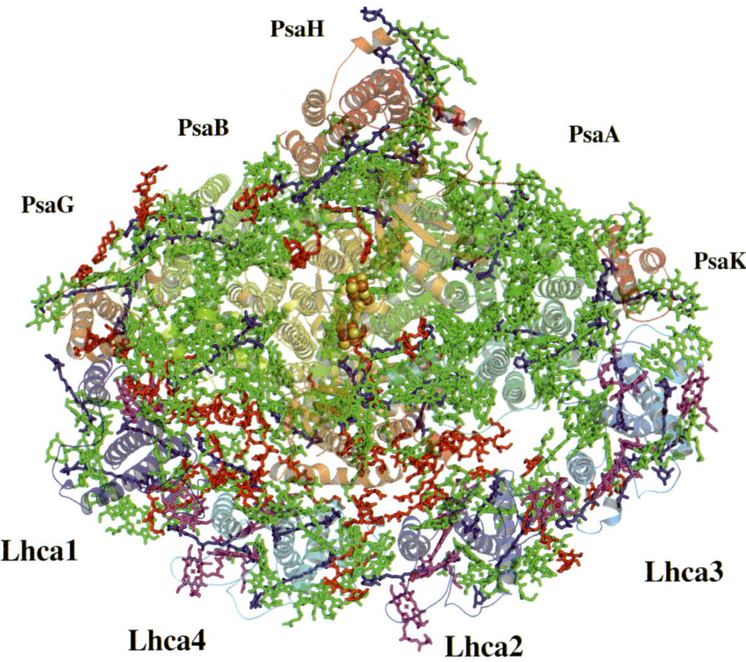

Figure 2. Plant PSI at 2.6 Å resolution. The model of plant PSI (Mazor *et al.*, 2017). Chlorophyll *a* in green, chlorophyll *b* in magenta, carotenoids in blue and lipids in red.

Because PSI contains three iron–sulfur clusters, data were collected at the iron edge (1.74 Å). Then, systematic errors were minimized by including measurements from multiple crystals, oriented differently in the beamline. The amount of data collected from each crystal represented a compromise, because radiation damage was the tradeoff for high resolution and the need to obtain a sufficient number of common reflections to be able to perform the subsequent scaling steps. Surprisingly, when PHASER software was run with the SHELXD substructure, 47 correct sulfur or iron atom positions were identified out of 110 atoms in the final model. The phases from PHASER were combined with the intensities from the high-resolution data set for use in the refinement step with PHENIX software. Refinement with the phenix-refine program was followed by successive rounds of manual model building, with the COOT program, and real space refinement with the phenix.real_space_refine tool [Emsley *et al.*, 2010]. As a bonus, anomalous signals were detected and identified as phosphates attached to some of the phospholipids within the complex. Figure 3 shows the position of the detected anomalous signals superimposed onto the plant PSI structure.

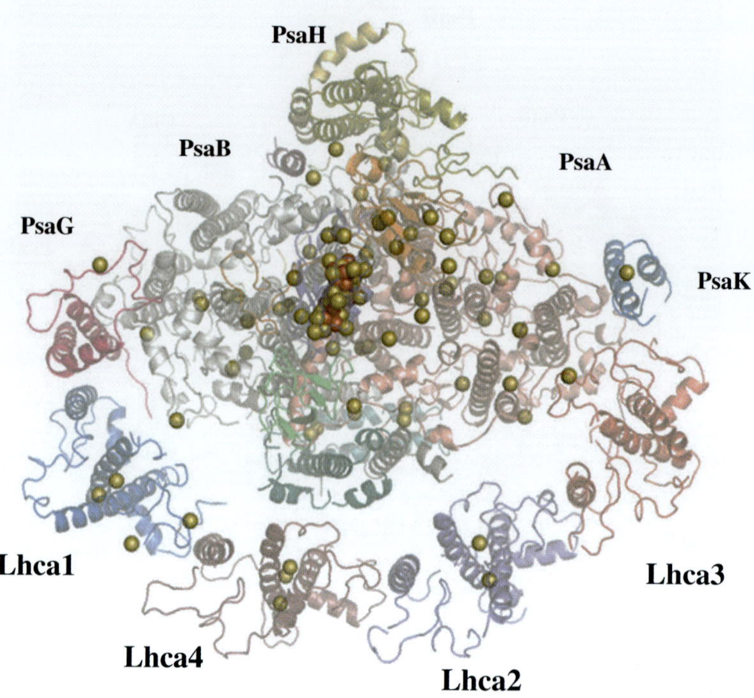

Figure 3. Anomalous signals in Plant PSI. Position of detected anomalous signals in PSI. Anomalous signals in gold on the background of PSI polypeptides.

3. The Light Harvesting Complex of Plant PSI

Our PSI structure includes the fully modeled LHCI belt, with nearly complete structures for all four Lhca proteins. These structures revealed several essential features, including the specific interactions between each Lhca protein and the core; the red chlorophyll assembly present in Lhca4 and Lhca3; and a previously unknown pigment binding site, which is the probable site for the recently discovered non-photochemical quenching at the luminal gap region of LHCI [de Bianchi *et al.*, 2010; Ballottari *et al.*, 2014]. The light harvesting chlorophyll *a* and *b* binding proteins that constitute the peripheral antennas of PSI and PSII (LHCII) show sequence and structural homology; nevertheless, their oligomeric states vary considerably [Nelson and Ben-Shem, 2004; Boekema *et al.*, 2001]. The light harvesting proteins that associate with PSII form either trimers or monomers, as minor antenna members CP24, CP26, and CP29 [Wei *et al.*, 2016]. Lhca1-4 and LHCII bind to approximately 13 chlorophyll *a* + *b* molecules each, and they possess the LHCII general fold. In contrast, the LHCI proteins assemble into dimers. Dimerization in LHCI is mediated by relatively small contact surfaces at the luminal side, near the C-terminus, and at the stromal side of the N-terminal domain of the Lhca proteins [Qin *et al.*, 2015; Mazor *et al.*, 2015, 2017]. A high degree of flexibility may be important for LHCI proteins to assemble into the half-moon shape. This flexibility may also contribute to the efficient energy-transfer along the LHCI belt and between it and the reaction center.

We have presented the most complete plant PSI-LHCI structure obtained to date. An analysis of this structure revealed the locations of and interactions among its protein subunits and more than 200 non-covalently bound photochemical cofactors. With this new crystal structure, we could examine the network of contacts among the protein subunits, from a structural perspective, which has provided a basis for elucidating the functional organization of the complex.

Acknowledgements

The authors would like to thank the ESRF, SLS and BESSYII synchrotrons for beam time and the staff scientists for excellent guide and assistance. This work is supported by grant no. 293579 — HOPSEP from the European Research Council, The Israel Science Foundation through grant no. 71/14 and by the I-CORE Program of the Planning and Budgeting Committee and The Israel Science Foundation (grant no. 1775/12).

References

Abrahams, J.P., Leslie, A.G., Lutter, R. and Walker, J.E. (1994). Structure at 2.8 Å resolution of F1-ATPase from bovine heart mitochondria, *Nature*, 370, 621–628.

Amunts, A., Ben-Shem, A. and Nelson, N. (2005). Solving the structure of plant photosystem I — biochemistry is vital, *Photochem. Photobiol. Sci.*, 4, 1011–1015.

Amunts, A., Drory, O. and Nelson, N. (2007). The structure of a plant photosystem I supercomplex at 3.4 Å resolution, *Nature*, 447, 58–63.

Amunts, A. and Nelson, N. (2009). Plant photosystem I design in light of evolution, *Structure*, 17, 637–650.

Amunts, A., Toporik, H., Borovikova, A. and Nelson, N. (2010). Structure determination and improved model of plant photosystem I, *J. Biol. Chem.*, 285, 3478–3486.

Anderson, J.M. and Boardman, N.K. (1966). Fractionation of the photochemical systems of photosynthesis I. Chlorophyll contents and photochemical activities of particles isolated from spinach chloroplasts, *Biochim. Biophys. Acta*, 112, 403–421.

Ballottari, M., *et al.* (2014). Regulation of photosystem I light harvesting by zeaxanthin, *Proc. Natl. Acad. Sci. USA*, 111, 2431–2438.

Barber, J. (2004). Engine of life and big bang of evolution: a personal perspective, *Photosynth. Res.*, 80, 137–155.

Ben-Shem, A., Frolow, F. and Nelson, N. (2003). The crystal structure of plant photosystem I, *Nature*, 426, 630–635.

Boekema, E.J., Jensen, P.E., van Breemen, J.F.L., van Roon, H., Scheller, H.V. and Dekker, J.P. (2001). Green plant photosystem I binds light-harvesting complex I on one side of the complex, *Biochemistry*, 40, 1029–1036.

Croce, R. and van Amerongen, H. (2013). Light-harvesting in photosystem I, *Photosynth. Res.*, 116, 153–166.

de Bianchi, S., Ballottari, M., Dall'osto, L. and Bassi, R. (2010). Regulation of plant light harvesting by thermal dissipation of excess energy, *Biochem. Soc. Trans.*, 38, 651–660.

Emsley, P., Lohkamp, B., Scott, W.G. and Cowtan, K. (2010). Features and development of Coot, *Acta Crystallogr. D Biol. Crystallogr.*, 66(Pt 4), 486–501.

Ferreira, K.N., Iverson, T.M., Maghlaoui, K., Barber, J. and Iwata, S. (2004). Architecture of the photosynthetic oxygen-evolving center, *Science*, 303, 1831–1838.

Jordan, P. *et al.* (2001). Three-dimensional structure of cyanobacterial photosystem I at 2.5 Å resolution, *Nature*, 411, 909–917.

Kargul, J. *et al.* (2005). Light-harvesting complex II protein CP29 binds to photosystem I of *Chlamydomonas reinhardtii* under State 2 conditions, *FEBS J.*, 272, 4797–4806.

Kouril, R. *et al.* (2005). Structural characterization of a complex of photosystem I and light-harvesting complex II of *Arabidopsis thaliana*, *Biochemistry*, 44, 10935–10940.

Kurisu, G., Zhang, H., Smith, J.L. and Cramer, W.A. (2003). Structure of the cytochrome b6f complex of oxygenic photosynthesis: tuning the cavity, *Science*, 302, 1009–1014.

Loll, B., Kern, J., Saenger, W., Zouni, A. and Biesiadka, J. (2005). Towards complete cofactor arrangement in the 3.0 Å resolution structure of photosystem II, *Nature*, 438, 1040–1044.

Lunde, C.P., Jensen, P.E., Haldrup, A., Knoetzel, J. and Scheller, H.V. (2000). The PSI-H subunit of photosystem I is essential for state transitions in plant photosynthesis, *Nature*, 408, 613–615.

Mazor, Y., Greenberg, I., Toporik, H., Beja, O. and Nelson, N. (2012). The evolution of photosystem I in light of phage-encoded reaction centers, *Phil. Trans. R. Soc. B*, 367, 3400–3405.

Mazor, Y., Nataf, D., Toporik, H. and Nelson, N. (2014). Crystal structures of virus-like photosystem I complexes from the mesophilic cyanobacterium, *Synechocystis* PCC6803, *eLife 2014*, 3, p. e01496. DOI: 10.7554/eLife.01496.

Mazor, Y., Borovikova, A. and Nelson, N. (2015). The structure of plant photosystem I super-complex at 2.8 Å resolution, *eLife 2015*, 4, p. e07433. DOI: 10.7554/eLife.07433.

Mazor, Y., Borovikova, A., Caspy, I. and Nelson, N. (2017). Structure of the plant photosystem I supercomplex at 2.6 Å resolution, *Nat Plants*, 3, p. 17014. DOI: 10.1038/nplants.2017.14.

Nelson, N. and Ben-Shem, A. (2004). The complex architecture of oxygenic photosynthesis, *Nat. Rev. Mol. Cell Biol.*, 5, 971–982.

Nelson, N. and Yocum, C. (2006). Structure and function of photosystems I and II, *Annu. Rev. Plant Biol.*, 57, 521–565.

Nelson, N. (2009). Plant Photosystem I — The most efficient nano-photochemical machine, *J. Nanosci. Nanotechnol.*, 9, 1709–1713.

Nelson, N. (2013). Evolution of photosystem I and the control of global enthalpy in an oxidizing world, *Photosynth. Res.*, 116, 145–151.

Nelson, N. and Junge, W. (2015). Structure and energy transfer in photosystems of oxygenic photosynthesis, *Annu. Rev. Biochem.*, 84, 659–683.

Page, C.C., Moser, C.C., Chen, X. and Dutton, P.L. (1999). Natural engineering principles of electron tunnelling in biological oxidation-reduction, *Nature*, 402, 47–52.

Qin, X., Suga, M., Kuang, T. and Shen, J.R. (2015). Photosynthesis. Structural basis for energy transfer pathways in the plant PSI–LHCI supercomplex, *Science*, 348, 989–995.

van Raaij, M.J., Abrahams, J.P., Leslie, A.G. and Walker, J.E. (1996). The structure of bovine F1-ATPase complexed with the antibiotic inhibitor aurovertin B, *Proc. Natl. Acad. Sci. USA*, 93, 6913–6917.

Rochaix, J.D. (2011). Regulation of photosynthetic electron transport, *Biochim. Biophys. Acta*, 1807, 375–383.

Rochaix, J.D., Lemeille, S., Shapiguzov, A., Samol, I., Fucile, G., *et al.* (2012). Protein kinases and phosphatases involved in the acclimation of the photosynthetic apparatus to a changing light environment, *Philos. Trans. R. Soc. Lond. B Biol. Sci.*, 367, 3466–3474.

Scheller, H.V., Jensen, P.E., Haldrup, A., Lunde, C. and Knoetzel, J. (2001). Role of subunits in eukaryotic Photosystem I, *Biochim. Biophys. Acta*, 1507, 41–60.

Standfuss, J., Terwisscha van Scheltinga, A.C., Lamborghini, M. and Kuhlbrandt, W. (2005). Mechanisms of photoprotection and nonphotochemical quenching in pea light-harvesting complex at 2.5 Å resolution, *EMBO J.*, 24, 919–928.

Umena, Y., Kawakami, K., Shen, J-R. and Kamiya, N. (2011). Crystal structure of oxygen-evolving photosystem II at a resolution of 1.9 Å, *Nature*, 473, 55–60.

Walker, J.E., Saraste, M., Runswick, M.J. and Gay, N.J. (1982). Distantly related sequences in the alpha- and beta-subunits of ATP synthase, myosin, kinases and other ATP-requiring enzymes and a common nucleotide binding fold, *EMBO J.*, 1, 945–951.

Walker, J.E., Saraste, M. and Gay, N.J. (1984). The unc operon. Nucleotide sequence, regulation and structure of ATP-synthase, *Biochim. Biophys. Acta.*, 768, 164–200.

Wei, X. *et al.* (2016). Structure of spinach photosystem II-LHCII supercomplex at 3.2 Å resolution, *Nature*, 534, 69–74.

Chapter 8

Rubisco Activase: The Molecular Chiropractor of the World's Most Abundant Protein

Devendra Shivhare and Oliver Mueller-Cajar*

*School of Biological Sciences, Nanyang Technological University,
60 Nanyang Drive, Singapore 637551*
**cajar@ntu.edu.sg*

The photosynthetic CO_2 fixing enzyme ribulose 1,5-bisphosphate carboxylase/oxygenase (Rubisco) is slow and non-specific. In order to maintain sufficient flux through the Calvin-Benson cycle, the enzyme is extremely abundant in photosynthetic tissue. In addition to its catalytic frailties, Rubisco forms dead-end inhibited complexes with its substrate ribulose-1,5-bisphosphate (RuBP) and other sugar phosphates. In order to remove these complexes, photoautotrophs have recruited multiple molecular chaperones of the AAA+ clade (ATPases associated with diverse cellular activities) known as the Rubisco activases. Here we outline a current perspective on the properties of Rubisco and summarize our knowledge of the green group of Rubisco activases, which is found in all higher plants. We acknowledge the rapid increase of mechanistic understanding currently developing for the molecular motors of the AAA+ class. Integrating this knowledge towards a detailed mechanistic understanding of Rubisco activase's mechanism has the potential to contribute to the long-standing goal of enhancing photosynthetic efficiency.

1. Introduction

Photosynthesis provides humanity with energy, food and materials. Almost all significant CO_2 assimilation occurs *via* the Calvin–Benson cycle, and uses the

"

world's most abundant protein, ribulose 1,5-bisphosphate carboxylase/oxygenase (Rubisco), as the catalyst. In spite of the crucial importance of this process to the organism, Rubisco is a sluggish and non-specific enzyme [Andersson and Backlund, 2008; Whitney *et al.*, 2011]. Its tendency to become inhibited is an additional short-coming that requires servicing by the helper protein Rubisco activase [Portis, 2003]. As a consequence of Rubisco's abundance and low catalytic prowess, the CO_2-fixing step of photosynthesis is accepted to be the rate-limiting step of the process under most conditions, and significantly contributes to the very low efficiency of light energy capture and conversion of plants and algae [Blankenship *et al.*, 2011]. Consequently improvements in the CO_2-capturing step are predicted to lead to increases in crop yield and their potential impact towards human agriculture and energy economy is hard to overstate [Parry *et al.*, 2007, 2013]. Here we focus our attention on the type of Rubisco enzyme found in higher plants. We outline our understanding of the catalyst, and in particular its unusual property of forming inhibited complexes with its substrate and other sugar phosphates. We then describe our current perspective regarding a molecular chaperone of the AAA+ class that has specialized to conformationally remodel the carboxylase. This enzyme, Rubisco activase, is a critical component of photosynthetic carbon fixation. Understanding its mechanistic subtleties and consequently modifying its properties is a promising target for crop improvement efforts.

2. The Most Abundant Protein on Earth is a Poor Catalyst

Ribulose 1,5-bisphosphate carboxylase/oxygenase, also known as Rubisco, (EC.4.1.1.39) is believed to be the most abundant protein on earth [Ellis, 1979] and is the key enzyme of the dark reactions. It often makes up 50% of the total soluble protein found in plant leaf tissue and accounts for the majority of the protein in some phototrophic microbes [Spreitzer and Salvucci, 2002; Andersson and Backlund, 2008]. It catalyses the initial steps of photosynthetic carbon fixation by converting ribulose 1,5-bisphosphate (RuBP) and CO_2 to two molecules of 3-phosphoglycerate (3PG). Rubisco has slow catalytic properties (with higher plant k_{cat} numbers falling between 3–6 s^{-1}) and due to its poor ability to differentiate between CO_2 and O_2, it also catalyses the oxygenation of RuBP which leads to formation of toxic 2-phosphoglycolate (2PG) [Andersson and Backlund, 2008; Whitney *et al.*, 2011]. This compound has to be recycled to 3PG at the cost of metabolic energy and CO_2, which reduces the efficiency of photosynthesis [Bauwe *et al.*, 2010]. In addition, Rubisco is prone to inhibition by its own substrate and other inhibitory sugar phosphates, some of them generated by its own active site, leading to a phenomenon called fallover [Edmondson *et al.*, 1990; Parry *et al.*, 2008]. The manipulation of the enzyme's kinetic properties has been an elusive objective for many years as it would lead to an enhancement of photosynthetic efficiency in higher plants [Andrews and Whitney, 2003; Parry *et al.*, 2007]. This hope is supported by the fact that Rubisco from different species possess greatly diverse kinetics and substrate specificity [Jordan and Ogren, 1981]. For instance, Rubisco from red algae such as *Griffithsia monilis* shows double the CO_2/O_2 specificity ($S_{C/O}$) compared to typical plant Rubiscos [Read and Tabita, 1994; Uemura *et al.*, 1997; Whitney *et al.*, 2001].

2.1. *The structure and diversity of Rubisco*

The Rubisco enzyme found in higher plants consists of ~52 kDa large subunits (RbcL) and ~15 kDa small subunits (RbcS) in a hexadecameric stoichiometry (L_8S_8) [Tabita, 1999]. However, it is only one example of a diverse protein family. Rubisco is classified into four major forms namely form I–IV, of which only the first three catalyze RuBP-dependent carbon fixation reactions (carboxylation/ oxygenation) [Badger and Bek, 2008; Tabita *et al.*, 2008]. Form IV enzymes are designated as Rubisco-like proteins (RLP) and do not catalyse the CO_2 fixation reaction [Tabita *et al.*, 2007]. Out of the four types, only Form I enzymes possess the small subunit. This is the most common form of Rubisco and is further classified into four subgroups: Form IA–ID. Form IA and Form IB are commonly referred to as green-type Rubisco [Tabita *et al.*, 2008] found in plants, green algae, cyanobacteria and proteobacteria. In plants and green algae the RbcL is encoded by chloroplast whereas RbcS is encoded by the nuclear genome [Blair and Ellis, 1973]. Form IC and form ID are referred as red-type Rubisco [Tabita *et al.*, 2008] in which RbcL and RbcS are encoded in the same operon [Hartman and Harpel, 1994] and is found in some proteobacteria, red algae and other non-green algae.

The overall fold of the Rubisco large subunit (RbcL) monomer is well-conserved in all forms of Rubisco. It is made up of a N-terminal $\alpha+\beta$ domain and a larger C-terminal α/β TIM barrel domain ($\beta_8\alpha_8$) (Figure 1A). The active site is at the intradimer (RbcL$_2$) interface between the N-domain of one subunit and the C-domain of the adjacent subunit. Consequently, the functional unit of Rubisco is an L$_2$ dimer with two active sites per dimer (Figure 1B) [Andersson and Taylor, 2003]. In Form I Rubiscos four dimers are assembled by the association of four small subunits to the top and bottom of the L$_8$ core to give the L$_8$S$_8$ hexadecamer (Figure 1C).

Although the Form I RbcL amino acid sequences are highly conserved, the Rubisco small subunit (RbcS) sequences are more diverse with only 30–40% homology between different species [Hauser *et al.*, 2015]. Even though not strictly required for enzymatic function, the small subunits are important for maximal catalytic rates and substrate specificity [Spreitzer, 2003, 2005; Ishikawa *et al.*,

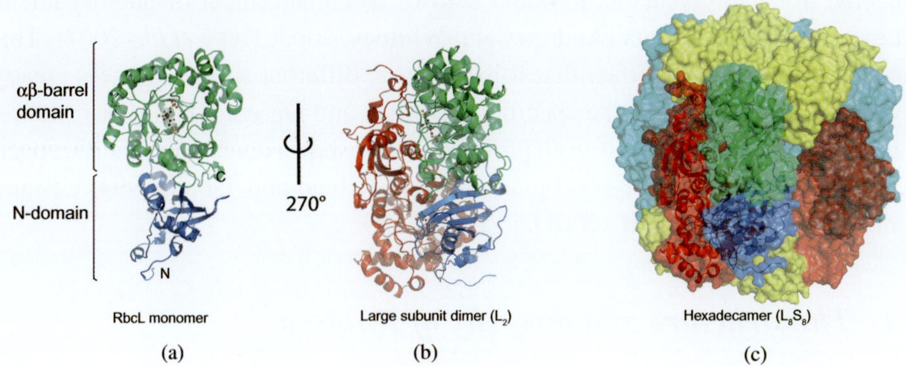

Figure 1.   The structure of higher plant (Form IB) Rubisco. (A) The large subunit consists of a N-terminal domain containing a mixed beta sheet and a C-terminal triose isomerase (TIM) barrel domain, which harbours the active site. (B) The functional unit of Rubisco is an anti-parallel large subunit dimer. (C) Four L$_2$ dimers are capped by four small subunits (in yellow) at the top and bottom of the L$_8$ core. The structure shown is from spinach, PDB:8RUC [Andersson, 1996]. This figure was prepared using pymol molecular visualization software.

2011]. The green and red type Rubisco enzymes differ in several structural features, the most notable being an extension in the C-terminal domains of the small subunits in the red-type enzymes. The C-terminal extensions form a small barrel that protrudes into the solvent channel of the holoenzyme [Joshi *et al.*, 2015]. It is possible that the structural variance between the two subclasses could explain the higher substrate specificity for CO_2 over O_2 of some red-type Rubiscos as compared to green-type enzymes.

Form II Rubisco found in proteobacteria and dinoflagellates is active as RbcL$_2$ or higher oligomers of the dimer. In many chemoautotrophs Form II often co-exists with Form I Rubisco with different numbers of RbcL dimeric pairs, (L$_2$)n. Although overall tertiary structure, as well as the residues involved in catalysis are conserved, the sequence homology between Form I and Form II large subunits is only 30% [Tabita, 1999; Tabita *et al.*, 2008].

Form III Rubisco exists in some thermophilic archaea, which lack the Calvin-Benson cycle but catalyze the carboxylation reaction of RuBP, which is generated *via* an alternate pathway [Sato *et al.*, 2007]. They exist as dimers, however pentamers of dimers (RbcL$_{2-5}$) have also been found as active form [Ezaki *et al.*, 1999; Maeda *et al.*, 1999; Watson *et al.*, 1999; Kitano *et al.*, 2001; Alonso *et al.*, 2009].

Form IV Rubisco are a very diverse group of proteins known as Rubisco-like proteins (RLP) which form antiparallel dimers of large subunits similar to bona-fide Rubiscos [Li *et al.*, 2005; Imker *et al.*, 2007]. Form IV Rubiscos are generally missing a number of conserved active site residues and consequently do not catalyse CO_2 fixation [Hanson and Tabita, 2001; Tabita *et al.*, 2007]. The RLPs from *Bacillus subtilis* and *Geobacillus kaustophilus* were reported to be involved in a methionine salvage pathway by performing an enolase-type function similar to Rubisco on a different substrate [Ashida *et al.*, 2003; Imker *et al.*, 2007]. The RLP from green sulphur bacteria *Chlorobium tepidum* is involved in thiosulphate oxidation [Hanson and Tabita, 2001].

2.2. *Reaction mechanism: co-factor binding and catalysis*

In spite of the diversity in primary and quaternary structure discussed above, all genuine Rubiscos (Form I–III) follow identical principles with regards to their activation mechanism and catalysis. The Rubisco apo form (E) requires two co-factors, CO_2 and Mg^{2+}, to bind at the active site to perform its catalytic function [Lorimer *et al.*, 1976]. The active site for carboxylation is present at the interface between the N-terminal domain of one subunit and the C-terminal domain of the adjacent subunit of the antiparallel RbcL2 [Schneider *et al.*, 1986]. Activation takes place by the binding of a non-substrate CO_2 molecule with the ε-amino group of lysine 201 (using spinach RbcL numbering) located at the catalytic sites to form a carbamate (EC). Once the carbamate is formed, it is stabilized by the addition of Mg^{2+} to form the active holoenzyme (ECM) [Lorimer *et al.*, 1976; Cleland *et al.*, 1998]. These spontaneously occurring, reversible reactions are only possible in the "open state" of the enzyme, which allow solvent access to the binding pockets. The addition of Mg^{2+} at the Rubisco active site is important for the positioning of RuBP for electrophilic attack of the substrate CO_2 molecule [Andersson and Backlund, 2008]. The substrate RuBP can then proceed to bind at the active site, triggering the closure of α/β TIM-barrel loop 6 over the active site. The extended conformation of loop 6 is then fixed in position by the flexible C-terminal tail of RbcL, excluding the active site from solvent [Schreuder *et al.*, 1993; Taylor and Andersson, 1996]. Rubisco then proceeds with the carboxylation of RuBP. The active site opens and closes during every catalytic cycle, mostly involving movement of loop 6 of the $\alpha\beta$-barrel [Duff *et al.*, 2000].

During catalysis, in the "closed state", loop 6 forms a latch-like structure on top of the substrate and reaction intermediates that are deeply buried in the active site of the complex. Carboxylation is initiated when the carbamylated K201 abstracts a proton from RuBP to form the enediol isomer of RuBP (enolisation) at the active site (Figure 2) [Cleland *et al.*, 1998; Spreitzer and Salvucci, 2002]. This enediol form makes a nucleophilic attack on the substrate CO_2 molecule to form a ketonic acid intermediate. The unstable 6-carbon product intermediate of carboxy-lation is subsequently hydrated and cleaved to produce the two molecules of 3PG [Andrews and Lorimer, 1987]. The oxygenation reaction takes place when O_2

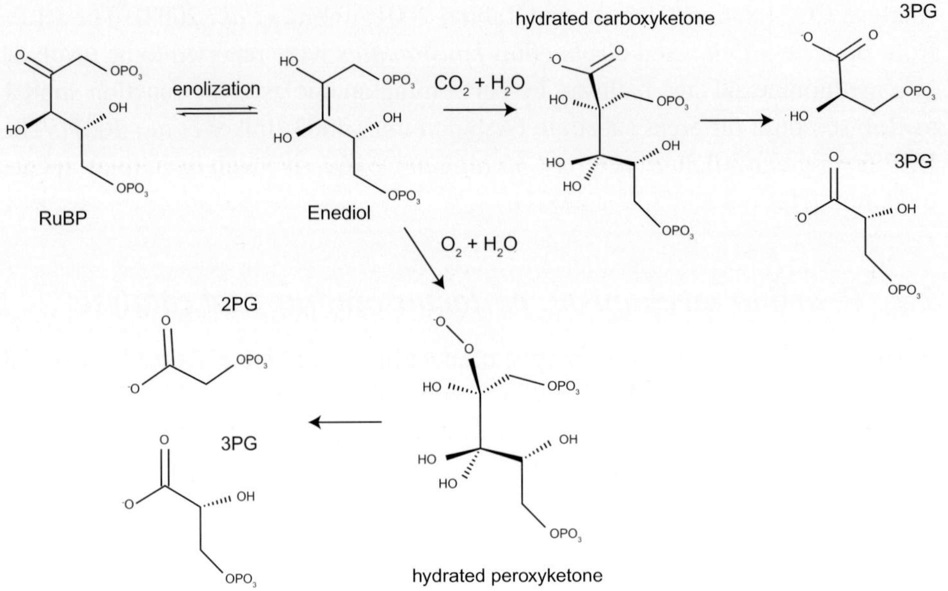

Figure 2. Carboxylation and oxygenation reactions catalyzed by Rubisco. Carboxylase activity: RuBP forms an enediol intermediate. CO_2 is polarized by the amino group of Lys-334 and undergoes nucleophilic attack by the enediolate, producing a branched six-carbon intermediate. C-3 of this intermediate is hydroxylated by a water molecule forming a β-keto acid intermediate. Two molecules of 3-phosphoglycerate (3PG) are formed as a result of aldol cleavage and stereospecific protonation. Oxygenase activity: Addition of O_2 to the common enediol results in an unstable intermediate that splits into two molecules, 2-phosphoglycolate (2PG) and 3PG.

molecules compete at the active site instead of CO_2 molecules. The mechanism is similar to CO_2 fixation where the enediol isomer makes a nucleophilic attack on the O_2 molecule resulting in an intermediate, which is hydrolysed to form one 3PG and one 2PG molecule [Andrews and Lorimer, 1987]. The energy released from the bond cleavage between the two phosphate substituents is likely required to permit the active site to open, allowing release of the products at the end of catalysis [Andrews, 1996].

2.3. *Rubisco is prone to inhibition by sugar phosphates*

The Rubisco active site has an unusual property, in that it is highly susceptible to becoming occupied by inhibitory sugar phosphates. Binding of the enzyme's substrate RuBP to the decarbamylated form of Rubisco results in an inhibited complex known as ER which is catalytically inactive [Jordan and Chollet, 1983]. Additionally, several misfire products generated during catalysis bind to non-carbamylated Rubisco and prevent catalysis [Parry *et al.*, 2008]. In many plants this tendency appears to have been recruited as a regulatory mechanism, since sugar phosphates specifically synthesized by the plant, such as the nocturnal inhibitor 2-carboxy-arabinitol-1-phosphate (CA1P) also bind to the carbamylated form to form stable inactive complexes at night (Figure 3) [Berry *et al.*, 1987; Andralojc *et al.*, 2012]. In this case locking the active sites into the closed, inhibited conformation when they are not required may provide protection against proteolysis [Khan *et al.*, 1999].

The affinity of Rubisco for RuBP is low in the carbamylated state in order to facilitate cleavage and release at the active site, while in the decarbamylated state the affinity is high leading to inhibition [Jordan *et al.*, 1983; McNevin *et al.*, 2006]. Different inhibitors bind to Rubisco in the carbamylated and non-carbamylated state with different affinities. RuBP and its misfire products like D-xylulose 1,5-bisphosphate (XuBP), glycero-2,3-pentodiulose-1,5-bisphosphate (PDBP) favour binding to the non-carbamylated enzyme (E) to form the non-carbamylated inhibited complex (E.I). Other inhibitors such as CA1P and the non-physiological transition state analogue 2-carboxy-D-arabinitol 1,5-bisphosphate (CABP) preferably bind to the Rubisco holoenzyme (E.C.M) and form a carbamylated inhibited complex (E.C.M.I) (Figure 3) [Pierce *et al.*, 1980; Seemann *et al.*, 1985].

Rubiscos with ligands occupying the substrate binding sites in the carbamylated and uncarbamylated state are structurally similar, indicating that the co-factors

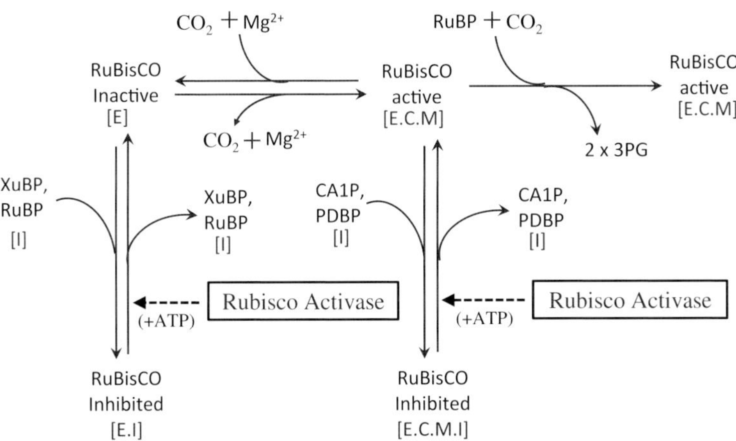

Figure 3.   Decarbamylated Rubisco (E) binds to CO$_2$ and Mg^{2+} to become the active holoenzyme (E.C.M) which actively fixes CO$_2$ by carboxylation of its substrate RuBP. E form can also bind substrate RuBP or inhibitors to form a dead inhibited complex (E.I). E.C.M can also become inhibited by its nocturnal inhibitor 2-carboxy-arabinitol-1-phosphate (CA1P) or other substrate analogues, such as D-glycero-2,3-pentodiulose-1,5-bisphosphate (PDBP) and 3-ketoarabinitol-1,5-bisphosphate (KABP) to form inhibited complex (E.C.M.I). The helper enzyme Rubisco activase remodels the Rubisco in both situations by removing the inhibitory sugar-phosphates.

do not play a critical role in substrate binding to the active sites [Knight *et al.*, 1990; Newman and Gutteridge, 1994; Andersson and Backlund, 2008]. Once E.I or E.C.M.I has formed, the most favourable route for opening and release of product no longer exists. This path, which involves cleavage of the substrate to produce 3PG, is only available for the substrate RuBP bound to the carbamylated and Mg^{2+}-bound active site [Andrews, 1996]. Different inhibitors have greatly varying affinities for the enzyme [Pearce, 2006], but to give an extreme example, a half-life of 530 days was determined for the release of CABP from carbamylated Rubisco in spinach, making the binding almost irreversible [Pierce *et al.*, 1980; Schloss, 1988].

The inhibition of Rubisco activity due to binding of sugar phosphates in the plant is part of the physiological regulation of the enzyme *in vivo*; however activity of activated Rubisco also declines in a non-linear manner under *in vitro* assay conditions. This phenomenon is known as "fallover" [Andrews and Hatch, 1971; Robinson and Portis, 1989; Edmondson *et al.*, 1990; Zhu and Jensen, 1991]. It is the outcome of the accumulation of misfire products during catalysis of RuBP (oxygenation or carboxylation) which block the active sites and inhibit Rubisco activity. Although these misfire products are present in much smaller quantities compared to RuBP, they can form slow tightly bound inhibited complexes [Paech *et al.*, 1978; Zhu and Jensen, 1991; Zhu *et al.*, 1998]. Therefore when active photosynthesis takes place these inactive Rubisco complexes would accumulate and obstruct CO_2 fixation. The fallover is less severe at high temperature due to looser binding of inhibitors in spite of a higher rate of formation of misfire products [Schrader *et al.*, 2006]. This could be the outcome of enhanced flexibility of the Rubisco active site at higher temperature leading to an increased rate of release of inhibitors from the active site.

The tendency of inhibition by such mechanisms differs between Rubisco from phylogenetically diverse sources [Uemura *et al.*, 1998; Pearce and Andrews, 2003; Pearce, 2006]. Generally there exists a compromise between substrate specificity ($S_{c/o}$) and catalytic rate of the Rubisco in different species [Zhu *et al.*, 2004; Tcherkez *et al.*, 2006]. According to the prevailing model, Rubisco enzyme exhibiting higher CO_2/O_2 specificity favours tight binding of transition state intermediates of carboxylation and consequently the cleavage of tightly bound reaction intermediates becomes slower. Although this tight binding favours carboxylation over oxygenation but the slower rate unfortunately also favours aggravated misfire product formation [Tcherkez *et al.*, 2006]. Also highly selective enzymes form slow-tight binding complexes with inhibitors, indicating that the active sites of these enzymes are more rigid. For instance, plant Rubiscos possessing high specificity factors form inhibited complexes with RuBP and display fallover. Highly specific red algal Rubiscos do not exhibit fallover inhibition, but when inhibited complexes are formed, they are extremely stable [Pearce and Andrews, 2003; Pearce, 2006]. Recently, a red algal Rubisco from *Cyanidioschyzon merolae* was reported to form a stable inhibited complex only at high temperature (45°C) [Loganathan *et al.*, 2016] that did not reactivate spontaneously at 25°C. Rubisco from green algae with intermediate specificity forms the inhibited complex with RuBP but did not exhibit fallover when assayed [Uemura *et al.*, 1998]. Less specific Rubisco from cyanobacteria neither form inhibited ER complexes nor display fallover [Andrews and Abel, 1981; Lee *et al.*, 1991; Uemura *et al.*, 1998; Pearce, 2006]. Thus highly specific Rubiscos exhibiting slower catalytic rates are particularly prone to form highly stable dead-end inhibited complexes.

In order to overcome this problem of inhibition, during the course of evolution a molecular chiropractor of Rubisco was recruited from the molecular chaperone pool of the host, called Rubisco activase (Rca) [Salvucci *et al.*, 1985; Robinson *et al.*, 1988]. Rca utilizes energy from ATP hydrolysis to remodel inhibited Rubisco active sites to trigger inhibitor release; subsequently specific phosphatases assist in degradation of non-RuBP inhibitors [Robinson and Portis, 1989; Andralojc *et al.*, 2012; Bracher *et al.*, 2015].

3. Rubisco Activase, the Molecular Chiropractor

Rubisco activase was discovered serendipitously, as part of large scale screens for *Arabidopsis* mutants that required high CO_2 for growth in an effort to delineate the photorespiratory pathway. It was observed that the Rca deletion strain was unable to grow at ambient CO_2 level and accumulated RuBP due to lack of activation of the Rubisco, but survived at high CO_2 levels [Somerville *et al.*, 1982; Portis Jr and Salvucci, 2002]. The gene responsible was therefore designated as *rca* for Rubisco activation. Later, it was found that two polypeptide bands were missing on a two dimensional polyacrylamide gel electrophoresis in the *rca* mutant during analysis of soluble polypeptides from the *Arabidopsis* mutants and wild-type strains. Biochemical and genetic analysis confirmed that the two missing polypeptides were responsible for the *rca* mutant phenotype and this protein was later named as Rubisco activase [Salvucci *et al.*, 1985].

Identifying the presence of Rca in soluble protein extracts of various C3 and C4 plants species confirmed that Rca-dependent regulation of Rubisco was pervasive in higher plants [Salvucci *et al.*, 1987]. In the same year, it was also discovered that the Rca is an ATP-dependent remodeler of Rubisco [Streusand and Portis, 1987] which promotes the release of bound sugar phosphates from the active site of both carbamylated and decarbamylated forms of Rubisco [Robinson *et al.*, 1988; Robinson and Portis, 1989]. Addition of Rca to uncarbamylated Rubisco showed no activity loss or to carbamylated Rubisco showed no fallover.

Thus, it applies to Rubisco inhibited by CA1P or other inhibitors that are formed during catalysis apart from RuBP alone. However, Rubisco activase is not directly involved in carbamylation or binding of magnesium ion at the active site [Robinson *et al.*, 1988].

For a number of decades it was thought that the phenomenon of protein-mediated Rubisco activation was restricted to the green-type Form I enzyme found in higher land plants. However, 25 years after the discovery of green-type Rca in plants, a convergently evolved red-type Rubisco activase (CbbX) from α-proteobacteria was reported [Mueller-Cajar *et al.*, 2011]. Four years later a third class of activase, the CbbQO complex (purple-type) was shown to be functional as Rubisco activase in chemoautotrophic bacteria [Tsai *et al.*, 2015]. Homologues of the red-type activase had long been found in all red-lineage phytoplankton inspected [Oudot-Le Secq *et al.*, 2007; Hovde *et al.*, 2015], and their function has now also been confirmed. A hetero-oligomeric red-type Rubisco activase was described in a thermophilic red alga *C. merolae* that operates as an hetero-oligomeric complex of a plastid and nuclear-encoded CbbX isoform (CmNP) [Loganathan *et al.*, 2016]. Although all three classes of Rubisco activases appear to employ distinct mechanisms, the energy required to remodel inhibited Rubisco complexes is generated by an AAA+ protein module. Therefore, in order to generate a foundation for functional understanding of Rubisco activase, it is necessary to appreciate the basic structural and functional aspects of AAA+ proteins in different systems.

3.1. *Rubisco activases are members of the AAA+ family of proteins*

A sequence comparison with related ATPases revealed earlier on that Rca belongs to the family of AAA+ (ATPases associated with diverse cellular activities) ring-shaped p-loop ATPase motor proteins [Neuwald *et al.*, 1999]. AAA+ proteins as the name suggests, have been described to perform an array of functions at the cellular level and are ubiquitously found across kingdoms of life. The overarching theme in the function of AAA+ proteins involves conformational remodelling of their target substrate (a macromolecule such as protein or polynucleotides) by utilizing energy from ATP hydrolysis [Vale, 2000; Hanson and Whiteheart, 2005]. They have an active role in various cellular processes such as protein folding and degradation, DNA replication membrane fusion, vesicle trafficking, transcriptional activation, motor activity, *etc.* [Vale, 2000; Ogura and Wilkinson, 2001; Hanson and Whiteheart, 2005]. Class I AAA+ proteins (*e.g.* ClpA, ClpB, ClpC and HSP104) contain two highly conserved AAA domains, referred to as AAA-1 (or D1) and AAA-2 (or D2), separated by a variable length linker sequence.

Class II proteins (*e.g.* ClpX, HslU), including all known Rubisco activases, contain only a single AAA domain [Hanson and Whiteheart, 2005; Erzberger and Berger, 2006].

AAA+ proteins consist of a distinct set of conserved sequences and are characterized by nucleotide binding domain consisting of a N-terminal α/β Rossmann fold and a less conserved C-terminal 4-helix bundle subdomain [Hanson and Whiteheart, 2005]. An α/β Rossman fold forms the core of the N-terminal subdomain with a β-sheet of parallel strands, arranged in a $\beta5$–$\beta1$–$\beta4$–$\beta3$–$\beta2$ fashion at its core (Figure 4). This large AAA+ subdomain comprises the signature Walker A, B and sensor I motifs [Wendler *et al.*, 2012] essential for nucleotide binding and hydrolysis. The nucleotide binding domain also contains conserved arginine residues called arginine fingers for intersubunit communication which affects ATP hydrolysis and subunit oligomerization [Ogura *et al.*, 2004]. Another typical feature of the AAA+ proteins are the pore loops facing the central cavity that is lined by residues from the individual subunits [Hanson and Whiteheart, 2005]. The underlying mechanism of action often involves protein threading, where the substrate protein is translocated through a central pore formed by a ring shaped oligomer of AAA+ protein subunits [Yamada-Inagawa *et al.*, 2003; Siddiqui *et al.*, 2004; Martin *et al.*, 2008]. In the α-helical subdomain, a conserved sequence motif called sensor II, characterized by a conserved arginine, is essential for nucleotide binding and hydrolysis [Ogura *et al.*, 2004; Wendler *et al.*, 2012]. In many cases AAA+ ATPases act on a large number of substrate proteins, such as the general protein disaggregation machine Hsp104/ClpB or the motor of the bacterial degradation machinery ClpX. Generally, AAA+ proteins form homohexameric arrangements with a central pore, stabilised by the presence of nucleotides. The AAA+ modules of class I proteins therefore form two stacked hexameric rings [Davies *et al.*, 2008; Wang *et al.*, 2011], while class II proteins form single hexameric ring [Vale, 2000; Roll-Mecak and Vale, 2008].

To date, three plant Rca crystal structures have been published. A 3.0 Å model of the tobacco AAA+ domain (residues 68–360 out of 383 total) [Stotz *et al.*, 2011], a 1.9 Å model of the creosote bush C-terminal small domain (residues 250–351 out of 379 total) [Henderson *et al.*, 2011] and a 2.9 Å model of the arabidopsis Rca (residues 65–362 out of 416 total) [Hasse *et al.*, 2015]. These structures reveal the organisation of the apo state and architecture of the AAA+ domain.

Figure 4 shows the crystal structure of Rubisco activase from tobacco in apo state [Stotz *et al.*, 2011]. The structure represents a typical AAA+ topology with large AAA+ domain comprising the nucleotide binding α/β subdomain and small α-helical domain characterized by a four helix bundle. Rca also comprises the N-terminal domain (67 residues) and a C-terminal extension (23 residues) which were not resolved in the structure due to flexibility. Also, four flexible loop regions

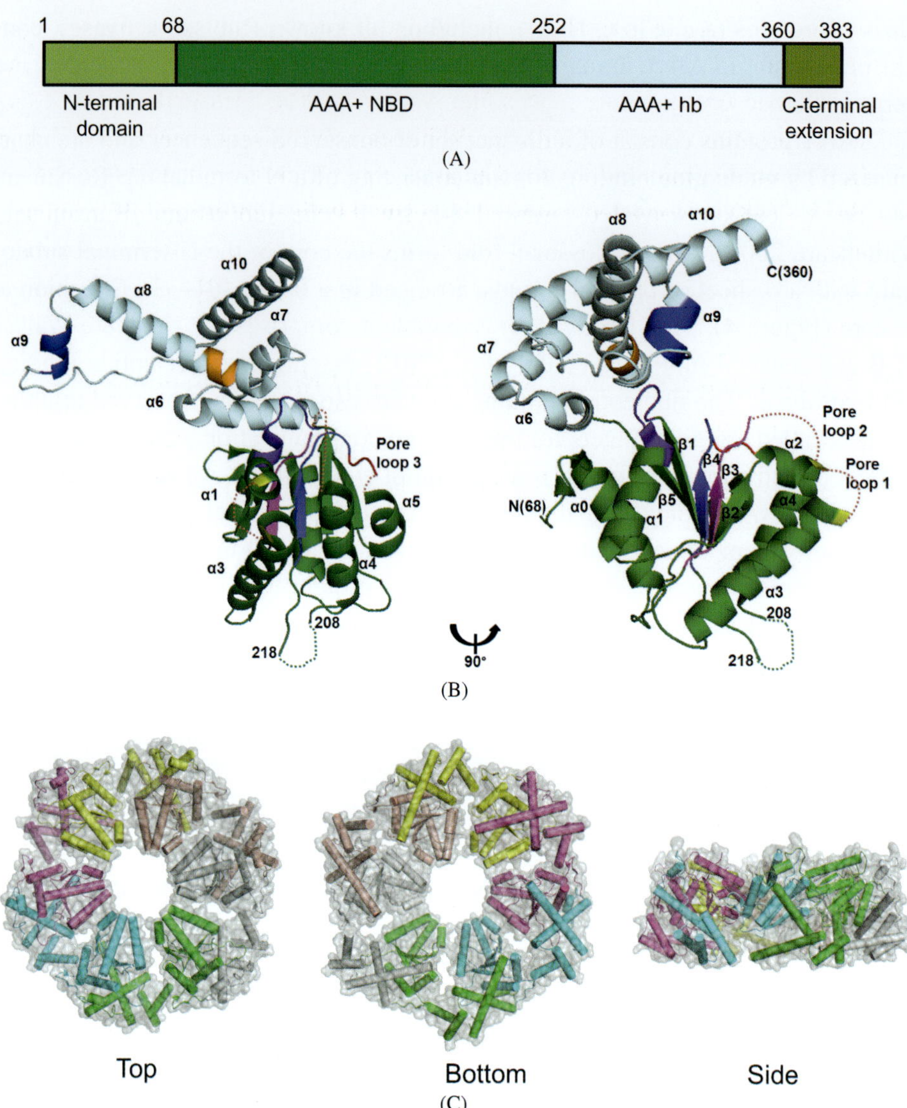

Figure 4. Structure and domain architecture of Rca from *Nicotiana tabacum*. (A) Domain architecture of tobacco activase (NtRca) showing the N-terminal domain, AAA+ nucleotide binding α/β sub-domain, α helical small subdomain (hb) and C-terminal extension. NBD: nucleotide binding domain, hb: α-helical subdomain. (B) Ribbon representation of the crystal structure of NtRca (3T15). Disordered loops are indicated by dotted lines. Two views related by 90° are shown. The α/β and the α-helical subdomains are indicated in green and pale cyan, respectively. The canonical AAA+ structural motifs are indicated as follows: Walker A (purple), Walker B (magenta), sensor I (tv blue) and sensor II (orange). The disordered pore loops are indicated as light red dots. The specificity helix (αH9) is shown in blue. Another disordered loop (residue 208–218) and chain termini are indicated. (C) A hexameric model of NtRca (3ZW6).Top, bottom and side view of the hexameric ring, helices are shown as cylinders. Each subunit is shown in a different colour. The images were generated using pymol molecular visualization software.

were not resolved in the structure and were extrapolated using dotted lines. The α/β subdomain is comprised of a Rossmann-fold typical for the AAA+ proteins and contains the key signature motifs like Walker A, B and sensor 1 region essential for ATPase function [Hanson and Whiteheart, 2005]. The small C-terminal domain with 4-helix bundle ($\alpha 6$–10) fold at its core differs from the typical AAA+ fold. Helix 8 ($\alpha 8$) is elongated followed by a small helical insertion ($\alpha 9$) responsible for substrate specificity between Rubisco of the Solanaceae and non-Solanaceae family of plants [Li *et al.*, 2005; Mueller-Cajar *et al.*, 2014].

3.2. *Oligomeric state and mechanism of the Rubisco activase*

The structural organisation of Rubisco activase oligomers has been a controversial issue. The published crystal structures do not provide a true functional oligomeric form of Rca. Purified Rubisco activase proteins from diverse species are variedly polydisperse in nature. The oligomeric state of Rca in solution was observed to be varied in a concentration and nucleotide dependent manner. Various oligomeric sizes (50-600 kDa) from 1–16 subunits were reported with larger species observed at higher enzyme concentrations [Wang *et al.*, 1993; Barta *et al.*, 2010; Keown *et al.*, 2013]. The specific ATPase, Rubisco activation activity and the oligomeric size increased with increasing concentration and in the presence of macromolecular crowding agents such as polyethylene glycol [Salvucci, 1992].

It has been reported that a complex of not more than two to four subunits is required for biological activity of tobacco Rca [Keown *et al.*, 2013] in a study using small angle X-ray scattering (SAXS) and analytical ultracentrifugation. However, this study was conducted in the absence of nucleotides under various dilutions. It has also been hypothesized that Rca may form open spiral structures in solution [Chakraborty *et al.*, 2012; Henderson *et al.*, 2013; Keown *et al.*, 2013] rather than a closed hexamer. It has also been interpreted that a large oligomer of more than 8 subunits and up to 16 subunits is the active state of activase [Buchen-Osmond *et al.*, 1992; Wang *et al.*, 1993; Barta *et al.*, 2010]. However, this appears inconsistent with the canonical active functional oligomeric form (hexamer) of most AAA+ proteins [Erzberger and Berger, 2006]. Therefore it is possible that transient hexamers may exist in solution in the presence of nucleotide, however formation of hexamers and their dissociation is too rapid to be detected by the reported methods and therefore a range of oligomers were observed.

Biochemical analysis of interface mutants in tobacco revealed 2.9–3.7 and 3.5–5.4 as minimal subunits necessary for ATPase and Rubisco activation activity respectively [Stotz *et al.*, 2011] providing evidence for the hexamer as minimal functional unit. Indeed mutating the Arginine 294 (R294) residue of tobacco Rca resulted in the formation of stable hexamers in the presence of Mg.ATPγS as observed by mass spectroscopy [Blayney *et al.*, 2011], negative stain EM [Stotz *et al.*, 2011] and analytical ultracentrifugation [Keown and Pearce, 2014]. The biochemical analysis of this variant (R294V) revealed it to be fully functional, further providing a strong evidence in support of the hexamer as active species. Also recently, α isoform and equimolar mixture of α and β isoform of spinach was observed to form a stable hexamer in presence of Mg.ATPγS as the first instance for any wild type Rca [Keown and Pearce, 2014]. The non-homologous bacterial red-type activase CbbX was observed to change its oligomeric form in response to the presence of an allosteric regulator [Mueller-Cajar *et al.*, 2011]. Both the red-type activase and the CbbQO-type activase are confirmed to form hexameric assemblies in their functional state [Mueller-Cajar *et al.*, 2011; Tsai *et al.*, 2015]. Thus it is quite possible that green-type Rubisco activase may acquire variable oligomeric states in solution but form transient hexamers as an active species in presence of nucleotide, especially when engaging inhibited substrate Rubisco.

Although we are confident that green-type Rca will function as a hexameric ring-shaped particle, and utilize the conformational changes brought about by ATP hydrolysis to remodel inhibited Rubisco active sites, the actual mechanism of plant Rubisco activase function remains elusive. The red-type Rca CbbX threads the C-terminal extension of the Rubisco large subunit through its central pore [Mueller-Cajar *et al.*, 2011], but this mode of action appears unavailable to higher plant Rca [Mueller-Cajar *et al.*, 2014]. Currently our mechanistic model is mostly limited to a confirmed interaction between the surface-exposed βC-βD loop of the Rubisco large subunit's N-terminal domain with the small helical insertion of Rca's α-helical subdomain [Ott *et al.*, 2000; Li *et al.*, 2005]. In addition loss of Rca's N-terminal domain (residues 1–66 in tobacco) and mutation of selected pore loop residues also eliminates activase, but not ATPase function [Stotz *et al.*, 2011]. A low affinity of Rca for its substrate Rubisco has complicated attempts to obtain Rca-Rubisco protein complexes for further analysis. It is critical to engage in a more systematic approach by testing many structure-guided mutations to throw light on this critical topic.

3.3. *Regulation of Rubisco activase*

In most plants, including arabidopsis, spinach and rice [Werneke *et al.*, 1988, 1989; To *et al.*, 1999], Rubisco activase occurs as two isoforms often as a result of alternative splicing of a single transcript. Consequently, the isoforms are essentially identical with the α isoform having a ~40 residue C-terminal extension, which contains two conserved cysteine residues that are essential for its regulation [Zhang and Portis, 1999; Portis *et al.*, 2008]. However, in some plants like cotton, two isoforms are generated by different genes rather than splicing [Salvucci *et al.*,

2003]. The activase is constitutively ATPase active and in contrast to many other AAA+ proteins the ATPase activity of activase is not stimulated by its protein substrate [Robinson and Portis, 1989; Hazra *et al.*, 2015].

The ATPase activity of Rubisco activase is inhibited by ADP and regulated by the ATP/ADP ratio in the chloroplast stroma. This modulation of ATPase activity of activase consequently regulates Rubisco activity in response to light fluctuations [Streusand and Portis, 1987; Shen *et al.*, 1991]. It has been shown that the α isoform is more sensitive to ADP inhibition than the β isoform [Shen *et al.*, 1991; Zhang and Portis, 1999].

The α isoform is also regulated by changes in the redox state of two cysteine residues located at its C-terminal extension by thioredoxin-f [Zhang and Portis, 1999]. Thioredoxins are a class of small redox proteins, which are involved in several regulatory processes in plants [Schurmann and Jacquot, 2000]. Apart from Rubisco activase, thioredoxin-f regulates the activity of several other Calvin-Benson cycle enzymes such as fructose1,6-bisphosphatase, sedoheptulose 1,7-bisphosphatase, phosphoribulokinase, and glyceraldehyde-3-phosphate dehydrogenase [Schurmann and Jacquot, 2000; Balmer and Buchanan, 2002]. Trx-f is reduced in the light *via* ferredoxin and ferredoxin-thioredoxin reductase (FTR) using electrons originating from the action of photosystem I [Dai *et al.*, 2007]. Reduced thioredoxin-f then diffuses throughout the stroma and reduces oxidized target proteins in the light leading to their activation.

In vitro, the redox regulation of Rca can be mediated by thioredoxin-f, along with a DTT electron source, which makes the ATP/ADP ratio much less inhibitory, causing an increase in ATPase and Rubisco activation activities of the α isoform. A similar effect was not observed in the presence of DTT and thioredoxin-m. This effect is reversible when DTT is replaced by oxidized glutathione leading to a reduction in activity [Zhang and Portis, 1999]. Substituting either of these two cysteine residues from the C-terminal extension to alanine reduces the ATP/ADP sensitivity of the activase and the light responsiveness of Rubisco activity *in vivo* and *in vitro* [Zhang and Portis, 1999; Zhang *et al.*, 2002].

Although the α isoform is critical for redox regulation and important in regulating Rubisco activation in response to fluctuating light irradiance, interestingly some species like tobacco lack an α isoform. Plants expressing only the β isoform did not show any downregulation in Rubisco activity during a light–dark transition [Zhang *et al.*, 2002] and the β isoform was not observed to be redox sensitive when assayed alone. The regulation of the β isoform by the ADP/ATP ratio varies among different species. For instance, the β isoform of tobacco is reportedly more prone to inhibition by ADP than the β isoform of *Arabidopsis* [Carmo-Silva and Salvucci, 2013]. Subunit mixing experiments between α and β isoforms under *in vitro* assay condition validated a regulatory effect of the α isoform on the β isoform in arabidopsis. At equimolar ratio, the β isoform is fully inhibited by the presence of the α isoform at an ATP/ADP ratio of 1:3 under oxidizing conditions [Zhang *et al.*, 2001], indicating that the α isoform allosterically modulates the activity of β isoforms.

Under *in vivo* conditions, reducing equivalents increase with light illumination, hence it was assumed that redox chemistry of the α isoform modulates activase activity in response to light and regulates Rubisco activity. This notion was confirmed by a study expressing just a single isoform *in vivo*. Arabidopsis Rca mutants expressing only the α isoform respond to light fluctuations and Rubisco activation was downregulated under limiting condition, while the expression of just β isoform did not downregulate Rubisco activity in response to light [Zhang *et al.*, 2002]. Based on the results of mutational and cross-linking experiments, the C-terminal extension of the α isoform in the oxidized state folds back towards the nucleotide binding pocket, perturbing its activity [Zhang *et al.*, 2001; Wang and Portis, 2006]. Furthermore, there is fluctuation in the diurnal expression of activase in plants with maximum expression at the onset of the photocycle and lowest at night [Rundle and Zielinski, 1991], which closely mimics the circadian rhythm of Rubisco activation.

3.4. *The role of Rubisco activase in regulating photosynthesis at elevated temperatures*

Since Rubisco activase is critically responsible for maintaining the activation state of Rubisco *in vivo*, the mechanism and regulation of Rca are the focus of current research. However, Rca is a thermolabile enzyme and inactivation of Rca often limits photosynthetic efficiency at moderately elevated temperatures [Crafts-Brandner and Salvucci, 2000]. Therefore activase with improved thermal stability is believed to be result in plants with improved photosynthetic performance and growth phenotypes under moderate heat stress [Salvucci and Crafts-Brandner, 2004; Kurek *et al.*, 2007; Sage *et al.*, 2008]. This hypothesis has been strongly supported by expressing more thermostable variants of Rca in arabidopsis, which resulted in plants with higher rate of photosynthesis and growth at elevated temperature [Kurek *et al.*, 2007; Kumar *et al.*, 2009].

It has also been argued that deactivation of Rubisco at elevated temperature may be a regulatory response to protect the plant during heat stress [Cen and Sage, 2005; Sharkey, 2005; Sage and Kubien, 2007]. Photosystem II (PSII) was considered as the major cause of thermal damage in the past [Berry and Bjorkman, 1980] however, PSII was reported to be unaffected at temperature that inhibits CO_2 fixation [Havaux and Tardy, 1996; Law and Crafts-Brandner, 1999]. Other electron transport limitations such as RuBP regeneration, stability of the thylakoid membrane and ATP/ADP levels were also supposed to constrain photosynthesis at high temperature [Wise *et al.*, 2004; Sharkey, 2005; Yamori *et al.*, 2014]. Also, the solubility of CO_2 and Rubisco specificity ($S_{c/o}$) reduces at high temperature favouring the oxygenation reaction leading to production of more misfire products [Jordan and Ogren, 1984; Schrader *et al.*, 2006]. Therefore, it was proposed that activase may function as a fuse to reduce Calvin–Benson cycle flux as a regulatory mechanism to protect Rubisco and the photosynthetic apparatus at high temperature [Sharkey, 2005].

It has been hypothesized that conformational flexibility of Rca is required for it to remodel Rubisco conformation and this could be the reason why Rca cannot be thermostable [Parry *et al.*, 2013]. Furthermore, research comparing a warm-season plant and a cool season plant also showed that there was a trade-off between Rca thermostability and its activity at low temperature [Salvucci and Crafts-Brandner, 2004; Carmo-Silva and Salvucci, 2013; Parry *et al.*, 2013]. The analysis of different genetic variants of rice yielded similar results where a rice species (*Oryza australiensis*) endemic to hot and arid zone displayed higher thermotolerance than the crop species (*Oryza sativa*) at the cost of catalytic activity [Scafaro *et al.*, 2016].

The stability of Rca varies between different species, with the α isoform reported to be more thermostable than the β isoform in spinach but not in cotton and arabidopsis [Kallis *et al.*, 2000; Salvucci *et al.*, 2003] where both isoforms have similar thermostability. In spinach, the α isoform also confers its thermostability to the β isoform [Crafts-Brandner *et al.*, 1997; Keown and Pearce, 2014]. Rca shows enhanced thermostability in presence of nucleotide, with ADP and ATP guarding the β isoform of arabidopsis and cotton against heat inactivation [Barta *et al.*, 2010; Henderson *et al.*, 2013]. Improved thermotolerance of the α and the β isoform was observed in spinach in presence of ADP, while the α isoform of spinach showed a dramatic increase in its thermal stability in presence of ATPγS [Crafts-Brandner *et al.*, 1997; Keown and Pearce, 2014]. The molecular mechanism underlying the thermolability of photosynthesis is of particular pertinence due to the predicted effects of global climate change [Sage *et al.*, 2008]. Further insights from structural and biochemical work may generate the possibility to discover and engineer more thermostable variants. Depending on whether the activase is in fact a temperature switch, this engenders the possibility of engineering crops that can maintain carbon fixation at supra-optimal temperatures.

4. Outlook

Rubisco activase received abundant attention from biochemists in the two decades prior to detailed structural information becoming available [Portis, 2003; Stotz *et al.*,

2011]. However, in spite of the current existence of a number of atomic models, the new data has not yet been taken advantage of by more extensive mutational studies. In order to resolve the many questions that remain regarding this important chaperone it is critical that our new and rapidly developing understanding of the vast AAA+ protein family [Sysoeva, 2016] is integrated with Rca structural data and a plethora of biochemical experiments. An extensive molecular toolbox comprising site-directed mutants of all relevant Rca functional aspects such as ATPase activity, protein–protein interaction, and regulation needs to be constructed. The ability of Rca to rapidly exchange subunits, make this system highly amenable to powerful subunit doping experiments that have recently been extensively exploited in the study of the disaggregase Hsp104 [DeSantis *et al.*, 2012] and a range of other AAA+ proteins [Moreau *et al.*, 2007; Werbeck *et al.*, 2008].

A series of systematic experiments can then be initiated to carefully dissect the relative contribution of different structural elements in Rca function. This approach will permit question of general interest to molecular chaperone biology to be asked. For example the relative ability to tolerate ATPase inactive subunits reveals whether a concerted or stochastic mechanism is employed in the function [Moreau *et al.*, 2007]. These questions are of major interest to the AAA+ ATPase community in general and are fiercely debated [Martin *et al.*, 2005; Lyubimov *et al.*, 2011]. Here it is also worth noting that whereas most other AAA+ ATPases utilize a range of substrates and rely on complex functional assays based on protein degradation or refolding, Rubisco activase has a well-defined substrate, and has rapid and sensitive biochemical assays permitting rapid functional dissection [Lan and Mott, 1991; Loganathan *et al.*, 2016]. One could therefore argue that Rubisco activase and the convergently evolved red-type and CbbQO-type Rcas are highly suitable workhorses to expand our appreciation of AAA+ protein mechanism.

From a more pragmatic point of view, strong evidence pointing to Rubisco activase as a key thermal lesion of photosynthesis calls for careful physiological characterization of plants expressing more thermostable activase variants. The published work indicated improved photosynthetic performance and biomass accumulation of such *Arabidopsis* plants at moderately elevated temperatures [Kurek *et al.*, 2007; Kumar *et al.*, 2009]. However, it did not seek to characterize the fitness of such plants under stressed or highly variable conditions. It will be critical to rigorously test the validity of the thermal fuse hypothesis [Sharkey, 2005], which predicts that Rca thermolability is a rapid mechanism to shut down Calvin cycle flux. Clearly this would then permit ATP and reducing equivalents to be utilized for more urgent tasks than biomass accumulation. Finally it should be noted that although activase thermolability may turn out to enhance survival of plants under natural conditions, this may be a switch that can be safely overridden under highly controlled agricultural conditions, such as those often found in modern production systems. Therefore this helper protein of the world's most abundant enzyme, may well form part of the solution towards the necessary task of dramatically enhancing agricultural productivity in the coming decades [Ort *et al.*, 2015].

References

Alonso, H., Blayney, M.J., Beck, J.L. and Whitney S.M. (2009). Substrate-induced assembly of *Methanococcoides burtonii* d-ribulose-1,5-bisphosphate carboxylase/oxygenase dimers into decamers, *J. Biol. Chem.*, 284, 33876–33882.

Andersson, I. (1996). Large structures at high resolution: The 1.6 Å crystal structure of spinach ribulose-1,5-bisphosphate carboxylase/oxygenase complexed with 2-carboxyarabinitol bisphosphate, *J. Mol. Biol.*, 259, 160–174.

Andersson, I. and Backlund, A. (2008). Structure and function of Rubisco, *Plant Physiol. Biochem.*, 46, 275–291.

Andersson, I. and Taylor, T.C. (2003). Structural framework for catalysis and regulation in ribulose-1,5-bisphosphate carboxylase/oxygenase, *Arch. Biochem. Biophys.*, 414, 130–140.

Andralojc, P.J., Madgwick, P.J., Tao, Y., Keys, A., Ward, J.L., Beale, M.H., Loveland, J.E., Jackson, P.J., Willis, A.C., Gutteridge, S. and Parry, M.A.J. (2012). 2-Carboxy-D-arabinitol 1-phosphate (CA1P) phosphatase: evidence for a wider role in plant Rubisco regulation, *Biochem. J.*, 442, 733–742.

Andrews, T. and Hatch, M. (1971). Activity and properties of ribulosediphosphate carboxylase from plants with the C_4-dicarboxylic acid pathway of photosynthesis, *Phytochemistry*, 10, 9–15.

Andrews, T.J. (1996). The bait in the Rubisco mousetrap, *Nat. Struct. Mol. Biol.*, 3, 3–7.

Andrews, T.J. and Abel, K.M. (1981). Kinetics and subunit interactions of ribulose bisphosphate carboxylase-oxygenase from the cyanobacterium, *Synechococcus* sp., *J. Biol. Chem.*, 256, 8445–8451.

Andrews, T.J. and Lorimer, G.H. (1987). Rubisco: structure, mechanisms and prospects for improvement. In *The Biochemistry of Plants*, 10, Hatch, M.D. and Boardman N.K., eds. (New York: Academic Press), pp. 131–218.

Andrews, T.J. and Whitney, S.M. (2003). Manipulating ribulose bisphosphate carboxylase/ oxygenase in the chloroplasts of higher plants, *Arch. Biochem. Biophys.*, 414, 159–169.

Ashida, H., Saito, Y., Kojima, C., Kobayashi, K., Ogasawara, N. and Yokota, A. (2003). A functional link between RuBisCO-like protein of *Bacillus* and photosynthetic RuBisCO, *Science*, 302, 286–290.

Badger, M.R. and Bek, E.J. (2008). Multiple Rubisco forms in proteobacteria: their functional significance in relation to CO_2 acquisition by the CBB cycle, *J. Exp. Bot.*, 59, 1525–1541.

Balmer, Y. and Buchanan, B.B. (2002). Yet another plant thioredoxin, *Trends Plant Sci.*, 7, 191–193.

Barta, C., Dunkle, A.M., Wachter, R.M. and Salvucci, M.E. (2010). Structural changes associated with the acute thermal instability of Rubisco activase, *Arch. Biochem. Biophys.*, 499, 17–25.

Bauwe, H., Hagemann, M. and Fernie, A.R. (2010). Photorespiration: players, partners and origin, *Trends Plant Sci.*, 15, 330–336.

Berry, J. and Bjorkman, O. (1980). Photosynthetic response and adaptation to temperature in higher plants, *Annu. Rev. Plant Physiol.*, 31, 491–543.

Berry, J.A., Lorimer, G.H., Pierce, J.,. Seemann, J.R., Meek, J. and Freas, S. (1987). Isolation, identification, and synthesis of 2-carboxyarabinitol 1-phosphate, a diurnal regulator of ribulose-bisphosphate carboxylase activity, *Proc. Natl. Acad. Sci. USA*, 84, 734–738.

Blair, G.E. and Ellis, R.J. (1973). Protein-synthesis in chloroplasts.1. Light-driven synthesis of large subunit of fraction I protein by isolated pea chloroplasts, *Biochim. Biophys. Acta.*, 319, 223–234.

Blankenship, R.E., Tiede, D.M., Barber, J., Brudvig, G.W., Fleming, G., Ghirardi, M., Gunner, M.R., Junge, W., Kramer, D.M., Melis, A. Moore, T.A., Moser, C.C., Nocera, D.G., Nozik, A.J., Ort, D.R., Parson, W.W., Prince, R.C. and Sayre, R.T. (2011). Comparing photosynthetic and photovoltaic efficiencies and recognizing the potential for improvement, *Science*, 332, 805–809.

Blayney, M.J., Whitney, S.M. and Beck, J.L. (2011). NanoESI mass spectrometry of Rubisco and Rubisco activase structures and their interactions with nucleotides and sugar phosphates, *J. Am. Soc. Mass. Spectrom.*, 22, 1588–1601.

Bracher, A., Sharma, A., Starling-Windhof, A., Hartl, F.U. and Hayer-Hartl, M. (2015). Degradation of potent Rubisco inhibitor by selective sugar phosphatase, *Nat. Plants*, 1, p. 14002.

Buchen-Osmond, C., Portis, A.R., Jr. and Andrews, J. (1992). Rubisco activase modifies the appearance of Rubisco in the electron microscope. In *Research in Photosynthesis, Vol. III; Proceedings of the 9th International Congress on Photosynthesis*, Nagoya, Japan, August 30–September 4, 1992 (ed. Murata), pp. 635–638. Kluwer Academic Publishers, Dordrecht, The Netherlands..

Carmo-Silva, A.E. and Salvucci, M.E. (2013). The regulatory properties of Rubisco activase differ among species and affect photosynthetic induction during light transitions, *Plant Physiol.*, 161, 1645–1655.

Cen, Y.P. and Sage, R.F. (2005). The regulation of Rubisco activity in response to variation in temperature and atmospheric CO_2 partial pressure in sweet potato, *Plant Physiol.*, 139, 979–990.

Chakraborty, M., Kuriata, A.M., Nathan Henderson, J., Salvucci, M.E., Wachter, R.M. and Levitus, M. (2012). Protein oligomerization monitored by fluorescence fluctuation spectroscopy: self-assembly of Rubisco activase, *Biophys. J.*, 103, 949–958.

Cleland, W.W., Andrews, T.J., Gutteridge, S., Hartman, F.C. and Lorimer, G.H. (1998). Mechanism of Rubisco: the carbamate as general base, *Chem. Rev.*, 98, 549–561.

Crafts-Brandner, S.J. and Salvucci, M.E. (2000). Rubisco activase constrains the photosynthetic potential of leaves at high temperature and CO_2, *Proc. Natl. Acad. Sci. USA*, 97, 13430–13435.

Crafts-Brandner, S.J., Van De Loo, F.J. and Salvucci, M.E. (1997). The two forms of ribulose-1,5-bisphosphate carboxylase/oxygenase activase differ in sensitivity to elevated temperature, *Plant Physiol.*, 114, 439–444.

Dai, S., Friemann, R., Glauser, D.A., Bourquin, F., Manieri, W., Schurmann, P. and Eklund, H. (2007). Structural snapshots along the reaction pathway of ferredoxin-thioredoxin reductase, *Nature*, 448, 92–96.

Davies, J.M., Brunger, A.T. and Weis, W.I. (2008). Improved structures of full-length p97, an AAA ATPase: implications for mechanisms of nucleotide-dependent conformational change, *Structure*, 16, 715–726.

DeSantis, M.E., Leung, E.H., Sweeny, E.A., Jackrel, M.E., Cushman-Nick, M., Neuhaus-Follini, A., Vashist, S., Sochor, M.A., Knight, M.N. and Shorter, J. (2012). Operational plasticity enables Hsp104 to disaggregate diverse amyloid and nonamyloid clients, *Cell*, 151, 778–793.

Duff, A.P., Andrews, T.J. and Curmi, P.M.G. (2000). The transition between the open and closed states of Rubisco is triggered by the inter-phosphate distance of the bound bisphosphate, *J. Mol. Biol.*, 298, 903–916.

Edmondson, D.L., Badger M.R. and Andrews, T.J. (1990). Slow inactivation of ribulosebisphosphate carboxylase during catalysis is caused by accumulation of a slow, tight-binding inhibitor at the catalytic site, *Plant Physiol.*, 93, 1390–1397.

Edmondson, D.L., Kane, H.J. and Andrews, T.J. (1990). Substrate isomerization inhibits ribulosebisphospate carboxylase-oxygenase during catalysis, *FEBS Lett.*, 260, 62–66.

Ellis, R.J. (1979). The most abundant protein in the world, *Trends Biochem. Sci.*, 4, 241–244.

Erzberger, J.P. and Berger, J.M. (2006). Evolutionary relationships and structural mechanisms of AAA+ proteins, *Annu. Rev. Biophys. Biomol. Struct.*, 35, 93–114.

Ezaki, S., Maeda, N., Kishimoto, T., Atomi, H. and Imanaka, T. (1999). Presence of a structurally novel type ribulose-bisphosphate carboxylase/oxygenase in the hyperthermophilic archaeon, *Pyrococcus kodakaraensis* KOD1, *J. Biol. Chem.*, 274, 5078–5082.

Hanson, T.E. and Tabita, F.R. (2001). A ribulose-1,5-bisphosphate carboxylase/oxygenase (RubisCO)-like protein from *Chlorobium tepidum* that is involved with sulfur metabolism and the response to oxidative stress, *Proc. Natl. Acad. Sci. USA*, 98, 4397–4402.

Hanson, P.I. and Whiteheart, S.W. (2005). AAA+ proteins: have engine, will work, *Nat. Rev. Mol. Cell Biol.*, 6, 519–529.

Hartman, F.C. and Harpel, M.R. (1994). Structure, function, regulation, and assembly of D-ribulose-1,5-bisphosphatecarboxylase oxygenase, *Annu. Rev. Biochem.*, 63, 197–234.

Hasse, D., Larsson, A.M. and Andersson, I. (2015). Structure of *Arabidopsis thaliana* Rubisco activase, *Acta Crystallogr. D Biol. Crystallogr.*, 71, 800–808.

Hauser, T., Popilka, L., Hartl, F.U. and Hayer-Hartl, M. (2015). Role of auxiliary proteins in Rubisco biogenesis and function, *Nat. Plants*, 1, p. 15065.

Havaux, M. and Tardy, F. (1996). Temperature-dependent adjustment of the thermal stability of photosystem II *in vivo*: possible involvement of xanthophyll-cycle pigments, *Planta*, 198, 324–333.

Hazra, S., Henderson, J.N., Liles, K., Hilton, M.T. and Wachter, R.M. (2015). Regulation of ribulose-1,5-bisphosphate carboxylase/oxygenase (Rubisco) activase: product inhibition, cooperativity, and magnesium activation, *J. Biol. Chem.*, 290, 24222–24236.

Henderson, J.N., Hazra, S., Dunkle, A.M., Salvucci, M.E. and Wachter, R.M. (2013). Biophysical characterization of higher plant Rubisco activase, *Biochim. Biophys. Acta*, 1834, 87–97.

Henderson, J.N., Kuriata, A.M., Fromme, R., Salvucci, M.E. and Wachter, R.M. (2011). Atomic resolution x-ray structure of the substrate recognition domain of higher plant ribulose-bisphosphate carboxylase/oxygenase (Rubisco) activase, *J. Biol. Chem.*, 286, 35683–35688.

Hovde, B.T., Deodato, C.R., Hunsperger, H.M., Ryken, S.A., Yost, W., Jha, R.K., Patterson, J., Monnat, R.J., Jr., Barlow, S.B., Starkenburg, S.R. and Cattolico, R.A. (2015). Genome sequence and transcriptome analyses of *Chrysochromulina tobin*: metabolic tools for enhanced algal fitness in the prominent order Prymnesiales (Haptophyceae), *PLOS Genet.*, 11, p. e1005469.

Imker, H.J., Fedorov, A.A., Fedorov, E.V., Almo, S.C. and Gerlt, J.A. (2007). Mechanistic diversity in the RuBisCO superfamily: the "Enolase" in the methionine salvage pathway in *Geobacillus kaustophilus, Biochemistry*, 46, 4077–4089.

Ishikawa, C., Hatanaka, T., Misoo, S., Miyake, C. and Fukayama, H. (2011). Functional incorporation of sorghum small subunit increases the catalytic turnover rate of Rubisco in transgenic rice, *Plant Physiol.*, 156, 1603–1611.

Jordan, D.B. and Chollet, R. (1983). Inhibition of ribulose bisphosphate carboxylase by substrate ribulose 1,5-bisphosphate, *J. Biol. Chem.*, 258, 13752–13758.

Jordan, D.B., Chollet, R. and Ogren, W.L. (1983). Binding of phosphorylated effectors by active and inactive forms of ribulose 1,5-bisphosphate carboxylase, *Biochemistry*, 22, 3410–3418.

Jordan, D.B. and Ogren, W.L. (1981). Species variation in the specificity of ribulose bisphosphate carboxylase/oxygenase, *Nature*, 291, 513–515.

Jordan, D.B. and Ogren, W.L. (1984). The CO_2/O_2 specificity of ribulose 1,5-bisphosphate carboxylase/oxygenase, *Planta*, 161, 308–313.

Joshi, J., Mueller-Cajar, O., Tsai, Y.C.C., Hartl, F.U. and Hayer-Hartl, M. (2015). Role of small subunit in mediating assembly of red-type form I Rubisco, *J. Biol. Chem.*, 290, 1066–1074.

Kallis, R.P., Ewy, R.G. and Portis, A.R., Jr. (2000). Alteration of the adenine nucleotide response and increased Rubisco activation activity of *Arabidopsis* Rubisco activase by site-directed mutagenesis, *Plant Physiol.*, 123, 1077–1086.

Keown, J.R., Griffin, M.D., Mertens, H.D. and Pearce, F.G. (2013). Small oligomers of ribulose-bisphosphate carboxylase/oxygenase (Rubisco) activase are required for biological activity, *J. Biol. Chem.*, 288, 20607–20615.

Keown, J.R. and Pearce, F.G. (2014). Characterization of spinach ribulose-1,5-bisphosphate carboxylase/oxygenase activase isoforms reveals hexameric assemblies with increased thermal stability, *Biochem. J.*, 464, 413–423.

Khan, S., Andralojc, P.J., Lea, P.J. and Parry, M.A. (1999). 2′-carboxy-d-arabitinol 1-phosphate protects ribulose 1,5-bisphosphate carboxylase/oxygenase against proteolytic breakdown, *Eur. J. Biochem.*, 266, 840–847.

Kitano, K., Maeda, N., Fukui, T., Atomi, H., Imanaka, T. and Miki, K. (2001). Crystal structure of a novel-type archaeal Rubisco with pentagonal symmetry, *Structure*, 9, 473–481.

Knight, S., Andersson, I. and Brändén, C.I. (1990). Crystallographic analysis of ribulose 1,5-bisphosphate carboxylase from spinach at 2.4 Å resolution: subunit interactions and active site, *J. Mol. Biol.*, 215, 113–160.

Kumar, A., Li, C. and Portis, A.R., Jr. (2009). Arabidopsis thaliana expressing a thermostable chimeric Rubisco activase exhibits enhanced growth and higher rates of photosynthesis at moderately high temperatures, *Photosynth. Res.*, 100, 143–153.

Kurek, I., Chang, T.K., Bertain, S.M., Madrigal, A., Liu, L., Lassner, M.W. and Zhu, G. (2007). Enhanced thermostability of Arabidopsis Rubisco activase improves photosynthesis and growth rates under moderate heat stress, *Plant Cell*, 19, 3230–3241.

Lan, Y. and Mott, K.A. (1991). Determination of apparent Km values for ribulose 1,5-bisphosphate carboxylase oxygenase (Rubisco) activase using the spectrophotometric assay of Rubisco activity, *Plant Physiol.*, 95, 604–609.

Law, R.D. and Crafts-Brandner, S.J. (1999). Inhibition and acclimation of photosynthesis to heat stress is closely correlated with activation of ribulose-1,5-bisphosphate carboxylase/oxygenase, *Plant Physiol.*, 120, 173–182.

Lee, B.G., Read, B.A. and Tabita, F.R. (1991). Catalytic properties of recombinant octameric, hexadecameric, and heterologous cyanobacterial/bacterial ribulose-1,5-bisphosphate carboxylase/oxygenase, *Arch. Biochem. Biophys.*, 291, 263–269.

Li, C., Salvucci, M.E. and Portis, A.R., Jr., (2005). Two residues of Rubisco activase involved in recognition of the Rubisco substrate, *J. Biol. Chem.*, 280, 24864–24869.

Li, H., Sawaya, M.R., Tabita, F.R. and Eisenberg, D. (2005). Crystal structure of a RuBisCO-like protein from the green sulfur bacterium *Chlorobium tepidum*, *Structure*, 13, 779–789.

Loganathan, N., Tsai, Y.-C.C. and Mueller-Cajar, O. (2016). Characterization of the heterooligomeric red-type Rubisco activase from red algae, *Proc. Natl. Acad. Sci. USA*, 113, 14019–14024.

Lorimer, G.H., Badger, M.R. and Andrews, T.J. (1976). The activation of ribulose-1,5-bisphosphate carboxylase by carbon dioxide and magnesium ions. Equilibria, kinetics, a suggested mechanism, and physiological implications, *Biochemistry*, 15, 529–536.

Lyubimov, A.Y., Strycharska, M. and Berger, J.M. (2011). The nuts and bolts of ring-translocase structure and mechanism, *Curr. Opin. Struct. Biol.*, 21, 240–248.

Maeda, N., Kitano, K., Fukui, T., Ezaki, S., Atomi, H., Miki, K. and Imanaka, T. (1999). Ribulose bisphosphate carboxylase/oxygenase from the hyperthermophilic archaeon *Pyrococcus kodakaraensis* KOD1 is composed solely of large subunits and forms a pentagonal structure, *J. Mol. Biol.*, 293, 57–66.

Martin, A., Baker, T.A. and Sauer, R.T. (2005). Rebuilt AAA plus motors reveal operating principles for ATP-fuelled machines, *Nature*, 437, 1115–1120.

Martin, A., Baker, T.A. and Sauer, R.T. (2008). Pore loops of the AAA+ ClpX machine grip substrates to drive translocation and unfolding, *Nat. Struct. Mol. Biol.*, 15, 1147–1151.

McNevin, D., Von Caemmerer, S. and Farquhar, G. (2006). Determining RuBisCO activation kinetics and other rate and equilibrium constants by simultaneous multiple non-linear regression of a kinetic model, *J. Exp. Bot.*, 57, 3883–3900.

Moreau, M.J., McGeoch, A.T., Lowe, A.R., Itzhaki, L.S. and Bell, S.D. (2007). ATPase site architecture and helicase mechanism of an archaeal MCM, *Mol. Cell.*, 28, 304–314.

Mueller-Cajar, O., Stotz, M. and Bracher, A. (2014). Maintaining photosynthetic CO_2 fixation *via* protein remodelling: the Rubisco activases, *Photosynth. Res.*, 119, 191–201.

Mueller-Cajar, O., Stotz, M., Wendler, P., Hartl, F.U., Bracher, A. and Hayer-Hartl, M. (2011). Structure and function of the AAA(+) protein CbbX, a red-type Rubisco activase, *Nature*, 479, 194–199.

Neuwald, A.F., Aravind, L., Spouge, J.L. and Koonin, E.V. (1999). AAA+: a class of chaperone-like ATPases associated with the assembly, operation, and disassembly of protein complexes, *Genome Res.*, 9, 27–43.

Newman, J. and Gutteridge, S. (1994). Structure of an effector-induced inactivated state of ribulose 1,5-bisphosphate carboxylase/oxygenase: the binary complex between enzyme and xylulose 1,5-bisphosphate, *Structure*, 2, 495–502.

Ogura, T., Whiteheart, S.W. and Wilkinson, A.J. (2004). Conserved arginine residues implicated in ATP hydrolysis, nucleotide-sensing, and inter-subunit interactions in AAA and AAA+ ATPases, *J. Struct. Biol.*, 146, 106–112.

Ogura, T. and Wilkinson, A.J. (2001). AAA+ superfamily ATPases: common structure–diverse function, *Genes Cells*, 6, 575–597.

Ort, D.R., Merchant, S.S., Alric, J., Barkan, A., Blankenship, R.E., Bock, R., Croce, R., Hanson, M.R., Hibberd, J.M., Long, S.P., Moore, T.A., Moroney, J., Niyogi, K.K., Parry, M.A., Peralta-Yahya, P.P., Prince, R.C., Redding, K.E., Spalding, M.H., van Wijk, K.J., Vermaas, W.F., von Caemmerer, S., Weber, A.P., Yeates, T.O., Yuan, J.S. and Zhu, X.G. (2015). Redesigning photosynthesis to sustainably meet global food and bioenergy demand, *Proc. Natl. Acad. Sci. USA*, 112, 8529–8536.

Ott, C.M., Smith, B.D., Portis, A.R., Jr. and Spreitzer, R.J. (2000) Activase region on chloroplast ribulose-1,5-bisphosphate carboxylase/oxygenase. Nonconservative substitution in the large subunit alters species specificity of protein interaction, *J. Biol. Chem.*, 275, 26241–26244.

Oudot-Le Secq, M.P., Grimwood, J., Shapiro, H., Armbrust, E.V., Bowler, C. and Green, B.R. (2007). Chloroplast genomes of the diatoms *Phaeodactylum tricornutum* and *Thalassiosira pseudonana*: comparison with other plastid genomes of the red lineage, *Mol. Genet. Genomics*, 277, 427–439.

Paech, C., Pierce, J., McCurry, S.D. and Tolbert, N. (1978). Inhibition of ribulose-1,5-bisphosphate carboxylase/oxygenase by ribulose-1,5-bisphosphate epimerization and degradation products, *Biochem. Biophys. Res. Commun.*, 83, 1084–1092.

Parry, M.A., Andralojc, P.J., Scales, J.C., Salvucci, M.E., Carmo-Silva, A.E., Alonso, H. and Whitney, S.M. (2013). Rubisco activity and regulation as targets for crop improvement, *J. Exp. Bot.*, 64, 717–730.

Parry, M.A., Keys, A.J., Madgwick, P.J., Carmo-Silva, A.E. and Andralojc, P.J. (2008). Rubisco regulation: a role for inhibitors, *J. Exp. Bot.*, 59, 1569–1580.

Parry, M.A.J., Madgwick, P.J., Carvalho, J.F.C. and Andralojc, P.J. (2007). Prospects for increasing photosynthesis by overcoming the limitations of Rubisco, *J. Agric. Sci.*, 145, 31–43.

Pearce, F.G. (2006). Catalytic by-product formation and ligand binding by ribulose bisphosphate carboxylases from different phylogenies, *Biochem. J.*, 399, 525–534.

Pearce, F.G. and Andrews, T.J. (2003). The relationship between side reactions and slow inhibition of ribulose-bisphosphate carboxylase revealed by a loop 6 mutant of the tobacco enzyme, *J. Biol. Chem.*, 278, 32526–32536.

Pierce, J., Tolbert, N. and Barker, R. (1980). Interaction of ribulosebisphosphate carboxylase/oxygenase with transition-state analogs, *Biochemistry*, 19, 934–942.

Portis, A.R., Jr. (2003). Rubisco activase-Rubisco's catalytic chaperone, *Photosynth. Res.*, 75, 11–27.

Portis, A.R., Jr., Li, C., Wang, D. and Salvucci, M.E. (2008). Regulation of Rubisco activase and its interaction with Rubisco, *J. Exp. Bot.*, 59, 1597–1604.

Portis, A.R., Jr. and Salvucci, M.E. (2002). The discovery of Rubisco activase — yet another story of serendipity, *Photosynth. Res.*, 73(1–3), 257–264.

Read, B.A. and Tabita, F.R. (1994). High substrate specificity factor ribulose bisphosphate carboxylase/oxygenase from eukaryotic marine algae and properties of recombinant cyanobacterial Rubisco containing "algal" residue modifications, *Arch. Biochem. Biophys.*, 312, 210–218.

Robinson, S.P. and Portis, A.R., Jr. (1989a) Adenosine triphosphate hydrolysis by purified Rubisco activase, *Arch. Biochem. Biophys.*, 268, 93–99.

Robinson, S.P. and Portis, A.R., Jr. (1989b). Ribulose-1,5-bisphosphate carboxylase/oxygenase activase protein prevents the *in vitro* decline in activity of ribulose-1,5-bisphosphate carboxylase/oxygenase, *Plant Physiol.*, 90, 968–971.

Robinson, S.P., Streusand, V.J., Chatfield, J.M. and Portis, A.R., Jr. (1988). Purification and assay of Rubisco activase from leaves, *Plant Physiol.*, 88, 1008–1014.

Roll-Mecak, A. and Vale, R.D. (2008). Structural basis of microtubule severing by the hereditary spastic paraplegia protein spastin, *Nature*, 451, 363–367.

Rundle, S.J. and Zielinski, R.E. (1991). Alterations in barley ribulose-1,5-bisphosphate carboxylase oxygenase activase gene-expression during development and in response to illumination, *J. Biol. Chem.*, 266, 14802–14807.

Sage, R.F. and Kubien, D.S. (2007). The temperature response of C3 and C4 photosynthesis, *Plant Cell Environ.*, 30, 1086–1106.

Sage, R.F., Way, D.A. and Kubien, D.S. (2008). Rubisco, Rubisco activase, and global climate change, *J. Exp. Bot.*, 59, 1581–1595.

Salvucci, M.E. (1992). Subunit interactions of Rubisco activase: polyethylene glycol promotes self-association, stimulates ATPase and activation activities, and enhances interactions with Rubisco, *Arch. Biochem. Biophys.*, 298, 688–696.

Salvucci, M.E. and Crafts-Brandner, S.J. (2004). Relationship between the heat tolerance of photosynthesis and the thermal stability of Rubisco activase in plants from contrasting thermal environments, *Plant Physiol.*, 134, 1460–1470.

Salvucci, M.E., Portis, A.R., Jr. and Ogren, W.L. (1985). A soluble chloroplast protein catalyzes ribulosebisphosphate carboxylase/oxygenase activation *in vivo*, *Photosynth. Res.*, 7, 193–201.

Salvucci, M.E., van de Loo, F.J. and Stecher, D. (2003). Two isoforms of Rubisco activase in cotton, the products of separate genes not alternative splicing, *Planta*, 216, 736–744.

Salvucci, M.E., Werneke, J.M., Ogren, W.L. and Portis, A.R., Jr. (1987). Purification and species distribution of Rubisco activase, *Plant Physiol.*, 84, 930–936.

Sato, T., Atomi, H. and Imanaka, T. (2007). Archaeal type III RuBisCOs function in a pathway for AMP metabolism, *Science*, 315, 1003–1006.

Scafaro, A.P., Galle, A., Van Rie, J., Carmo-Silva, E., Salvucci, M.E. and Atwell, B.J. (2016). Heat tolerance in a wild Oryza species is attributed to maintenance of Rubisco activation by a thermally stable Rubisco activase ortholog, *New Phytol.*, 211, 899–911.

Schloss, J. (1988). Comparative affinities of the epimeric reaction-intermediate analogs 2- and 4-carboxy-D-arabinitol 1,5-bisphosphate for spinach ribulose 1,5-bisphosphate carboxylase, *J. Biol. Chem.*, 263, 4145–4150.

Schneider, G., Lindqvist, Y., Branden, C.I. and Lorimer, G. (1986). 3-dimensional structure of ribulose-1,5-bisphosphate carboxylase-oxygenase from *Rhodospirillum rubrum* at 2.9 Å resolution, *EMBO J.*, 5, 3409–3415.

Schrader, S.M., Kane, H.J., Sharkey, T.D. and von Caemmerer, S. (2006). High temperature enhances inhibitor production but reduces fallover in tobacco Rubisco, *Funct. Plant Biol.*, 33, 921–929.

Schreuder, H.A., Curmi, P.M., Cascio, D., Eisenberg, D., Andersson, I., Knight, S., Brändén, C.I. and Sweet, R.M. (1993). Crystal structure of activated tobacco Rubisco complexed with the reaction-intermediate analogue 2-carboxy-arabinitol 1,5-bisphosphate, *Protein Sci.*, 2, 1136–1146.

Schurmann, P. and Jacquot, J.P. (2000). Plant thioredoxin systems revisited, *Annu. Rev. Plant Physiol. Plant Mol. Biol.*, 51. 371–400.

Seemann, J.R., Berry, J.A., Freas, S.M. and Krump, M.A. (1985). Regulation of ribulose bisphosphate carboxylase activity *in vivo* by a light-modulated inhibitor of catalysis, *Proc. Natl. Acad. Sci. USA*, 82, 8024–8028.

Sharkey, T.D. (2005). Effects of moderate heat stress on photosynthesis: importance of thylakoid reactions, Rubisco deactivation, reactive oxygen species, and thermotolerance provided by isoprene, *Plant Cell Environ.*, 28, 269–277.

Shen, J.B., Orozco, E.M. and Ogren, W.L. (1991). Expression of the two isoforms of spinach ribulose 1,5-bisphosphate carboxylase activase and essentiality of the conserved lysine in the consensus nucleotide-binding domain, *J. Biol. Chem.*, 266, 8963–8968.

Siddiqui, S.M., Sauer, R.T. and Baker, T.A. (2004). Role of the processing pore of the ClpX AAA+ ATPase in the recognition and engagement of specific protein substrates. *Genes Dev.*, 18, 369–374.

Somerville, C.R., Portis, A.R., Jr. and Ogren, W.L. (1982). A mutant of *Arabidopsis thaliana* which lacks activation of RuBP carboxylase *in vivo*, *Plant Physiol.*, 70, 381–387.

Spreitzer, R.J. and Salvucci, M.E. (2002). Rubisco: structure, regulatory interactions, and possibilities for a better enzyme, *Annu. Rev. Plant Biol.*, 53, 449–475.

Spreitzer, R.J. (2003). Role of the small subunit in ribulose-1,5-bisphosphate carboxylase/ oxygenase, *Arch. Biochem. Biophys.*, 414, 141–149.

Spreitzer, R.J., Peddi, S.R. and Satagopan, S. (2005). Phylogenetic engineering at an interface between large and small subunits imparts land-plant kinetic properties to algal Rubisco, *Proc. Natl. Acad. Sci. USA*, 102, 17225–17230.

Stotz, M., Mueller-Cajar, O., Ciniawsky, S., Wendler, P., Hartl, F.U., Bracher, A. and Hayer-Hartl, M. (2011). Structure of green-type Rubisco activase from tobacco, *Nat. Struct. Mol Biol.* 18, 1366–1370.

Streusand, V.J. and Portis, A.R., Jr. (1987). Rubisco activase mediates ATP-dependent activation of ribulose bisphosphate carboxylase, *Plant Physiol.*, 85, 152–154.

Sysoeva, T.A. (2016). Assessing heterogeneity in oligomeric AAA+ machines, *Cell. Mol. Life Sci.* 74, 1001–1018.

Tabita, F.R. (1999). Microbial ribulose 1,5-bisphosphate carboxylase/oxygenase: a different perspective, *Photosynth. Res.*, 60, 1–28.

Tabita, F.R., Hanson, T.E., Li, H.Y., Satagopan, S., Singh, J. and Chan, S. (2007). Function, structure, and evolution of the RubisCO-like proteins and their RubisCO homologs, *Microbiol. Mol. Biol. Rev.*, 71, 576–599.

Tabita, F.R., Hanson, T.E., Satagopan, S., Witte, B.H. and Kreel, N.E. (2008). Phylogenetic and evolutionary relationships of RubisCO and the RubisCO-like proteins and the functional lessons provided by diverse molecular forms, *Philos. Trans. R. Soc. Lond. B Biol. Sci.*, 363, 2629–2640.

Tabita, F.R., Satagopan, S., Hanson, T.E., Kreel, N.E. and Scott, S.S. (2008). Distinct form I, II, III, and IV Rubisco proteins from the three kingdoms of life provide clues about Rubisco evolution and structure/function relationships, *J. Exp. Bot.*, 59, 1515–1524.

Taylor, T.C. and Andersson, I. (1996). Structural transitions during activation and ligand binding in hexadecameric Rubisco inferred from the crystal structure of the activated unliganded spinach enzyme, *Nat. Struct. Biol.*, 3, 95–101.

Tcherkez, G.G., Farquhar, G.D. and Andrews, T.J. (2006). Despite slow catalysis and confused substrate specificity, all ribulose bisphosphate carboxylases may be nearly perfectly optimized, *Proc. Natl. Acad. Sci. USA*, 103, 7246–7251.

To, K.Y., Suen, D.F. and Chen, S.C.G. (1999). Molecular characterization of ribulose-1,5-bisphosphate carboxylase/oxygenase activase in rice leaves, *Planta*, 209, 66–76.

Tsai, Y.C., Lapina, M.C., Bhushan, S. and Mueller-Cajar, O. (2015). Identification and characterization of multiple Rubisco activases in chemoautotrophic bacteria, *Nat. Commun.*, 6, p. 8883.

Uemura, K., Miyachi, S. and Yokota, A. (1997). Ribulose-1,5-bisphosphate carboxylase/ oxygenase from thermophilic red algae with a strong specificity for CO_2 fixation, *Biochem. Biophys. Res. Commun.*, 233, 568–571.

Uemura, K., Tokai, H., Higuchi, T., Murayama, H., Yamamoto, H., Enomoto, Y., Fujiwara, S., Hamada, J. and Yokota, A. (1998). Distribution of fallover in the carboxylase reaction and fallover-inducible sites among ribulose 1,5-bisphosphate carboxylase/oxygenases of photosynthetic organisms, *Plant Cell Physiol.*, 39, 212–219.

Vale, R.D. (2000). AAA proteins. Lords of the ring, *J. Cell. Biol.*, 150, 13–19.

Wang, F., Mei, Z., Qi, Y., Yan, C., Hu, Q., Wang, J. and Shi, Y. (2011). Structure and mechanism of the hexameric MecA–ClpC molecular machine, *Nature*, 471, 331–335.

Wang, D. and Portis, A.R., Jr. (2006). Increased sensitivity of oxidized large isoform of ribulose-1,5-bisphosphate carboxylase/oxygenase (Rubisco) activase to ADP inhibition is due to an interaction between its carboxyl extension and nucleotide-binding pocket, *J. Biol. Chem.*, 281, 25241–25249.

Wang, Z.Y., Ramage, R.T. and Portis, A.R., Jr. (1993). Mg^{2+} and ATP or adenosine 5′-[γ-thio]-triphosphate (ATPγS) enhances intrinsic fluorescence and induces aggregation which increases the activity of spinach Rubisco activase, *Biochim. Biophys. Acta*, 1202, 47–55.

Watson, G.M.F., Yu, J.P. and Tabita, F.R. (1999). Unusual ribulose 1,5-bisphosphate carboxylase/oxygenase of anoxic archaea, *J. Bacteriol.*, 181, 1569–1575.

Wendler, P., Ciniawsky, S., Kock, M. and Kube, S. (2012). Structure and function of the AAA+ nucleotide binding pocket, *Biochim. Biophys. Acta*, 1823, 2–14.

Werbeck, N.D., Schlee, S. and Reinstein, J. (2008). Coupling and dynamics of subunits in the hexameric AAA+ chaperone ClpB, *J. Mol. Biol.*, 378, 178–190.

Werneke, J.M., Chatfield, J.M. and Ogren, W.L. (1989). Alternative mRNA splicing generates the two ribulosebisphosphate carboxylase/oxygenase activase polypeptides in spinach and Arabidopsis, *Plant Cell*, 1, 815–825.

Werneke, J.M., Zielinski, R.E. and Ogren, W.L. (1988). Structure and expression of spinach leaf cDNA encoding ribulosebisphosphate carboxylase/oxygenase activase, *Proc. Natl. Acad. Sci. USA*, 85, 787–791.

Whitney, S.M., Baldett, P., Hudson, G.S. and Andrews, T.J. (2001). Form I Rubiscos from non-green algae are expressed abundantly but not assembled in tobacco chloroplasts, *Plant J.*, 26, 535–547.

Whitney, S.M., Houtz, R.L. and Alonso, H. (2011). Advancing our understanding and capacity to engineer nature's CO_2-sequestering enzyme, Rubisco, *Plant Physiol.*, 155, 27–35.

Wise, R.R., Olson, A.J., Schrader, S.M. and Sharkey, T.D. (2004). Electron transport is the functional limitation of photosynthesis in field-grown Pima cotton plants at high temperature, *Plant Cell Environ.*, 27, 717–724.

Yamada-Inagawa, T., Okuno, T., Karata, K., Yamanaka, K. and Ogura, T. (2003). Conserved pore residues in the AAA protease FtsH are important for proteolysis and its coupling to ATP hydrolysis, *J. Biol. Chem.*, 278, 50182–50187.

Yamori, W., Hikosaka, K. and Way, D.A. (2014). Temperature response of photosynthesis in C3, C4, and CAM plants: temperature acclimation and temperature adaptation, *Photosynth. Res.*, 119, 101–117.

Zhang, N., Kallis, R.P., Ewy, R.G. and Portis, A.R., Jr. (2002). Light modulation of Rubisco in Arabidopsis requires a capacity for redox regulation of the larger Rubisco activase isoform, *Proc. Natl. Acad. Sci. USA*, 99, 3330–3334.

Zhang, N. and Portis, A.R., Jr. (1999). Mechanism of light regulation of Rubisco: a specific role for the larger Rubisco activase isoform involving reductive activation by thioredoxin-f, *Proc. Natl. Acad. Sci. USA*, 96, 9438–9443.

Zhang, N., Schürmann, P. and Portis, A., Jr. (2001). Characterization of the regulatory function of the 46-kDa isoform of Rubisco activase from Arabidopsis, *Photosynth. Res.*, 68, 29–37.

Zhu, G., Bohnert, H.J., Jensen, R.G. and Wildner, G.F. (1998). Formation of the tight-binding inhibitor, 3-ketoarabinitol-1,5-bisphosphate by ribulose-1,5-bisphosphate carboxylase/oxygenase is O_2-dependent, *Photosynth. Res.*, 55, 67–74.

Zhu, G. and Jensen, R.G. (1991). Xylulose 1,5-bisphosphate synthesized by ribulose 1,5-bisphosphate carboxylase/oxygenase during catalysis binds to decarbamylated enzyme, *Plant Physiol.*, 97, 1348–1353.

Zhu, X.G., Portis, A.R., Jr. and Long, S. (2004). Would transformation of C3 crop plants with foreign Rubisco increase productivity? A computational analysis extrapolating from kinetic properties to canopy photosynthesis, *Plant Cell Environ.*, 27, 155–165.

Chapter 9

Adaptive Reorganisation of the Light Harvesting Antenna

Alexander V. Ruban

Queen Mary University of London,
The Mile End Road E1 4NS, London, UK
a.ruban@qmul.ac.uk

The photosynthetic membrane is the most protein-rich biological membrane. It performs a number of energy transformation functions that include photon absorption, energy transfer between the photosynthetic light harvesting antenna complexes, excitation energy trapping by the reaction center complexes, charge separation and electron transport between a number of redox electron carriers, coupled to proton translocation across the membrane, reduction of NADP and synthesis of ATP. This sequence of coherently-coupled events is finely regulated, adjusting to changes in environmental and metabolic factors. Therefore, the study of these adaptive dynamics of the photosynthetic membrane and the factors that govern them fall increasingly in a focus of modern photosynthesis research. The light harvesting antenna of photosystem II (LHCII) of higher plants carries up to 70% of all protein in the photosynthetic membrane. It therefore inevitably governs membrane structure, dynamics and functions. Currently, a great deal of information exists on the organisation of the antennas of higher plants around photosystems and attempts are being made to interpret the significance of these structures for light harvesting. However, very little is said and actually known about the properties and processes involved in the short-term dynamics of light harvesting antenna underlying the adaptations to the environment. Here, I will present a view on the role of the composition and features of the LHCII antenna in enabling the dynamics of the photosynthetic membrane that makes its light harvesting function flexible even on a minute's timescale.

1. Introduction

Knowledge of the stages and functions of the light phase of photosynthesis emerged some decades before the structures responsible for them have been identified, characterised and now many of them solved at atomic resolution. The photosynthetic membrane is a unique example of a compact two-dimensional multienzyme association that transforms energy of photon into the energy and redox-potential of universal biological molecules, ATP and NADPH, respectively. The timescale of the light phase processes of photosynthesis that start from photon absorption and end with the synthesis of ATP and NADPH spans about 15 orders of magnitude, combining ultrafast dynamics with the energy transformations that take seconds and minutes [Ruban, 2012]. Remarkably, all these processes are well-synchronised, if not coherent, with the many dozens of proteins in multisubunit complexes interacting with each other in the sequence of photon-exciton-electron-proton transformation processes leading to their energy stabilisation in the form of chemical bonds. Whilst the significant progress in the functional and structural characterisation of the bioenergetics of the photosynthetic membrane has been made, the knowledge of the molecular mechanisms that enable these functions and structures to adapt to various environmental and metabolic factors remains incomplete.

The fact that the light phase of photosynthesis is a subject to adaptation has been known for some time. Classic experiments performed in the laboratories of Anderson, Barber and Boardman, for example, revealed the remarkable acclimative dynamics of the pigment protein complexes of higher plants [Boardman, 1977; Barber, 1983; Anderson, 1986]. Since acclimation requires alteration in gene expression it is a rather slow process and the term "dynamics" in its relation that was used at the time simply meant the fact that the photosynthetic membrane is not "set in concrete". Indeed, the ratios, the absolute amounts of the photosynthetic complexes were discovered to depend upon environmental light quantity as well as spectral quality. The term "dynamics" began to be used to reflect the long-term alterations in the composition of the photosynthetic machinery. At the same time, at the end of 1970s beginning of 1980s, along with the light regulation of gene expression, photoinhibiton and associated D1 protein turnover [Powles, 1984; Ohad *et al.*, 1984; Barber, 1994], and protein phosphorylation appeared as the most popular topics in the whole of the photosynthesis research [Anderson, 1986]. D1 protein repair, triggered by its phosphorylation, occurs on the hourly timescale, largely because it requires its *de novo* synthesis and photosystem II (PSII) complex assembly [Koivuniemi *et al.*, 1995; Nixon *et al.*, 2010]. The phosphorylation of the major light harvesting complex, LHCII, polypeptides revealed that a light adaptation of the photosynthetic membrane could occur on a much faster timescale than the D1 repair. Indeed, the process was suggested to underlie the molecular mechanism of state transitions [Bonaventura and Myers, 1969; Murata, 1969] that occurred on the timescale of minutes [Bennett, 1977]. As a consequence of phosphorylation a migration of LHCII complexes into stroma lamellae — a residence of photosystem I (PSI) — has been proposed [Kyle *et al.*, 1984]. The photosynthetic membrane began to reveal itself as constantly "alive" with the relatively fast adaptive movements and rearrangements of its components, mainly LHCII antenna complexes. The studies of the mechanism(s) of interactions between segregated in space photosystems and their dynamics have actually been at their most intense at that time [Andersson and Anderson, 1980; Barber, 1982, 1983; Anderson, 1986]. A range of membrane biochemical and biophysical techniques has been used those days to achieve this goal. The recent studies of the rapid adoptive membrane dynamics have been equipped with a number of novel progressive techniques that include various types of mutagenesis, single molecule spectroscopy, advanced biochemical methods, confocal, electron and atomic force microscopies [Mullineaux and Sarcina, 2002; Nevo *et al.*, 2012; Kirchhoff, 2014; Iwai *et al.*, 2014; Ruban and Johnson, 2015; Malý *et al.*, 2016]. The ambition was and remains to attempt to track in real time the adaptive dynamics of the photosynthetic complexes. The functional visualisation techniques are increasingly becoming more diverse with the ultimate goal of developing the "life" tracking super resolution of the protein movements and their structural dynamics in the intact membrane [Ruban and Johnson, 2015].

2. The Structure of LHCII Complex and the Landscapes of the Photosystems

Assemblies of thylakoids formed by the photosynthetic membrane often arrange into stacks, grana that contain mainly PSII. The formation of grana is promoted and stabilised by Mg^{2+} cations. Monovalent K^+ cations are also effective but at somewhat higher concentration [Barber, 1982]. Unstacked thylakoids contain mainly PSI. Grana and unstacked thylakoids contain all of the major components that run the light phase of photosynthesis. The special segregation of photosystems is one of the key features of the photosynthetic membrane of plants that actually requires it to be dynamic in order to regulate the electron transport. Long before the era of extensive biochemical studies of the photosynthetic membrane [Horton, 2014] it was believed that chlorophyll — a lipid like pigment was freely located in the membrane. Later it was shown that all chlorophyll is attached to three major classes of membrane protein — light harvesting antenna, PSI and PSII complexes [Blankenship, 2014]. The most resolved and studied is a structure of the major light harvesting complex, LHCII (Figure 1). The relatively small protein of about 27 kDa binds 14 chlorophyll molecules (8 Chl *a* and 6 Chl *b*) and four xanthophylls — neoxanthin, two luteins

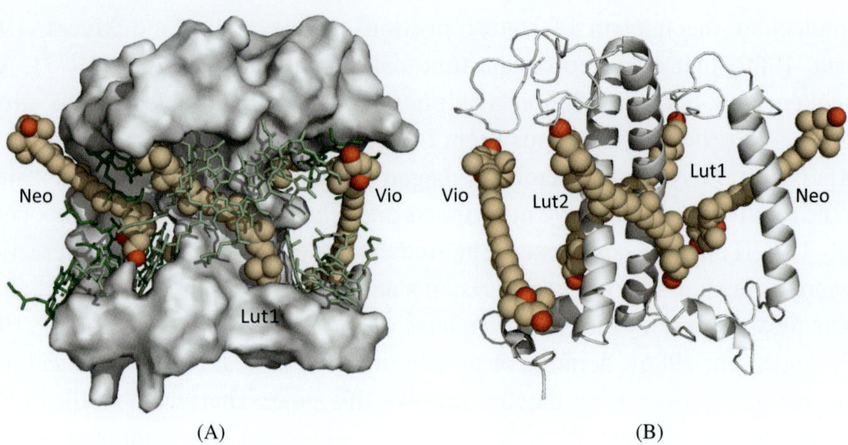

Figure 1. Atomic structure of the LHCII monomer. (A) The view of the chlorophylls and xanthophylls arranged around the hydrophobic core of the protein shown in the form of a van-der-Waals surface. (B) Three transmembrane alpha-helixes of LHCII monomer with four xanthophylls: Vio, violaxanthin; Neo, neoxanthin; Lut, lutein.

and violaxanthin. The latter is converted into zeaxanthin in light. Figure 1 shows how densely the pigments are surrounding the core of the protein (A) that is built of the three transmembrane helixes (B). The LHCII xanthophylls are essential not only for providing extra light energy to the chlorophylls but play a role in photoprotection and shape the LHCII tertiary and quaternary structure [Liu *et al.*, 2004]. Indeed, reconstitution experiments have clearly shown that for the effective assembly of the functional complex all xanthophylls are required, lutein in particular [Plumley and Schmidt, 1987; Paulsen *et al.*, 1993]. Whilst lutein 1 (Lut620) is essential for the monomer assembly, lutein 2 (Lut621) is important in stabilisation of the trimeric structure, that is the native state of the major LHCII complex in the membrane [Dall'Osto *et al.*, 2014]. Later the atomic structure of one of the minor LHCII complexes, CP29, has been solved [Pan *et al.*, 2011]. The structure has a very similar outline — three transmembrane helixes and three xanthophylls (instead of four of the major LHCII): lutein, violaxanthin and neoxanthin. Violaxanthin in CP29 replaced lutein 2 and, unlike violaxanthin that is bound in LHCII only by hydrophobic forces, is very tightly associated with the protein by hydrogen bonds.

Each photosystem was found to possess its own light harvesting antenna in the form of several related pigment-proteins [Jensen *et al.*, 2007; Kouřil *et al.*, 2012]. PSI contains at least four types of LHCI antenna polypeptides, called Lhca1–4. PSII has at least six types — Lhcb1–6. These polypeptides assemble into the major, trimeric complex, LHCIIb (Lhcb1–3) and three minor monomeric complexes LHCIIa (CP29), LHCIc (CP26) and LHCIId (CP24). These days, LHCIIb is called just LHCII and the minor antenna complexes — CP24, CP26, and CP29. The LHC units are organised around core complexes of photosystems (Figure 2). The atomic

Photosystem II

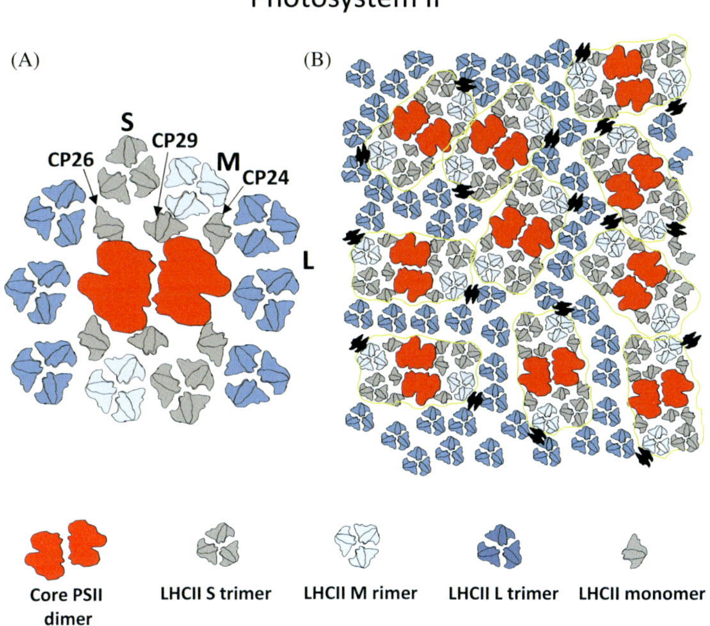

Photosystem I

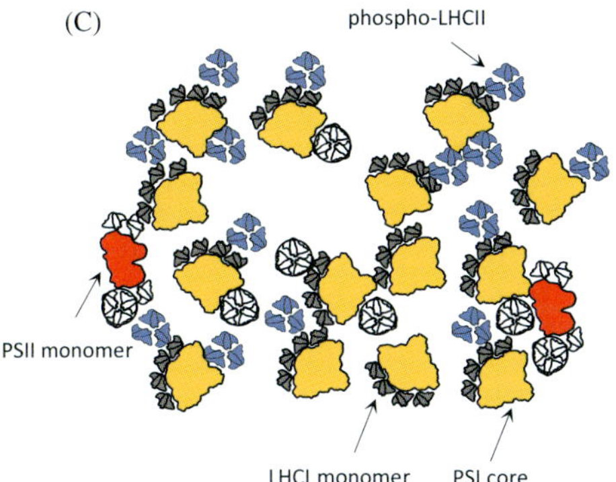

Figure 2. Organisation of the complexes of photosystems in the thylakoid membrane. (A) PSII $C_2S_2M_2L_6$ supercomplex. (B) The view of the grana membrane fragment showing PSII $C_2S_2M_2$ supercomplexes [Kouřil *et al.*, 2011] surrounded by L trimers at the trimer to RCII monomer ratio of 5 [Ruban, 2016]. (C) Redistribution of PSI supercomplexes containing four LHCI monomers each in the stroma thylakoid membrane [Ruban and Johnson, 2015]. Also shown monomeric PSII cores with some minor and trimeric LHCII as well as the phosphorylated LHCII interacting with PSI [Benson *et al.*, 2016].

 A. V. Ruban

structure of the whole PSI complex — a very stable particle — revealed the details of interactions and localisation of the four LHCI monomers (built of Lhca1–4 polypeptides) that are tightly associated with its core complex [Ben-Shem *et al.*, 2003; Qin *et al.*, 2015] (Figure 2C). The structure of individual LHCI complex again shows a very similar to LHCII three-transmembrane helical design. However, LHCI complexes are enriched in chlorophyll *a* and lack neoxanthin (in place of it they bind β-carotene). In addition, violaxanthin replaces lutein in lutein 2 position [Qin *et al.*, 2015]. PSI complex is also often found in association with LHCII trimers that are present in small amounts in the stroma lamellae but can increase upon LHCII phosphorylation prompting state transitions (see paragraphs 1 and 6) [Benson *et al.*, 2015]. The landscape of PSI is somewhat heterogeneous. Apart from the localisation in unstacked lamellae a fraction of PSI is localised in grana margins and plays an important role on the mentioned state transitions interacting with more than one phosphorylated LHCII trimer [Benson *et al.*, 2015; Ruban and Johnson, 2015]. Further, PSII monomers carrying photo-damaged RCII are found in the stroma lamellae, reflecting a stage of the repair process [Nixon *et al.*, 2010].

The knowledge of the arrangements, interactions and redistribution of the LHCII complexes around PSII is relatively new [Kouřil *et al.*, 2011; Ruban and Johnson, 2015]. The most important advance was made by the Barber's and Boekema's groups that discovered the supercomplex structure of the PSII dimer [Boekema *et al.*, 1995]. Figure 2A shows the structure of the PSII supercomplex $C_2S_2M_2$ with additional six loosely bound LHCII trimers (L). C stands for "core", S is a strongly-bound trimer and M is a medium-bound trimer. The $C_2S_2M_2$ particle has been obtained by Boekema's group [Kouřil *et al.*, 2011]. They also managed to see, albeit infrequently, some L-trimers associated with the supercomplex [Boekema *et al.*, 1999]. On average, 4–5 trimers serve a single PSII reaction center [Peter and Thornber, 1991], however, since more than 50% of them are L-trimers it proved so far to be impossible to prepare the PSII with the complete light harvesting antenna. The structural flexibility of PSII peripheral antenna (namely L-trimers of LHCII) binding can be an essential entropic giveaway in order to making the system more adaptable to the environmental factors (see next paragraphs). Hence, the particle shown on the Figure 2A — $C_2S_2M_2$ — surrounded by 6 L trimers is largely a hypothetical structure. The three minor antenna complexes, CP24, 26 and 29 have been found to play an important structural role linking the trimers to the core complex [Boekema *et al.*, 1999] that enables LHCII trimers to efficiently supply energy to RCII [Oort, *et al.*, 2010; Dall'Osto *et al.*, 2014]. Further, CP24 complex was found to be essential in binding LHCII M-trimer [Kovács *et al.*, 2006].

Staehelin's work pioneered freeze-fracture electron microscopy for obtaining visual information on the arrangement of photosystems in the membrane. It clearly visualized the spatial segregation of PSI and PSII complexes [Staehelin, 1976]. The great advantage of this microscopy is that it yields snapshots of the intact frozen membrane in different physiological states. However, it does not allow for high resolution (>5 nm). Therefore, the Boekema group used negative stain transmission electron microscopy on membrane fragments (rather than intact membranes) in early studies of the PSII supercomplex assemblies and arrangements. However, for the PSII analysis in such membrane fragments the supercomplexes had to be ordered in crystalline arrays [Kouřil *et al.*, 2011]. This work enabled, for the first time, the analysis of PSII with different amounts of LHCII trimers, but never more than two per one RCII core monomer. A large part of the light harvesting antenna was missing. Moreover, as it turned out the arrangement of PSII in the ordered arrays is relatively rare [Ruban and Johnson, 2015] (see the next paragraphs).

The next step in the investigation of the *in vivo* PSII organisation was undertaken by Kouřil and coworkers [Kouřil *et al.*, 2011], who applied cryoelectron microscopy to study several stacked grana membranes. The advantage of cryoelectron microscopy is that it does not use any artificial environment or staining procedures to fix the sample. Almost intact grana membranes have been isolated from thylakoids using a gentle treatment with digitonin. This enabled the identification of densities of the dimeric PSII core complexes. They were fairly randomly redistributed in the membrane (Figure 2B). Localisation and orientation of the PSII dimeric cores was used as initial information to model upon them the $C_2S_2M_2$ supercomplex structures (Figure 2B, outlined by yellow lines). These complexes are not only randomly oriented but well-spaced from each other. The spaces between the complexes are likely to be filled mostly with the rest of the LHCII L-trimers. Figure 2B incorporates LHCII trimers into empty spaces between $C_2S_2M_2$ complexes to give 5 to 1 LHCII/RCII ratio. The density of LHCII is high but sufficient for the mobility of L-trimers. The next paragraphs will address how and why the membrane mobility is governed by PSII structure, antenna composition and its functional states like repair of photodamaged RCII complexes, energy redistribution between PSII and PSI and photoprotection of RCII.

3. Evidence for the Fast Dynamics of the Photosynthetic Membrane

Although, the photosynthetic membrane is packed with proteins (Figure 2) and the grana structure provides additional constrain to the protein mobility such diffusion does take place [Kirchhoff, 2014]. This is not only vital for the functioning of the electron transport chain but for its regulation, repair and protection. Direct tracking of the movements of certain proteins, like LHCII, was successfully undertaken using antibody labelling technique [Consoli *et al.*, 2005]. However, the tagging methods are prone to artifacts as far as the grana membrane is concerned, since they prevent the natural mobility as well as partitioning of the labelled protein within the grana stacks [Goral *et al.*, 2010].

A more accurate approach, fluorescence recovery after photobleaching, FRAP, has been used to assess the photosynthetic membrane dynamics using the external fluorescence probes (to assess lipid diffusion) or autofluorescence of chlorophyll-containing complexes. This method is based on confocal microscopy combined with the selective wavelength bleaching lasers [Mullineaux, 2004]. Figure 3A shows relatively slow bleach recovery of the chlorophyll fluorescence in comparison to the BODIPY (lipid probe) diffusion. The protein bleach recovery can be stopped by a cross-linker, suggesting that freezing protein-protein interactions within the crowded membrane stops all diffusive processes within. Similar effect can be achieved by lowering the temperature. Figure 3B shows actual size of the mobile fraction of protein involved. It is relatively small in the control plants but getting even smaller in plants lacking the two types of kinases, stn7 and stn8 that phosphorylate LHCII and D1 polypeptides, respectively. Hence the process of phosphorylation enhances protein mobility as a fundamental prerequisite for LHCII and D1 protein relocalisation in the photosynthetic membrane. The latter is required for removal of damaged PSII into the stroma lamellae [Nixon *et al.*, 2010] and the former is a key dynamic adaptive process that enables balancing light energy input into photosystems — state transitions, as was already mentioned [Kyle *et al.*, 1984]. The observed relatively small mobile protein fraction of granal chlorophyll proteins is indicative of the ability of these proteins to not only be able to diffuse within the stacked membranes but diffuse through the connecting stroma lamellae connecting different granae. This fraction of chlorophyll protein is likely

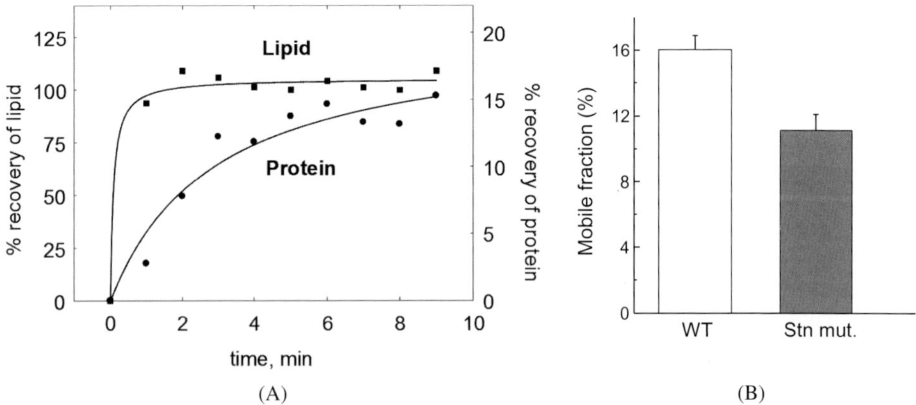

Figure 3. (A) Mobility of lipids and proteins in the photosynthetic membrane probed by Fluorescence Recovery After Photobleaching (FRAP) method. (B) Calculated mobile protein fraction in isolated PSII membranes of wild-type and *stn7asn8* double mutant of Arabidopsis.

to be enriched in the L-trimers of LHCII (see the previous paragraph) that carry on average 30% of the total PSII chlorophyll — a number that is close to the published predicted mobile fraction [Kirchhoff *et al.*, 2008]. Hence, potentially, this fraction can be higher than 16% shown in Figure 3B. The next paragraph is set to explore the variation of the size of the mobile fraction depending on the antenna protein composition.

4. LHCII Composition and Membrane Dynamics

The availability of a range of Lhcb antisense and eventually knock-out mutants of Arabidopsis prompted the study that related the dynamics of the thylakoid membrane to the PSII antenna landscapes and composition [Goral *et al.*, 2012].

The aim was also to see whether the mutations affected the percentage of PSII crystallinity. Figures 4A and 4B present freeze-fracture electron microscopy (FFEM) images of the thylakoid membranes from the wild type Arabidopsis chloroplasts. The PFs surfaces are crowded with LHCII (A) and some rare images show isles of PSII crystallinity (B). Comparison of LHCII clustering with the size of mobile fraction measured by FRAP is shown on the Figure 4C. The relationship obtained was found for the most cases to be reciprocal — the higher is the size of the mobile fraction, the lower is clustering. Hence, the principle that protein crowdedness limits mobility worked well for this data [Kirchhoff *et al.*, 2008]. The only one off case was the knock-out mutant lacking Lhcb3 protein, where very low LHCII clustering was accompanied by the somewhat lower size of the mobile fraction. This could be a consequence of the modified position of the M-trimer within the supercomplex that caries Lhcb3 and, as a result, alteration in the binding of L-trimers to the PSII, restricting their mobility [Goral *et al.*, 2012].

The extent of crystallinity was found to be highest in plants lacking Lhcb6 (CP24 complex apoprotein) that has a significant impact on PSII organisation with the appearance of more frequent C_2S_2 supercomplexes arranged in long rows [Goral *et al.*, 2012]. The likely explanation for this is that since CP24 is important in binding the M-trimer it is crucial in stabilisation of $C_2S_2M_2$ particle. As a result, the C_2S_2 produce more frequent crystallinity that in turn limits protein diffusion.

Another interesting observation was made on PsbS enriched (L17) or lacking (*npq4*) plants (for the more detailed function of PsbS see paragraph 6). Although it does not carry chlorophyll and hence does not contribute to FRAP measurement directly it was found to greatly affect the LHCII mobility and clustering (Figure 4C). Overexpression of the protein significantly enhanced the size of the mobile fraction to above 20%. This correlated with decreased clustering. The absence of this protein, on the other hand, led to a strong decrease in the mobile fraction and increased clustering to almost the same extent as in the Lhcb6 knock-out mutant.

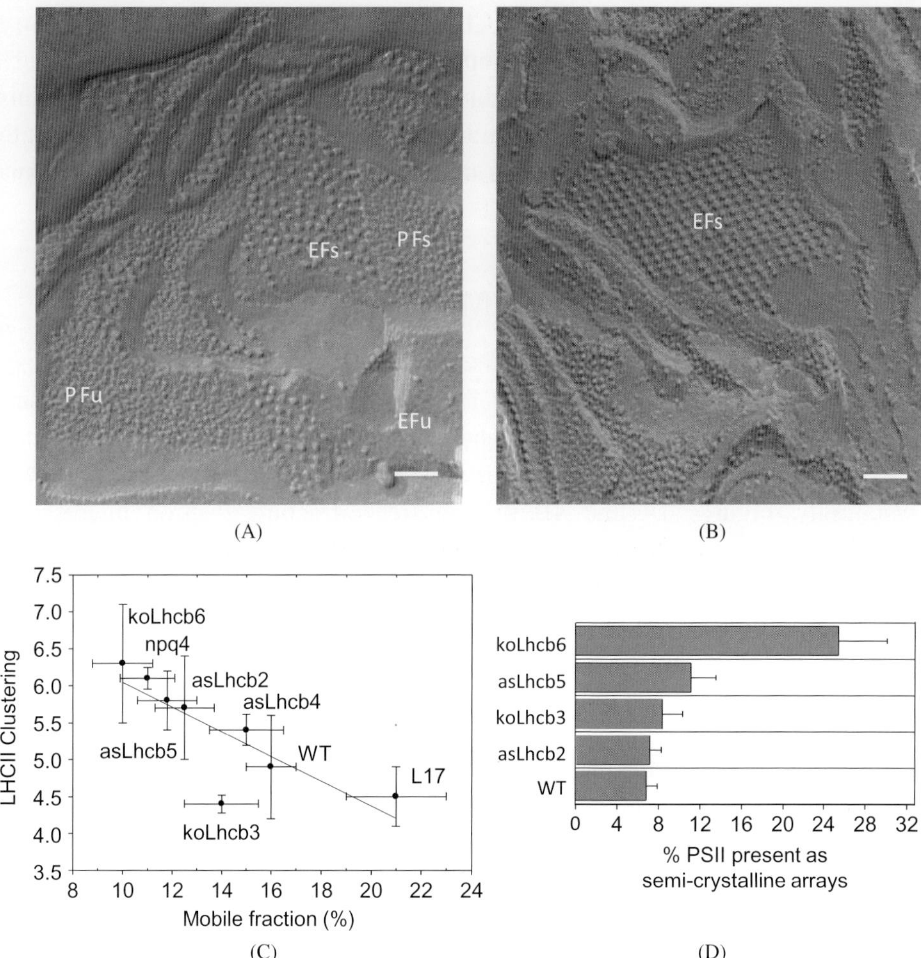

Figure 4. (A) Freeze-fracture micrograph of the thylakoid membrane from intact Arabidopsis chloroplasts showing four fracture surfaces, EF and PF, exoplasmic and periplasmic faces of grana, index "s" and stroma, index "u" membranes. (B) EFs surface of PSII units in semi-crystalline arrays. (C) Relationship between LHCII clustering (PFs surfaces) and the size of the mobile fraction in grana membranes. (D) Extent of PSII crystallinity in grana membranes in mutants lacking various lhcb proteins.

Hence, this non-chlorophyll binding protein was found to strongly affect the photosynthetic membrane dynamics, specifically the mobility of LHCII trimers.

5. State Transitions are Affected by PSII Antenna Composition

The reasons for grana formation, segregation of photosystems and their significance for photosynthetic performance were always a focus of photosynthesis research. Stacked thylakoids contain mainly PSII and its light harvesting antenna (LHCII). Unstacked membranes contain PSI with its own antenna, LHCI, and ATPase. Cytochrome b_6/f complexes tend to be located closer to the areas where thylakoids begin to stack. Cytochrome b_6/f and ATPase were also found in terminal thylakoids of grana and ATPase was identified on the grana margins. To expect the high efficiency of light energy utilisation along the whole sequence of reactions one may look for the existence of key forces or factors designed to integrate, conduct and coordinate light absorption, excitation energy redistribution between photosystems, electron transfer and proton translocation at conditions of spatial segregation between these components in the photosynthetic membrane. The control of the energy balance between the photosystems is effectively taken by state transitions (as mentioned in the paragraph 1). The essence of the molecular mechanism is relocation of phosphorylated LHCII (phospho-LHCII) from grana to stroma lamellae [Kyle *et al.*, 1984]. In fact, the phospho-LHCII is a shuttle that is ready to move to and from grana depending on the extent of plastoquinone reduction [Horton and Black, 1980]. Not only LHCII is capable of the long-range migration it incorporates in specific sites of PSI in the State II [Ruban and Johnson, 2009; Benson *et al.*, 2015]. Hence, phosphorylation/dephosphorylation should enhance membrane dynamics and as shown on the Figure 3B increase the concentration of mobile chlorophyll-carrying complexes in grana.

A couple of reports have been published on the kinetics and amplitude of the state transitions in Lhcb knock-out mutants. Kovács and co-workers reported strongly reduced but very fast state transition kinetics in Arabidopsis plants lacking CP24 minor antenna complex [Kovács *et al.*, 2006]. They interpreted this as the result of the change in the mutant PSII supercomplex structure that was lacking the M-trimer bound to it (see the previous paragraph). This promoted the formation of tight rows of C_2S_2 complexes that could have promoted the rigidity of the grana membrane. This is reflected in the strongly-enhanced PSII crystallinity, LHCII clustering and decrease in the size of the mobile protein fraction (Figure 4). Therefore, the reduced extent of state transitions is consistent with these independent observations. Another mutant devoid of Lhcb3 protein was reported to have an accelerated transition into State II [Damkjær *et al.*, 2009]. It can be due to a decrease in LHCII clustering in the mutant (Figure 4C).

However, it is the non-pigmented protein PsbS (Figure 2A) that has the most striking effect on the kinetics of the state transition (Figure 5). The lack of this protein slowed down the kinetics of State I-State II transition and its overexpression had a very strong acceleration effect. These observations are consistent with the fact that PsbS enhanced the mobile protein fraction in the grana with a strong reduction in the extent of LHCII clustering. Hence, PsbS appears as the factor that controls thylakoid membrane dynamics, exerting a regulatory role on the state of light harvesting antenna mobility.

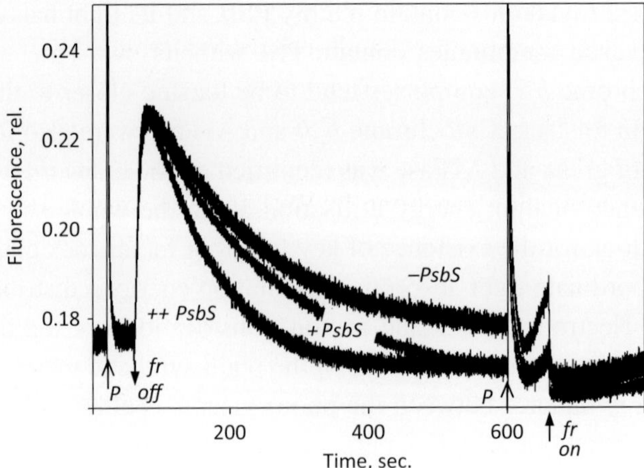

Figure 5. Kinetics of transition from State I to State II in the leaves of Arabidopsis plants possessing no (-PsbS), enhanced (++PsbS) or normal (+PsbS) levels of PsbS protein. P, saturating pulse; fr, far red light. The leaves were first illuminated with both, red (650 nm, PSII exciting) and far red (>700 nm, PSI exciting) lights in order to form State I. then the far red light was turned off immediately causing the imbalance of energy towards PSII and reduction of the plastoquinone pool. Fluorescence began to quench due to the migration of the phospho-LHCII towards PSI that triggered oxidation of plastoquinones, including the one in Q_A site of PSII complex that caused the fluorescence decrease.

6. Zeaxanthin and Non-Photochemical Quenching Modulate LHCII Dynamics

Non-photochemical chlorophyll fluorescence quenching (NPQ) reflects a process in which excess absorbed light energy is dissipated into heat. This process takes place in the photosynthetic membrane [Demmig-Adams *et al.*, 2014; Ruban, 2016]. This protective process is required because high light exposure causes rapid saturation of the photosynthetic reaction centers and their eventual closure, leading to a reduction in the fraction of energy utilized in photosynthesis and the subsequent build-up of harmful excess excitation energy in the photosynthetic membrane [Björkman and Demmig-Adams, 1995]. This energy can damage the most delicate part of the photosynthetic apparatus, the PSII reaction center (RCII). [Powles, 1984; Barber, 1995; Ohad *et al.,* 1984]. The recent evidence suggests that it is indeed protective and the effectiveness of this protection can now be assessed relatively strait forward and on intact plants [Ruban, 2017].

For a number of years several lines of evidence suggesting that the site of NPQ is localised in LHCII antenna and not in the PSII reaction center complex have been produced [Ruban *et al.*, 2012]. The quenching is associated with a significant decrease in antenna fluorescence when all reaction centers are open (Fo) [Horton and Ruban, 1993]. The process persists if samples are frozen to 77 K and is associated with quenched fluorescence bands originating from LHCII [Ruban *et al.*, 1991]. Time-resolved fluorescence data recorded for leaves are consistent with quenching in the antenna [Genty *et al.*, 1992]. Direct measurement of heat emission in NPQ state shows it to occur within 1.4 μsec, much faster than estimates for rates of the recombination reactions in PSII [Mullineaux *et al.*, 1994]. Cross-linkers block NPQ and transition of isolated LHCII complexes into a dissipative state [Ilioaia *et al.*, 2008]. NPQ and LHCII respond in the same way to a number of factors: antimycin A, tertiary amines and magnesium [Noctor *et al.*, 1993; Ruban *et al.*, 1994]. NPQ was discovered to be entirely dependent upon the presence of exclusively LHCII-bound xanthophylls, lutein and zeaxanthin [Niyogi *et al.*, 2001]. Hence, the origin of NPQ in the light harvesting antenna of PSII became widely accepted.

6.1. *LHCII clustering versus mobility*

Whilst a great deal of research has been performed on various isolated LHCII complexes in order to discover the mechanism of NPQ, the nature of the quencher in particular, very little has been done to study the state of light harvesting antenna in the grana membrane [Ruban *et al.*, 2012; Ruban, 2016]. Although, the ΔpH-induced alterations in grana geometry have been reported some time ago [Murakami and Packer, 1970] and Krause proposed subtle conformational changes that accompany NPQ [Krause, 1974], the more precise information on the state of LHCII antenna has been missing. More recently, on isolated grana membranes, derived from light-treated Arabidopsis leaves, a PsbS-dependent change in the distance between PSII core complexes was observed by electron microscopy (EM), implying that a reorganisation of the PSII-LHCII macrostructure may indeed occur during illumination [Betterle *et al.*, 2009]. This was supported by biochemical evidence that showed a fragment of the $C_2S_2M_2$ supercomplex, consisting of the LHCII M-trimer, CP24, and CP29 (B4C subcomplex), was dissociated by light treatment [Betterle *et al.*, 2009]. This implied physical disentanglement of the major trimeric complex from the minor antenna/core PSII complexes. The dissociation of the B4C subcomplex was also dependent on the presence of PsbS [Betterle *et al.*, 2009].

Generation of various Lhcb and PsbS protein mutants enabled the modulation of NPQ in order to relate the mobility to LHCII clustering in a similar way that is described in the previous paragraph in relation to the state transitions [Goral *et al.*, 2012]. Figure 6A shows that in the dark the size of the mobile protein fraction determines the extent of NPQ. The larger is this fraction in the dark, the larger NPQ can be obtained upon illumination. Hence the prerequisite of NPQ is the

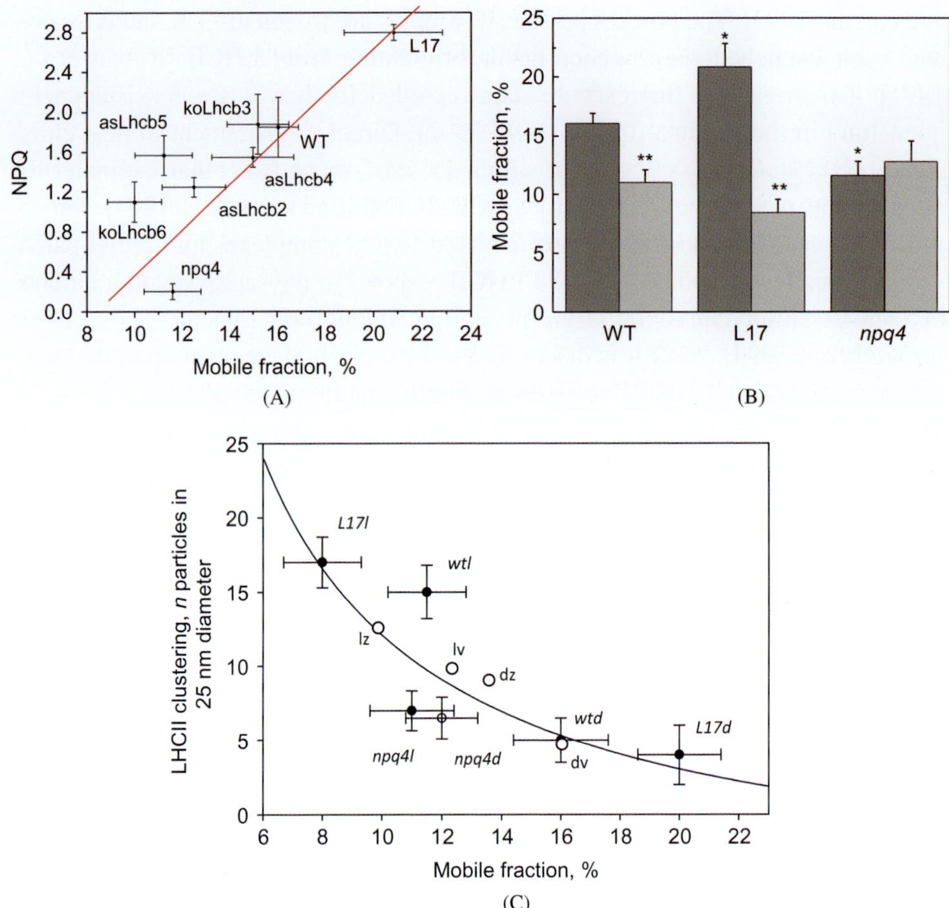

Figure 6.   Direct *in vivo* structural evidence of reorganisation of LHCII complexes and alteration of their mobility in NPQ state obtained by freeze-fracture electron microscopy and FRAP techniques. (A) Positive correlation of NPQ with the size of the mobile protein fraction in the dark in the grana membrane. (B) The size of the mobile fraction depends upon the level of PsbS protein in the dark (dark grey bars) and in NPQ states (light grey bars). L17, overexpressor of PsbS, *npq4*, mutant lacking PsbS. One and two stars indicate statistical difference from the wt in the dark and NPQ states, respectively. (C) The degree of LHCII clustering *versus* the size of the mobile protein fraction in the grana modulated by PsbS, NPQ and xanthophyll cycle (l and d, +NPQ and –NPQ states, respectively; v and z, violaxanthin and zeaxanthin-containing membranes, respectively). Closed symbols are Arabidopsis chloroplasts; open symbols are spinach chloroplasts.

ability of LHCII to move within the thylakoid membrane. PsbS strongly enhances this mobility. The mutations in the various LHCII antenna components, in particular deletions in CP24 and CP26 altered the described correlation so that the decrease in the mobility was not followed by appreciable decrease in NPQ. This could be explained by the altered supercomplex structure and PSII landscapes in these mutants that could have affected the LHCII binding, hence altering the relationship between mobility and clustering [Ruban *et al.*, 2003; Kovács *et al.*, 2006]. Whilst the mobility is enhanced by PsbS in the dark, it is more restricted in the NPQ state in comparison to the wild type (Figure 6B). In plants lacking PsbS protein, on the other hand, the mobility does not change much in the NPQ state. Therefore, in order to form NPQ, LHCII antenna (or part of it) should be relatively mobile in the dark to undergo a transition into a quenched state in light that leads to a decrease in the mobility accompanied by the increase in the LHCII clustering as shown on Figure 6C. Interestingly, whilst PsbS enhanced the mobile fraction, zeaxanthin caused its decrease. However, zeaxanthin also enhanced NPQ and LHCII clustering. Hence, the role of these two essential factors that modulate the quenching process is likely to be very different. Indeed, whilst both of them enhance and accelerate the formation of NPQ, they have an opposite effect upon quenching recovery. Whilst zeaxanthin slows it down, PsbS accelerates it [Crouchman *et al.*, 2006]. This view is also consistent with the recent ultrafast fluorescence kinetics study that concluded the distinct roles of PsbS and zeaxanthin in NPQ [Sylak-Glassman *et al.*, 2014]. Both the mentioned structural and fluorescence studies imply that PsbS is just a mobiliser that controls the concentration of LHCII's engaged in NPQ while zeaxanthin is associated with the actual quenching process, either directly or indirectly [Duffy and Ruban, 2015].

6.2. *NPQ control by LHCII antenna hydrophobicity*

The new developments in the time-resolved fluorescence enabled relatively common application of the measurements of the fluorescence lifetime/yield in the dark as well as NPQ states [van Amerongen and Croce, 2014]. This method gives certain advantages in comparison to pulse amplitude modulation (PAM) fluorometry. Unlike PAM time-resolved fluorescence provides absolute measurements of the fluorescence yield and has a spectral resolution too. This enabled the comparison of the fluorescence lifetimes of LHCII *in vitro* (in detergent micelle) and *in vivo* (in the intact thylakoid membrane) [Johnson and Ruban, 2009]. Figure 7A compares these fluorescence lifetimes as well as the lifetime at the NPQ state. Belgio and co-workers discovered that whilst isolated LHCII complex possess about 4 ns lifetime, the complex embedded in the membrane has a lifetime of only 2 ns [Belgio *et al.*, 2012]. This difference was explained *via* analysis of 77 K fluorescence spectral measurements. It appeared that the shortening of LHCII lifetime is due to its crowded or partially aggregated state in the thylakoid membrane. Indeed, LHCII aggregation induced *in vitro* causes very strong fluorescence quenching that is characterized by the appearance of a second fluorescence band around 700 nm [Ruban and Horton, 1992] (Figure 7A). Further, in the NPQ state *in vivo*

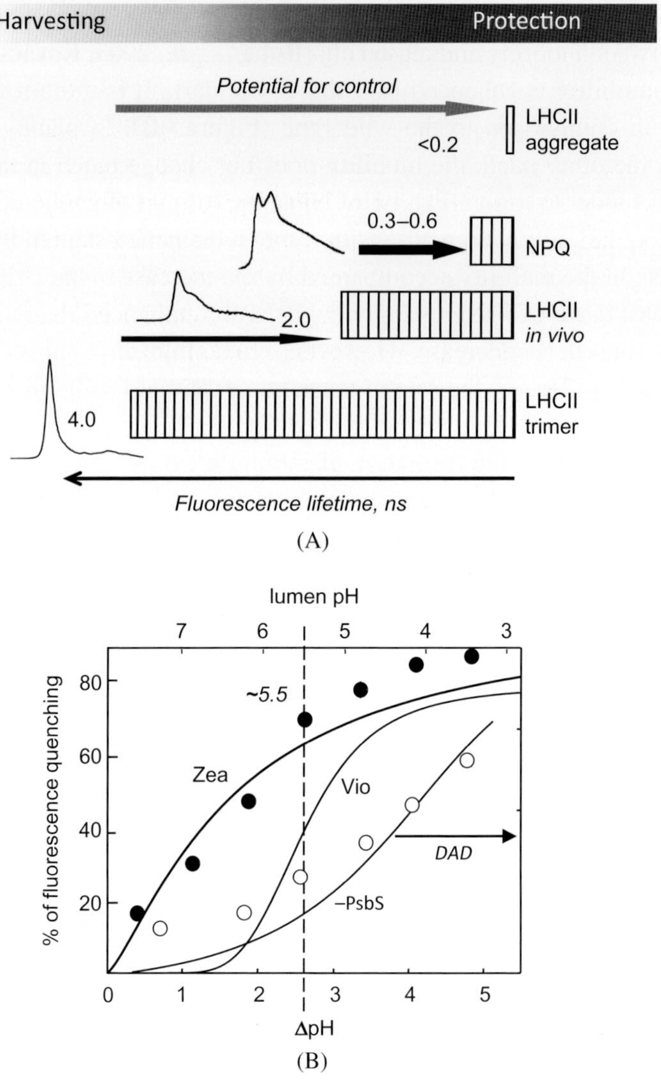

Figure 7. (A) Chlorophyll fluorescence lifetime of LHCII *in vitro* and *in vivo* in efficient harvesting and photoprotective states. Numbers indicate lifetimes in nanoseconds. Three traces on the left are 77 K fluorescence spectra of LHCII in different states. (B) pH titrations of fluorescence quenching in chloroplasts (solid lines) and LHCII solubilised with 20 μM (open symbols) or 6 μM of β-dodecylmaltoside (closed symbols). Zea and Vio are zeaxanthin or violaxanthin enriched chloroplasts. −PsbS are the chloroplasts from NPQ4 mutant.

the fluorescence lifetime is even shorter, within 300–500 ps, and accompanied by the strong increase in the 700 nm band [Johnson and Ruban, 2009; Ware *et al.*, 2015]. Hence, the early proposal of Horton and co-workers that LHCII aggregation may underlie NPQ seems to be reasonably confirmed by the recent *in vivo* experiments on the fluorescence lifetimes, 77 K fluorescence and electron microscopy (Figure 6) that shows increased clustering/aggregation of LHCII in the NPQ state [Johnson *et al.*, 2011].

Zeaxanthin was demonstrated to enhance the LHCII clustering (Figure 6C). What is the mechanism of this enhancement? It has been known for some time that zeaxanthin caused the amplification of NPQ because it increased its sensitivity to protons [Noctor *et al.*, 1991, 1993]. Specifically, it caused an increase in the pK of LHCII protonation from 5.0–5.5 to 6.5–7.0 (Figure 7B). Moreover, the fact that NPQ in plants devoid of zeaxanthin can be enhanced by lowering the acidity of the lumen using the artificial proton shuttle, DAD (Figure 7B), points out towards rather modulatory and not, at least, exclusive direct quenching role of zeaxanthin in NPQ [Johnson and Ruban, 2011]. At the first glance, both PsbS and zeaxanthin enhanced LHCII sensitivity to protons *in vivo* (Figure 7B). PsbS seems to possess somewhat stronger effect on NPQ than zeaxanthin as it does with the enhancement of LHCII clustering (Figure 6C). The *in vitro* quenching titrations of LHCII aggregation demonstrated that the partially aggregated complex becomes more sensitive to protonation, so that pK for quenching becomes higher. Figure 7B shows typical titrations of fluorescence quenching of LHCII in 20 μM of β-dodecylmaltoside (open circles) and 6 μM of the same detergent (closed circles). In the latter case LHCII was more aggregated and noticeably more quenched than in the former. On the other hand, completely separated LHCII trimers did not quench even at pH 4.0 [Petrou *et al.*, 2014]. These observations *in vitro* are consistent with the notion made in Section 6.1 that PsbS could simply control the populations of LHCII engaged in NPQ and that this control is exerted by the regulation of the LHCII exterior environment that promotes aggregation as shown in the Figure 7B.

Why do zeaxanthin and PsbS enhance LHCII sensitivity to protons? The answer could be found from the established fact that hydrophobicity of the environment of the proton-accepting aminoacids such as aspartate and glutamate is the key controller of their pKs [Li *et al.*, 2004; Mehler *et al.*, 2002; Thurlkill *et al.*, 2006]. In hydrophilic environment those residues are engaged in hydrogen bonding that greatly reduces the protonation hence lowering the pKs. Embedding in hydrophobic environment brings up the pKs by several pH units. Further, PsbS is one of the most hydrophobic protein in the thylakoid membrane and zeaxanthin is the most hydrophobic carotenoid of LHCII antenna. However, for the above mechanism to take place, first of all, interactions of zeaxanthin and PsbS with LHCII antenna complexes are required. It was believed for some time that most of the xanthophyll cycle carotenoids are free in the thylakoid membrane and only a small fraction is associated with the monomeric minor antenna complexes CP24, CP26 and CP29 [Bassi *et al.*, 1997]. The fact that the majority of PSII violaxanthin and zeaxanthin are actually bound to the LHCII trimers was established later in experiments that tested the affinity of xanthophyll binding to various LHC complexes [Ruban *et al.*, 1999]. The atomic structure of LHCII has confirmed the existence of the fourth xanthophyll binding site in the trimeric complex [Liu *et al.*, 2004]. Transformation of a very polar violaxanthin into hydrophobic zeaxanthin is likely to promote antenna oligomerisation by a similar mechanism to that recently observed on membrane lipids that transiently of permanently promote protein oligomerisation [Gupta *et al.*, 2017]. As far as interactions of PsbS with LHCII antenna are concerned, it has been proposed that PsbS exists in the form of a dimer in the dark and upon protonation monomerises [Bergantino *et al.*, 2003]. The monomeric protein gains high binding affinity to various, if not all, LHCII complexes as was recently found by magnetic bids pull-down assays [Sacharz *et al.*, 2017]. Further, zeaxanthin was found to enhance PsbS binding to the monomeric minor complexes, specifically CP29, prompting the whole antenna rearrangement into high molecular weight clusters, as was also reported by the electron microscopy study earlier [Johnson *et al.*, 2011].

Therefore, PsbS and zeaxanthin by promoting LHCII interactions could simply enhance its hydrophobicity. Figure 8 provides an additional *in vivo* proof showing the intensity of the ^{13}C-labelled DCCD (N,N′-dicyclohexylcarbodiimide) binding to LHCII polypeptides in the thylakoid membrane with or without

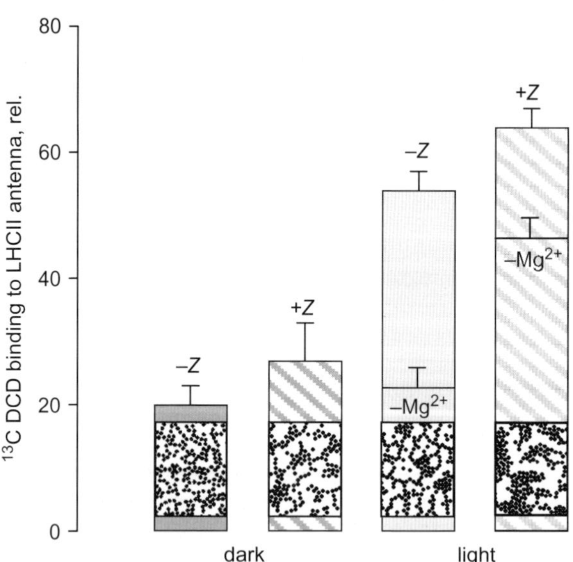

Figure 8. Intensity of ^{13}C DCCD binding to all Lhcb polypeptides in isolated spinach chloroplasts with (+Z) or without (–Z) zeaxanthin in the dark or after illumination (light) to form NPQ. The binding decreases in the absence in Mg^{2+} cations in the NPQ state. Insets in each bar represent the LHCII particles' co-ordinates on the PFs fracture surfaces of the freeze fracture images of the chloroplast membranes (see Duffy *et al.*, 2013).

zeaxanthin and the effect of NPQ ("light"). DCCD is known to bind to glutamate and aspartate residues hidden in hydrophobic domains of membrane proteins [Solios, 1984; Hassinen and Vuokila, 1993]. The binding is enhanced by increased concentration of such domains. Zeaxanthin increases DCCD binding in the dark and NPQ further bring it up, even without zeaxanthin, suggesting that the similar to zeaxanthin effect can be achieved by the LHCII clustering (promoted by PsbS in the NPQ state, see above). Interestingly, magnesium cations were needed to enhance DCCD binding in the NPQ state as well as NPQ itself [Noctor *et al.*, 1993]. Since magnesium is key in grana formation, it is likely that the granal structure itself is important in promoting the NPQ state.

7. Formation of the NPQ Quencher(s): Inner LHCII Complex Dynamics

The mechanistic insights of how LHCII antenna protonation and aggregation is promoted by hydrophobicity described in the previous paragraph show how the NPQ state is formed and what factors control it. However, they do not provide the answers to the questions "what is acting as an energy quencher" and "how is the quencher formed in LHCII aggregates?" The past witnessed a number of theories related to the identity of the NPQ quencher. Specific or non-specific chlorophyll molecules, zeaxanthin and/or lutein have been suggested as the quenchers (for review see [Duffy and Ruban, 2015]). Zeaxanthin in particular has been and is being considered by some groups to be a very likely candidate. One report proposed that PsbS brings the quencher zeaxanthin to LHCII antenna [Wilk *et al.*, 2013]. The other group suggested that this pigment binds in the lutein 2 site of the minor antenna complexes where it acts as a quencher (see Figure 1) [Ahn *et al.*, 2008; Fleming *et al.*, 2012]. However, neither knock-out of the minor antenna complexes nor inhibition of zeaxanthin formation greatly reduced NPQ (for review see [Ruban *et al.*, 2012]). Moreover, given large enough ΔpH NPQ could easily reach the wild type levels in plants lacking PsbS and zeaxanthin [Johnson and Ruban, 2011; Johnson *et al.*, 2012]. Chlorophyll molecules at high concentration could potentially become quenchers in the NPQ state [Beddard and Porter, 1976]. But why do they not quench in the dark while still being highly concentrated? Furthermore, the specific *Chl-Chl* quenching associates as candidates for the NPQ quenchers proposed by several groups [Ruban and Horton, 1992; Holzwarth *et al.*, 2009] were neither experimentally unambiguously identified nor theoretically justified [Duffy and Ruban, 2013]. All that was observed in the aggregated LHCII was the highly fluorescent red-shifted forms [Ruban and Horton, 1992; Ware *et al.*, 2015]. The mixt excitonic-charge transfer nature of these forms was recently proposed to be not originated from the quencher as such [Chmeliov *et al.*, 2016].

Eventually, the formation of the LHCII aggregates was found to be not a primary cause of quenching. High hydrostatic pressure could cause the quenching with all spectral features similar to the aggregation-related quenching in isolated LHCII [Connelly *et al.*, 1998; van Oort *et al.*, 2007]. The same quenching could be achieved in non-aggregated complexes embedded in the gel [Ilioaia *et al.*, 2008]. The crystals of LHCII, where the order of the trimers is such that they barely interact with each other have been found quenched too [Pascal *et al.*, 2005]. Finally, single molecule fluorescence spectroscopy (SMS) of LHCII trimers and monomers revealed their frequent transitions into various quenching conformations, called "blinking" [Krüger *et al.*, 2012, 2013, 2014]. These observations allowed to conclude that not only LHCII populations in the membrane but the inner structure of the monomer/trimer is dynamic too. The "blinking" dynamics was found to be modulated by NPQ factors like zeaxanthin and pH [Krüger *et al.*, 2013]. In addition, various spectroscopic studies revealed the changes in LHCII pigment configuration and orientation as well as the atomic details of LHCII apoprotein dynamics. Orientation of chlorophylls of the terminal emitter locus (*Chl* 610–612), changes in chlorophyll and xanthophyll conformation have been observed using linear and circular dichroism spectroscopy [Ruban *et al.*, 1997]. Twisting of neoxanthin and lutein in the quenched LHCII and NPQ states of isolated chloroplasts, have been observed using Resonance Raman spectroscopy [Ruban *et al.*, 2000, 2007]. Changes in interactions between chlorophylls and lutein in the quenched LHCII as well as signs of protein dynamics have been revealed using NMR spectroscopy [Pandit *et al.*, 2013; Duffy *et al.*, 2014]. Recent molecular dynamics stimulations showed details of several flexible domains within LHCII (lutein, neoxanthin and terminal chlorophyll emitter) that can bring about the formation of the quenching pigment associations [Liguori *et al.*, 2015].

Finally, it was discovered that also the inner LHCII-bound xanthophylls were effective modulators of NPQ *in vivo* and *in vitro* [Johnson *et al.*, 2010, 2012]. Indeed Figure 9 shows the relationship between the pK for the quenching and so-called xanthophyll H-parameter that reflects their hydrophobicity [Ruban, 2012; Dall'Osto *et al.*, 2014]. This parameter was derived from the experiments on solubility of LHCII xanthophylls in water/ethanol mixtures [Ruban *et al.*, 1993]. The plants/LHCII that contained most polar xanthophylls, neoxanthin and violaxanthin possessed the lowest pK for NPQ, about 4.5, indicative that since the estimated lumen pH was about 5.7 [Kramer *et al.*, 1999], they will never possess any significant amount of quenching, unless the ΔpH is enhanced by DAD (see Figure 7B). On the other hand, the plants that contained zeaxanthin only revealed very large pK for NPQ, which is about 6.8. These observations suggest that the LHCII xanthophylls control the inner complex conformational change that leads to NPQ. Most importantly, LHCII aggregation appear to be an event that underlies its

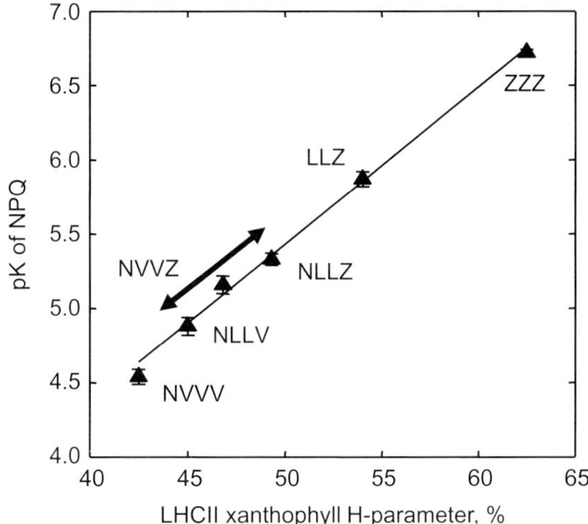

Figure 9. The relationship between pK for NPQ in isolated chloroplasts and LHCII xanthophyll hydrophobicity parameter.

co-operative character [Horton *et al.*, 2000]. Indeed, the NPQ ΔpH titration experiments showed that without zeaxanthin, for example, the Hill coefficient was about 4.5, whilst with zeaxanthin, that promotes LHCII aggregation it became about 1.6 [Ruban *et al.*, 2001]. In other words, for the formation of the quenching state interactions between LHCII units induced by PsbS and zeaxanthin are an essential facilitating event as in co-operative enzymatic reactions [Monod *et al.*, 1965].

Recently, the NPQ quencher has revealed its "economic" character, meaning that it is not fast enough to compete with the open RCII [Belgio *et al.*, 2014]. The quencher becomes effective when the centers are closed. Hence the protection is essential for the closed, not open RCII's. Therefore, the quenching pigment(s) cannot be of the RC-type as was proposed for chlorophyll dimers [Holzwarth *et al.*, 2009]. The trapping rate to the quencher has to be much slower than to the RCII. Time-resolved spectroscopy along with the theoretical modeling revealed that lutein is the agreeable candidate for such an NPQ quencher [Ruban *et al.*, 2007]. Indeed, it was found that the energy transfer rate from the nearest chlorophylls (*Chl* 610–612) to lutein 1 in LHCII is slow, within dozens of picoseconds [Chmeliov *et al.*, 2015; Ruban and Duffy, 2015]. Therefore, this pigment or in case of mutants, any other xanthophyll bound into the lutein 1 site is a good candidate for the NPQ quencher. Indeed, the nature combined two remarkable features of carotenoids: very weak coupling to the S_1 state of chlorophyll and very fast relaxation from their own S_1 state [Ruban and Duffy, 2015] — a perfect feature of an "economic" quencher.

8. Physiological Implications of the Dynamic Properties of LHCII

The work on the major photosynthetic antenna (LHCII) dynamics reviewed here shows that its outer (diffusion, rearrangements, clustering) and inner (conformational changes within an LHCII unit) plasticity are essential properties that enable *tunable, flexible, robust, economic* and *safe* control of the photosynthetic light harvesting process. Indeed, LHCII mobility in the membrane is an essential enabling factor for the state transitions. The diffusion of this membrane protein is somewhat limited by its crowded environment and is also determined by its binding properties. The latter are defined by the charge of the N-terminus that is controlled by phosphorylation — an effective reversible process that prescribes LHCII localisation within the photosystems, hence the energy redistribution between them. The D1 turnover process begins with a similar step — protein phosphorylation that, again, defines its destination into the stroma lamellae. It is not only phosphorylation but also apparently the assembly of the whole of PSII supercomplex that controls its function. Having the three types of interacting LHCII trimers: strongly-, medium- and loosely-bound defines not only the shape of PSII but also the yield and its dynamics *via* NPQ. The process is *tunable* by adjustment in the levels of zeaxanthin in the photosynthetic membrane that finely conditions the efficiency of light harvesting in a certain light environment (bright constant or fluctuating light or shade). Much slower epoxidation of zeaxanthin enables "light memory" by keeping the antenna in the partially dissipative state (Figure 10) and readiness to respond promptly to rapid increase in light intensity [Ruban, 2016]. PsbS levels are also under control [Zia *et al.*, 2011]. The protein enhances NPQ and enables it to track rapidly light fluctuations (Figure 10). Therefore, the combination of zeaxanthin and PsbS makes the light harvesting not only tunable but *flexible* too *via* the fine control over NPQ amplitude as well as kinetics [Ruban, 2012, 2016]. To enable these features evolution led to the formation of a crowded thylakoid membrane with LHCII complexes, some freely diffusing and interacting with each other and the PSII complex. Without zeaxanthin and PsbS, even under high light, the LHCII protonation is strongly inhibited (Figure 10) since the ΔpH levels are normally low (see above). The formation of the modulation factors starts in light. Even at zero zeaxanthin level, PsbS is constitutively present, can be readily protonated, monomerised and begins to interact with LHCII making its environment more hydrophobic that lowers its protonation pK (Figure 10). If the protonation happens LHCII complexes start to change conformation, aggregate and become even more embedded in hydrophobic environment. Formation of zeaxanthin picks up within minutes leading to a further enhancement of hydrophobicity that promotes further aggregation making LHCII

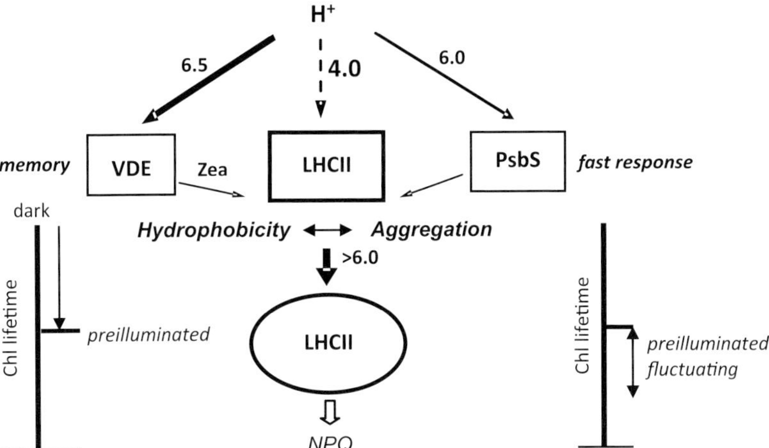

Figure 10. Model explaining the synergistic actions of PsbS and zeaxanthin on the pK of NPQ tuning the protonation of LHCII residues involved in the conformational change that leads to the quencher formation. The pK of LHCII is ~4.0, too low for NPQ activation by physiological lumen pH values (~5.8) [Kramer *et al.*, 1999]. PsbS and de-epoxidase (VDE), however, have a pK of ~6.0 and thus easily are activated by ΔpH. Together they trigger the aggregation of LHCII increasing the hydrophobicity of the environment of the NPQ-active residues and shifting the pK to ~6.0, hence activating NPQ at physiological lumen pH values. Thus, PsbS and zeaxanthin are modulators of LHCII antenna chlorophyll excited state lifetime, hence determine NPQ amplitude and kinetics and enable "light memory" (for more details see Ruban *et al.*, 2012].

antenna even more sensitive to protons so that more complexes change the unquenched conformation into quenched. With zeaxanthin LHCII antenna is partially quenched and conditioned to respond promptly to high light. PsbS just controls the size of the engaged into NPQ population of LHCII complexes.

The *robustness* of NPQ manifests itself in various mutants lacking different xanthophylls, PsbS or Lhcb complexes [Ruban, 2017]. Whilst it is possible to strongly reduce NPQ formation and recovery with mutagenesis, the quenching is hard to abolish completely and its reasonable protective power persists even in the mutants lacking PsbS or zeaxanthin [Ruban, 2017]. Hence, what the various deletion mutants show is alterations in the factors that modulate quenching. The major factor that enables NPQ robustness is LHCII antenna aggregation and the formation of the quencher(s) within. Indeed, downregulation of the levels of all Lhcb proteins was reported to abolish almost all NPQ [Jahns and Krause, 1994].

The *economic* character of NPQ is a remarkable evolutionary achievement, since it makes the quencher to protect only when the protection is needed, when the RCIIs are not able to process more excitations, in other words when they are saturated with the energy of light [Belgio *et al.*, 2014]. The evolution lead to the use of xanthophyll instead of chlorophyll for the NPQ quencher to enable this economic protection. Indeed, the trapping rate is very slow to xanthophyll as opposed to a quencher chlorophyll. Hence, the quencher cannot compete for energy with the working RCII. This feature enables high quantum efficiency of PSII in the low light environment even when NPQ is present but the RCIIs are not saturated with energy. It also ensures *safe* protection by not allowing to waste much energy to the quencher and on the other hand be strong enough under very high light to deal effectively with the damaging excess energy. Indeed, NPQ amplitude in some photosynthetic organisms is always sufficiently high to protect against any intensity of high light that can occur in specific environments on our planet [Ruban, 2015].

Acknowledgements

I dedicate this chapter to the memory of my friendly guide, Professor Jan Anderson. I would like to acknowledge the Royal Society for the Wolfson Research Merit Award.

References

Anderson, J. (1986). Photoregulation of the composition, function, and structure of thylakoid membranes, *Ann. Rev. Plant Physiol.*, 37, 93–136.

Anderson, B. and Anderson, J.M. (1980). Lateral heterogeneity in the distribution of chlorophyll–protein complexes of the thylakoid membranes of spinach chloroplasts, *Biochim. Biophys. Acta*, 593, 427–440.

Barber, J. (1982). Influence of surface charges on thylakoid structure and function, *Annu. Rev. Plant Physiol.*, 33, 261–295.

Barber, J. (1983). Photosynthetic electron transport in relation to thylakoid membrane composition and organization, *Plant Cell Environ.*, 6, 311–322.

Barber, J. (1995). Molecular-basis of the vulnerability of photosystem-II to damage by light, *Aust. J. Plant Physiol.*, 22, 201–208.

Bassi, R., Sandonà, D. and Croce, R. (1997). Novel aspects of chlorophyll *a/b*-binding proteins, *Physiol. Plant.*, 100, 769–779.

Beddard, G.S. and Porter, G. (1976). Concentration quenching in chlorophyll, *Nature*, 260, 366–367.

Belgio, E., Johnson, M.P., Jurić, S. and Ruban, A.V. (2012). Higher plant photosystem II light harvesting antenna, not the reaction center, determines the excited state lifetime—both the maximum and the non-photochemically quenched, *Biophys. J.*, 102, 2761–2771.

Belgio, E., Kapitonova, E., Chmeliov, E., Duffy, C.D.P., Ungerer, P., Valkunas, L. and Ruban, A.V. (2014). Economic photoprotection in photosystem II that retains a complete light harvesting system with slow energy traps, *Nat. Commun.*, 5, p. 4433.

Bennett, J. (1977). Phosphorylation of chloroplast membrane proteins, *Nature*, 269, 344–346.

Bennett, J. (1983). Regulation of photosynthesis by reversible phosphorylation of the light-harvesting chlorophyll a/b protein, *Biochem. J.*, 212, 1–13.

Ben-Shem, A., Frolow, F. and Nelson, N. (2003). Crystal structure of plant photosystem I, *Nature*, 426, 630–635.

Benson, S.L., Maheswaran, P., Ware, M.A., Hunter, C.N., Horton, P., Jansson, S., Ruban, A.V. and Johnson, M.P. (2015). An intact light harvesting complex I antenna system is required for complete state transitions in Arabidopsis, *Nat. Plants*, 1, p. 15176.

Bergantino, E., Segalia, A., Brunetta, A., Teardo, E., Rogoni, F., Giacometti, G.M. and Szabo, I. (2003). Light- and pH-dependent structural changes in the PsbS protein of photosystem II, *Proc. Natl. Acad. Sci. USA*, 100, 15265–15270.

Betterle, N., Ballottari, M., Zorzan, S., de Bianchi, S., Cazzaniga, S., Dall'osto, L., Morosinotto, T. and Bassi, R. (2009). Light-induced dissociation of an antenna hetero-oligomer is needed for nonphotochemical quenching induction, *J. Biol. Chem.* 284, 15255–15266.

Blankenship, R. (2002) *Molecular Mechanisms of Photosynthesis*, 2nd edn. (Blackwell Science, UK).

Boardman, N.K. (1977). Comparative photosynthesis of sun and shade plants, *Ann. Rev. Plant Physiol.*, 28, 355–377.

Boekema, E.J., van Roon, H., Calkoen, F., Bassi, R. and Dekker, J.P. (1999). Multiple types of association of photosystem II and its light-harvesting antenna in partially solubilized photosystem II membranes, *Biochemistry*, 38, 2233–2239.

Bonaventura, C. and Myers, J. (1969). Fluorescence and oxygen evolution from *Chlorella pyrenoidosa, Biochim. Biophys. Acta*, 189, 366–383.

Chmeliov, J., Bricker, W.P., Lo, C., Jouin, E., Valkunas, L., Ruban, A.V. and Duffy, C.D. (2015). An "all pigment" model of excitation quenching in LHCII, *Phys. Chem. Chem. Phys.*, 17, 15857–15867.

Chmeliov, J., Gelzinis, A., Songaila, E., Augulis, R., Duffy, C.D.P., Ruban, A.V. and Valkunas, L. (2016). The nature of self-regulation in photosynthetic light-harvesting antenna, *Nat. Plants*, 1, p. 16045.

Connelly, J.P., Ruban, A.V. and Horton, P. (1998). Pressure induced fluorescence quenching in plant light harvesting complex II. In *High Pressure Food Science, Bioscience and Chemistry*, ed. Isaacs, N.S., ed. (Cambridge: The Royal Society of Chemistry), pp. 417–422.

Consoli, E., Croce, R., Dunlap, D.D. and Finzi, L. (2005). Diffusion of light harvesting complex II in the thylakoid membranes, *EMBO Rep.*, 6, 782–786.

Crouchman, S., Ruban, A.V., and Horton, P. (2006). PsbS enhances nonphotochemical fluorescence quenching in the absence of zeaxanthin, *FEBS Lett.*, 580, 2053–2058.

Dall'Osto, L., Bassi, R. and Ruban, A. (2014). Plastid biology, advances in plant biology 5. In *Photoprotective Mechanisms: Carotenoids*, Theg, S.M. and Wollman, F.-A., eds. Chap. 15, (New York: Springer Science+Business Media), pp. 393–435.

Dall'Osto, L., Ünlü, C., Cazzaniga, S. and van Amerongen, H. (2014). Disturbed excitation energy transfer in *Arabidopsis thaliana* mutants lacking minor antenna complexes of photosystem II, *Biochim. Biophys. Acta*, 1837, 1981–1988.

Duffy, C.D.P., Pandit, A.P. and Ruban, A.V. (2014). Modelling the NMR signatures associated with the functional conformational switch in the major light-harvesting antenna of photosystem II in higher plants, *Phys. Chem. Chem. Phys.*, 16, 5571–5580.

Duffy, C.D.P. and Ruban, A.V. (2015). Dissipative pathways in the photosystem-II antenna in plants, *J. Photochem. Photobiol. B*, 152, 215–226.

Duffy, C.D.P., Valkunas, L. and Ruban, A.V. (2013). Light-harvesting processes in the dynamic photosynthetic antenna, *Phys. Chem. Chem. Phys.*, 15, 18752–18770.

Fleming, G.R., Schlau-Cohen, G.S., Amarnath, K. and Zaks, J. (2012). Design principles of photosynthetic light-harvesting, *Faraday Discuss.* 155, 27–41.

Genty, B., Goulas, Y., Dimon, B., Peltier, G., Briantais, J.M. and Moya, I. (1992). Research in Photosynthesis 4. In *Modulation of Efficiency of Primary Conversion in Leaves*, Murata, N., ed. (Dordrecht: Kluwer Academic Publishing), pp. 603–610.

Goral, T.K., Johnson, M.P., Brain, A.P., Kirchhoff, H. and Mullineaux, C.W. (2010). Visualizing the mobility and distribution of chlorophyll proteins in higher plant thylakoid membranes: effects of photoinhibition and protein phosphorylation, *Plant J.*, 62, 948–959.

Goral, T.K., Johnson, M.P., Duffy, C.D., Brain, A.P., Ruban, A.V. and Mullineaux, C.W. (2012). Light-harvesting antenna composition controls the macrostructure and dynamics of thylakoid membranes in Arabidopsis, *Plant J.*, 69, 289–301.

Gupta, K., Donlan, J.A.C., Hopper, J.T.S., Uzdavynis, P., Landreh, M., Struwe, W.B., Drew, D., Baldwin, A.J., Stansfeld, P.J. and Robinson, C.V. (2017). The role of lipids in stabilizing membrane protein oligomers, *Nature*, 541, 421–424.

Hassinen, I.E. and Vuokila, P.T. (1993). Reaction of dicyclohexylcarbodiimide with mitochondrial proteins, *Biochim. Biophys. Acta*, 1144, 107–124.

Holzwarth, A.R., Miloslavina, Y., Nilkens, M., and Jahns, P., (2009). Identification of two quenching sites active in the regulation of photosynthetic light-harvesting studied by time-resolved fluorescence, *Chem. Phys. Lett.*, 483, 262–267.

Horton, P. (2014). Developments in research on non-photochemical fluorescence quenching: emergence of key ideas, theories and experimental approaches. In *Non-Photochemical Quenching and Energy Dissipation in Plants, Algae and Cyanobacteria*Developments in Research on Non-Photochemical Fluorescence Quenching: Emergence of Key Ideas, Theories and Experimental Approaches, Demmig-Adams, B., Garab, G., Adams, W., III and Govindjee, eds. Chap. 3 (Dordrecht: Springer), pp. 73–95.

Horton, P. and Black, M. (1980). Activation of adenosine 5′-triphosphate induced quenching of chlorophyll fluorescence by reduced plastoquinone. The basis of state I–state II transitions in chloroplasts, *FEBS Lett.*, 119, 141–144.

Horton, P., and Ruban, A.V. (1993). Delta-pH-dependent quenching of the Fo-level of chlorophyll fluorescence in spinach leaves, *Biochim. Biophys. Acta*, 1142, 203–206.

Horton, P., Ruban, A.V. and Wentworth, M. (2000). Allosteric regulation of the light harvesting system of photosystem II, *Philos. Trans. R. Soc. Lond. B Biol. Sci.*, 355, 1361–1370.

Ilioaia, C., Johnson, M., Horton, P. and Ruban, A.V. (2008). Induction of efficient energy dissipation in the isolated light harvesting complex of photosystem II in the absence of protein aggregation, *J. Biol. Chem.*, 283, 29505–29512.

Iwai, M., Yokono, M. and Nakano, A. (2014). Visualizing structural dynamics of thylakoid membranes, *Sci. Rep.*, 4, p. 3768.

Jahns, P. and Krause, G. (1994). Xanthophyll cycle and energy-dependent fluorescence quenching in leaves from tea plants grown under intermittent light, *Planta*, 192, 176–182.

Jensen, P.E., Bassi, R., Boekema, E.J., Dekker, J.P., Jansson, S., Leister, D., Robinson, C. and Scheller, H.V. (2007). Structure, function and regulation of plant photosystem I, *Biochim. Biophys. Acta*, 1767, 335–352.

Johnson, M.P., Goral, T.K., Duffy, C.D.P., Brain, A.P.R., Mullineaux, C.W. and Ruban, A.V. (2011). Photoprotective energy dissipation involves the reorganization of photosystem II light harvesting complexes in the grana membranes of spinach chloroplasts, *Plant Cell*, 23, 1468–1479.

Johnson, M.P. and Ruban, A.V. (2009). Photoprotective energy dissipation in higher plants involves alteration of the excited state energy of the emitting chlorophyll in LHCII, *J. Biol. Chem.*, 284, 23592–23601.

Johnson, M.P. and Ruban, A.V. (2011). Restoration of rapidly reversible photoprotective energy dissipation in the absence of PsbS protein by enhanced pH, *J. Biol. Chem.*, 286, 19973–19981.

Johnson, M.P., Zia, A., Horton, P. and Ruban, A.V. (2010). Effect of xanthophyll composition on the chlorophyll excited state lifetime in plant leaves and isolated LHCII, *Chem. Phys.*, 373, 23–32.

Johnson, M.P., Zia, M.P. and Ruban, A.V. (2012). Elevated ΔpH restores rapidly reversible photoprotective energy dissipation in Arabidopsis chloroplasts deficient in lutein and xanthophyll cycle activity, *Planta*, 235, 193–204.

Kirchhoff, H. (2014). Diffusion of molecules and macromolecules in thylakoid membranes, *Biochim. Biophys. Acta*, 1837, 495–502.

Kirchhoff, H., Haferkamp, S., Allen, J.F., Epstein, D.B. and Mullineaux, C.W. (2008). Protein diffusion and macromolecular crowding in thylakoid membranes, *Plant Physiol.*, 146, 1571–1578.

Koivuniemi, A., Aro, E.-M. and Andersson, B. (1995). Degradation of the D1 and D2 proteins of photosystem II in higher plants is regulated by reversible phosphorylation, *Biochemistry*, 1995, 34, 16022–16029.

Kouřil, R., Dekker, J.P. and Boekema, E.J. (2011). Supramolecular organization of photosystem II in green plants, *Biochim. Biophys. Acta*, 1817, 2–12.

Kouřil, R., Ostergetel, J.P. and Boekema, E.J. (2012). Fine structure of granal thylakoid membrane organization using cryo-electron tomography, *Biochim. Biophys. Acta*, 1807, 368–374.

Kovács, L., Damkjær, J., Kereïche, S., Ilioaia, C., Ruban, A.V., Boekema, E.J., Jansson, S. and Horton, P. (2006) Lack of the light-harvesting complex CP24 affects the structure and function of the grana membranes of higher plant chloroplasts, *Plant Cell*, 18, 3106–3120.

Kramer, D.M., Sacksteder, C.A. and Cruz, J.A. (1999). How acidic is the lumen? *Photosynth. Res.*, 60, 151–163.

Krause, G.H. (1974). Changes in chlorophyll fluorescence in relation to light-dependent cation transfer across thylakoid membranes, *Biochim. Biophys. Acta*, 333, 301–313.

Krüger, T.P.J., Ilioaia, C., Johnson, M.P., Belgio, E., Ruban, A.V. and van Grondelle, R. (2013). The specificity of controlled protein disorder in the photoprotection of plants, *Biophys. J.*, 105, 1018–1026.

Krüger, T.P.J., Ilioaia, C., Johnson, M.P., Ruban, A.V., Papagiannakis, E., Horton, P. and van Grondelle, R. (2012). Controlled disorder in plant light harvesting complex II explains its photoprotective role, *Biophys. J.*, 102, 2669–2676.

Krüger, T.P.J., Ilioaia, C., Johnson, M.P., Ruban, A.V. and van Grondelle, R. (2014). Disentangling the low-energy states of the major light-harvesting complex of plants and their role in photoprotection, *Biochim. Biophys. Acta*, 1837, 1027–1038.

Kyle, D.J., Kuang, T.-Y, Watson, J.L. and Arntzen, C.J. (1984). Movement of a sub-population of the light harvesting complex (LHCII) from grana to stroma lamellae as a consequence of its phosphorylation, *Biochim. Biophys. Acta*, 765, 89–96.

Li, H., Robertson, A.D. and Jensen, J.H. (2004). The determinants of carboxyl pK(a) values in Turkey ovomucoid third domain, *Proteins*, 55, 689–704.

Liguori, N., Periole, X., Marrink, S.J. and Croce, R. (2015). From light-harvesting to photoprotection: structural basis of the dynamic switch of the major antenna complex of plants (LHCII), *Sci. Rep.*, 5, p. 15661.

Liu, Z., Yan, H., Wang, K., Kuang, T.Y., Zhang, J.P., Gui, L.L., An, X.M. and Chang, W.R. (2004). Crystal structure of spinach major light-harvesting complex at 2.72 Å resolution, *Nature*, 428, 287–292.

Malý, P., Gruber, J.M., van Grondelle, R. and Mančal, T. (2016). Single molecule spectroscopy of monomeric LHCII: experiment and theory, *Sci. Rep.*, 6, p. 26230.

Mehler, E.L., Fuxreiter, M., Simon, I. and Garcia-Moreno, E.B. (2002). The role of hydrophobic microenvironments in modulating pKa shifts in proteins, *Proteins*, 48, 283–292.

Monod, J., Wyman, J. and Changeux, J.-P. (1965). On the nature of allosteric transitions: a plausible model, *J. Mol. Biol.*, 12, 88–118.

Mullineaux, C.W. (2004). FRAP analysis of photosynthetic membranes, *J. Exp. Bot.*, 55, 1207–1211.

Mullineaux, C.W., Ruban, A.V., and Horton, P. (1994). Prompt heat release associated with delta-pH-dependent quenching in spinach thylakoid membranes, *Biochim. Biophys. Acta*, 1185, 119–123.

Mullineaux, C.W. and Sarcina, M. (2002). Probing the dynamics of photosynthetic membranes with fluorescence recovery after photobleaching, *Trends Plant Sci.*, 7, 237–240.

Murata, N. (1969). Control of excitation transfer in photosynthesis. I. Light-induced change of chlorophyll a fluorescence in *Porphyridium cruentum*, *Biochim. Biophys. Acta*, 172, 242–251.

Nevo, R., Charuvi, D., Tsabari, O. and Reich, Z. (2012). Composition, architecture and dynamics of the photosynthetic apparatus in higher plants, *Plant J.*, 70, 157–176.

Nixon, P.J., Michoux, F., Yu, J., Boehm, M. and Komenda, J. (2010). Recent advances in understanding the assembly and repair of photosystem II, *Ann. Bot.*, 106, 1–16.

Niyogi, K.K., Shih, C., Chow, W.S., Pogson, B.J., DellaPenna, D. and Björkman, O. (2001). Photoprotection in a zeaxanthin and lutein deficient double mutant of Arabidopsis, *Photosynth. Res.*, 67, 139–145.

Noctor, G., Rees, D., Young, A. and Horton, P. (1991). The relationship between zeaxanthin, energy-dependent quenching of chlorophyll fluorescence and the transthylakoid pH-gradient in isolated chloroplasts, *Biochim. Biophys. Acta*, 1057, 320–330.

Noctor, G., Ruban, A.V. and Horton, P. (1993). Interactions between the effects of potentiators and antagonists of DpH-dependent thermal dissipation of excitation energy in spinach thylakoids, *Biochim. Biophys. Acta*, 1183, 339–344.

Ohad, I., Kyle, D.J. and Arntzen, C.J. (1984). Membrane-protein damage and repair: removal and replacement of inactivated 32-kilodalton polypeptides in chloroplast membranes, *J. Cell Biol.*, 99, 481–485.

Oort, B., Alberts, M., de Bianchi, S., Dall'Osto, L., Bassi, R., Trinkunas, G., Croce, R. and van Amerongen, H. (2010). Effect of antenna-depletion in Photosystem II on excitation energy transfer in *Arabidopsis thaliana*, *Biophys. J.*, 98, 922–931.

Pan, X., Li, M., Wan, T., Wang, L., Jia, C., Hou, Z., Zhao, X., Zhang, J. and Chang, W. (2011). Structural insights into energy regulation of light-harvesting complex CP29 from spinach, *Nat. Struct. Mol. Biol.*, 18, 309–315.

Pandit, A., Reus, M., Morosinotto, T., Bassi, R., Holzwarth, A.R. and de Groot, H.J.M. (2013). An NMR comparison of the light-harvesting complex II (LHCII) in active and photoprotective states reveals subtle changes in the chlorophyll *a* ground-state electronic structures, *Biochim. Biophys. Acta*, 1827, 738–744.

Pascal, A.A., Liu, Z., Broess, K., van Oort, B., van Amerongen, H., Wang, C. Horton, P., Robert, B., Chang, W. and Ruban, A. (2005). Molecular basis of photoprotection and control of photosynthetic light-harvesting, *Nature*, 436, 134–137.

Paulsen, H., Finkenzeller, B. and Kühlein, N. (1993). Pigments induce folding of light-harvesting chlorophyll a/b-binding protein, *Eur. J. Biochem.*, 215, 809–816.

Peter, G.F. and Thornber, J.P. (1991). Biochemical composition and organization of higher plant photosystem II light-harvesting pigment-proteins, *J. Biol. Chem.*, 266, 16745–16754.

Petrou, K., Belgio, E. and Ruban, A.V. (2014). pH sensitivity of chlorophyll fluorescence quenching is determined by the detergent/protein ratio and the state of LHCII aggregation, *Biochim. Biophys. Acta*, 1837, 1533–1539.

Plumley, F.G. and Schmidt, G.W. (1987). Reconstitution of chlorophyll a/b light-harvesting complexes: xanthophyll-dependent assembly and energy transfer, *Proc. Natl Acad. Sci. USA*, 84, 146–150.

Powles, S.B. (1984). Photoinhibition of photosynthesis induced by visible light, *Annu. Rev. Plant Physiol.*, 35, 15–44.

Qin, X., Suga, M., Kuang, T. and Shen, J.-R. (2015). Structural basis for energy transfer pathways in the plant PSI-LHCI supercomplex, *Science*, 348, 989–995.

Ruban, A.V. (2012). *The Photosynthetic Membrane: Molecular Mechanisms and Biophysics of Light Harvesting*. (Wiley-Blackwell, UK).

Ruban, A.V. (2015). Evolution under the sun: optimising light harvesting in photosynthesis. *J. Exp. Bot.*, 66, 7–23.

Ruban, A.V. (2016). Non-photochemical chlorophyll fluorescence quenching: mechanism and effectiveness in protection against photodamage, *Plant Physiol.*, 170, 1903–1916.

Ruban, A.V. (2017). Quantifying the efficiency of photoprotection, *Philos. Trans. R. Soc. Lond. B Biol. Sci.*, in press.

Ruban, A.V., Berera, R., Ilioaia, C., van Stokkum, I.H.M., Kennis, J.T.M., Pascal, A.A., van Amerongen, H., Robert, B., Horton, P. and van Grondelle, R. (2007). Identification of a mechanism of photoprotective energy dissipation in higher plants, *Nature*, 450, 575–578.

Ruban, A.V., Calkoen, F., Kwa, S.L.S., van Grondelle, R., Horton, P. and Dekker, J.P. (1997). Characterisation of the aggregated state of the light harvesting complex of photosystem II by linear and circular dichroism spectroscopy, *Biochim. Biophys. Acta*, 1321, 61–70.

Ruban, A.V. and Horton, P. (1992). Mechanism of DpH-dependent dissipation of absorbed excitation energy by photosynthetic membranes. I. Spectroscopic analysis of isolated light harvesting complexes, *Biochim. Biophys. Acta*, 1102, 30–38.

Ruban, A.V., Horton, P. and Young, A.J. (1993). Aggregation of higher plant xanthophylls: differences in absorption spectra and in the dependency on solvent polarity, *J. Photochem. Photobiol. B: Biol.*, 21, 229–234.

Ruban, A.V. and Johnson, M.P. (2009). Dynamics of higher plant photosystem cross-section associated with state transitions, *Photosynth. Res.*, 99, 173–183.

Ruban, A.V. and Johnson, M.P. (2015). Towards visualisation of the dynamic structure of the plant photosynthetic membrane, *Nat. Plants*, 1, p. 15161.

Ruban, A.V., Johnson, M.P. and Duffy, C.D.P. (2012). Photoprotective molecular switch in photosystem II, *Biochim. Biophys. Acta*, 1817, 167–181.

Ruban, A.V., Lee, P.J., Wentworth, M., Young, A.J. and Horton, P. (1999). Determination of the stoichiometry and strength of binding of different xanthophylls to the photosystem II light harvesting complexes, *J. Biol. Chem.*, 274, 10458–10465.

Ruban, A.V., Pascal, A. and Robert, B. (2000). Xanthophylls of the major photosynthetic light-harvesting complex of plants: identification, conformation and dynamics, *FEBS Lett.*, 477, 181–185.

Ruban, A.V., Rees, D., Noctor, G.D., Young, A. and Horton, P. (1991). Long-wavelength chlorophyll species are associated with amplification of high-energy-state excitation quenching in higher-plants, *Biochim. Biophys. Acta*, 1059, 355–360.

Ruban, A.V., Wentworth, M. and Horton, P. (2001). Kinetic analysis of non-photochemical quenching of chlorophyll fluorescence. I. Isolated chloroplasts, *Biochemistry*, 40, 9896–9901.

Ruban, A.V., Wentworth, M., Yakushevska, A.E., Keegstra, W., Lee, P.J., Dekker, J.P., Jansson, S., Boekema, E. and Horton, P. (2003). Plants lacking the main light harvesting complex retain photosystem II macro-organisation, *Nature*, 421, 648–652.

Sacharz, J., Giovagnetti, V., Ungerer, P., Mastroianni, G. and Ruban, A.V. (2017). The xanthophyll cycle affects reversible interactions between PsbS and light-harvesting complex II to control non-photochemical quenching, *Nat. Plants*, 3, p. 16225.

Solioz, M. (1984). Dicyclohexylcarbodiimide as a probe for proton translocating enzymes, *Trends Biochem. Sci.*, 9, 309–312.

Staehelin, L.A. (1976). Reversible particle movements associated with unstacking and restacking of chloroplast membranes *in vitro*, *J. Cell Biol.*, 71, 136–158.

Sylak-Glassman, E.J., Malnoë, A., De Re, E., Brooks, M.D., Fischer, A.L., Niyogi, K.K. and Fleming, G.R. (2014). Distinct roles of the photosystem II protein PsbS and zea-xanthin in the regulation of light harvesting in plants revealed by fluorescence lifetime snapshots, *Proc. Natl. Acad. Sci. USA*, 111, 17498–17503.

Thurlkill, R.L., Grimsley, G.R., Scholtz, M. and Pace, C.N. (2006). Hydrogen bonding markedly reduces the pK of buried carboxyl grouPSIn proteins, *J. Mol. Biol.*, 362, 594–604.

Van Amerongen, H. and Croce, R. (2014). Light harvesting in photosystem II, *Photosynth. Res.*, 116, 251–263.

Ware, M.A., Giovagnetti, V., Belgio, E. and Ruban, A.V. (2015). PsbS protein modulates non-photochemical chlorophyll fluorescence quenching in membranes depleted from photosystems, *J. Photochem. Photobiol. B*, 152, 301–307.

Wilk, L., Grunwald, M., Liao, P.-N., Walla, P.J. and Kühlbrandt, W. (2013). Direct interaction of the major light-harvesting complex II and PsbS in nonphotochemical quenching, *Proc. Natl. Acad. Sci. USA*, 110, 5452–5456.

Zia, A., Johnson, M.P. and Ruban, A.V. (2011). Acclimation- and mutation-induced enhancement of PsbS levels increases photoprotective energy dissipation in the absence of antheraxanthin and zeaxanthin, *Planta*, 233, 1253–1264.

Chapter 10

Thylakoid Membrane Dynamics in Higher Plants

Haniyeh Koochak, Meng Li and Helmut Kirchhoff*

*Institute of Biological Chemistry, Washington State University,
PO Box 646340, Pullman, WA, 99164, USA
kirchhh@wsu.edu

The energy-converting thylakoid membrane system in higher plant chloroplasts dynamically respond to environmental cues. This chapter gives an update about architectural thylakoid membrane dynamics and its functional consequences. Two structural levels are addressed. The first is the mesoscopic or supramolecular level describes the collective behavior of protein ensembles. In this part, the switch from disordered to semi-crystalline mesoscopic reorganization of photosystem II supercomplexes in stacked grana thylakoids is addressed. Functional implications on different lateral diffusion processes are discussed. The second part gives a survey about architectural changes of the entire thylakoid membrane system that includes dynamic swelling/shrinkage of the thylakoid lumen or stacking/destacking of grana. Potential mechanisms that control these ultrastructural changes include thylakoid ion channels and protein phosphorylation. In summary, this chapter links structural features of thylakoid membranes to the functionality, regulation, and maintenance of the photosynthetic apparatus. It highlights that a combination of different structural levels is required for a holistic understanding of energy conversion.

1. Introduction

In higher plants, the photosynthetic conversion of sunlight into chemical energy is realized by nanometer (nm)-sized protein-supercomplexes that are embedded in the thylakoid membrane system inside chloroplasts [Nelson and Ben-Shem, 2004; Dekker and Boekema, 2005]. A structural hallmark of thylakoid membranes is the tight stacking part of the membranes to grana discs, cylindrical structures with a diameter of 400 nm to 600 nm and of variable height [Staehelin and van der Staay, 1996; Daum *et al.*, 2010; Nevo *et al.*, 2012]. Thylakoid membrane stacking leads to formation of three structurally distinct subdomains that form one continuous thylakoid membrane: stacked grana core, grana margins at the periphery of the grana cylinder, and unstacked stroma lamellae that connect grana stacks. Grana seem to have arisen in evolution after ancient photosynthetic organisms colonized the terrestrial habitats [Mullineaux, 2005]. Although many hypotheses try to explain why grana formed, the biological function of grana remains largely obscure. Grana membranes mainly host three protein supercomplexes: photosystem II (PSII), trimeric light harvesting complex II (LHCII), and the dimeric cytochrome $b_6 f$ (cyt *bf*) complex [Albertsson, 2001; Herbstová *et al.*, 2012; Puthiyaveetil *et al.*, 2014]. In contrast to the other granal protein complexes, PSII comes in different structural subspecies [Dekker and Boekema, 2005; Danielsson *et al.*, 2006; Caffarri *et al.*, 2009], ranging from a monomeric reaction center (C), over dimeric reaction center (C2), to PSII species that bind different numbers of LHCII, *i.e.* C2S2 (dimeric PSII with two strongly bound LHCII), C2S2M2 (C2S2 with two additional moderately bound LHCII). This diversity in PSII structures is likely a reflection of the high turnover of this complex due its vulnerability to photodamage [Ohad *et al.*, 1984; Melis, 1999; Chow *et al.*, 2005]. In recent years, good progress has been made in understanding the structure of the overall thylakoid system on a lower resolution μm length scale [Chow *et al.*, 2005; Chuartzman *et al.*, 2008; Kirchhoff, 2014b]. At high resolution, excellent structural data sets exist for the molecular architecture of individual protein supercomplexes on the nm down to the Å length scale.

In contrast to high resolution data on thylakoid proteins, an important gap in the knowledge base exists on the intermediate, mesoscopic length scale (supramolecular level) that encompasses the structural organization of larger protein ensembles. Supramolecular cooperation of nanometer sized protein complexes is a key element of biomembranes. As detailed below, in thylakoid membranes, the precise organization of protein nanomachines in stacked grana thylakoids on the mesoscopic level determines vital functions for the conversion of sunlight into chemical energy. This includes diffusion-dependent electron transport, harnessing of solar radiation by light-harvesting complexes (LHC), photoprotective high energy quenching, balancing energy distribution between the two photosystems (state transition), and protein repair processes. In short, almost all aspects of photosynthetic energy conversion depend on the mesoscopic protein ensemble architecture in grana thylakoids. The environment of the plant exerts strong control on the mesoscopic protein organization in grana. For example, changes in temperature, light, water content, or osmotic pressure all can switch the protein arrangement from random to a highly ordered semicrystalline state [Staehelin, 1986; Dekker and Boekema, 2005]. Thus, knowing how environmental factors feedback on mesoscopic membrane properties and how changes of different supramolecular arrangements determine photosynthetic energy conversion is critical for understanding the plasticity of the photosynthetic apparatus. The aspects of mesoscopic dynamics will be addressed in the first part of this book chapter.

Constant improvements in electron and atomic force microscopy and sample preparations revealed fascinating new insights in the overall organization of photosynthetic membranes. It becomes clearer that the whole thylakoid membrane structure responds dynamically to environmental stimuli, constantly changing its shape. These architectural changes are not well understood. In particular, it is unknown how they are controlled, *i.e.* what mechanisms lead to structural membrane alterations and how environmental signals are sensed and transduced into architectural dynamics. In this respect, recently discovered thylakoid ion channels could play an interesting role. Overall, membrane dynamics and the potential role of ion channels and reversible protein phosphorylation will be presented in the second part of this book chapter.

2. Supramolecular Dynamics

2.1. *Protein order and disorder in thylakoid membranes*

Structural features of thylakoid membranes on the mesoscopic length scale (about a few tens to 1000 nm), and their dynamics determine the efficiency, regulation, and maintenance of the photosynthetic apparatus. Small changes in the arrangement of photosynthetic supercomplexes in stacked grana membranes can switch the functionality of the whole system [Dekker and Boekema, 2005; Kirchhoff, 2008a; Kirchhoff, 2013a; Tietz *et al.*, 2015]. Therefore, the mesoscopic organization in grana thylakoids must be fine-tuned and regulated to adjust energy conversion to metabolic needs and dynamics in the plant environment. An impressive mesoscopic change in thylakoid membranes is the change from a disordered to a semicryalline state in stacked grana.

2.1.1. *Significance of mesoscopic dynamics for energy transformation and its regulation*

It turns out that on the mesoscopic scale, grana thylakoid membranes are highly dynamic. For example, it is well established that the arrangement of PSII can change between disordered and highly organized semicrystalline states [Staehelin, 1986; Dekker and Boekema, 2005, Kirchhoff, 2008a; Daum *et al.*, 2010; Sznee *et al.*, 2011; Tietz *et al.*, 2015]. Furthermore, large-scale rearrangements of PSII and LHCII were reported to be involved in photoprotective high-energy quenching processes [Betterle *et al.*, 2009; Johnson *et al.*, 2011]. Optimization of energy distribution to both photosystems under low light by a process called state transition also requires mesoscopic rearrangements in grana membranes by shifting LHCII from PSII to PSI and *vice versa* [Rochaix, 2007, 2014]. Finally, the repair of photodamaged PSII leads to alterations in the overall membrane organization of thylakoid membranes including grana [Herbstová *et al.*, 2012; Puthiyaveetil *et al.*, 2014]. We recently found indications that the mesoscopic organization into semicrystalline state in grana controls the traffic of photodamaged PSII between stacked and unstacked thylakoid regions, *i.e.* effects the PSII repair cycle [Tietz *et al.*, 2015]. These examples highlight that constant structural reorganizations on the mesoscopic length scale are essential for regulation, adaptation, and repair of the photosynthetic machinery.

Beyond the importance of the grana architecture for regulation and repair processes, supramolecular features of grana membranes are also central for even more fundamental functions of photosynthetic energy conversion. For example, the mobility of small hydrophobic molecules, like the electron carrier plastoquinone (PQ), is sensitive to the supramolecular protein organization. PQ shuttles electrons between PSII and cyt *bf* complexes by diffusion through the lipid bilayer in the grana thylakoid membrane. Although PQ is the smallest component of the photosynthetic electron transport chain, it has an eminent position because besides shuttling electrons, it controls kinase activity involved in the regulation of light harvesting and PSII degradation, determines the rate of photodamage of PSII, and acts as a signal for differential expression of photosynthetic genes [Kirchhoff, 2008b]. It was recognized that the dense protein packing in grana membranes could severely impact PQ mobility [Lavergne *et al.*, 1992; Kirchhoff *et al*, 2000]. The mesoscopic organization in crowded grana has a strong impact on the efficiency of PQ diffusion [Tremmel *et al.*, 2003; Kirchhoff *et al.*, 2004; Tietz *et al.*, 2015]. Another fundamental energy converting process that is sensitive to the mesoscopic protein organization in grana is light-harvesting by LHCII connected to PSII [Lavergne and Trissl, 1995; Kirchhoff *et al.*, 2004; Bennett *et al.*, 2013]. This is because efficient radiation-less energy transfer between chlorophylls (Chl) is strongly dependent on the exact pigment separation and orientation, *i.e.* on the supramolecular arrangement of LHCII and PSII [Bennett *et al.*, 2013; Caffarri *et al.*, 2011].

2.1.2. *Types of semi-crystalline protein arrays in grana thylakoids*

Detailed electron microscopic image analysis revealed that different types of semicrystalline arrays exist (different lattice constants, [Dekker and Boekema,

2005]). Semicrystalline arrays can be made of C2S2M2, C2S2M, or C2S2 supercomplexes that determine the type of the crystal [Dekker and Boekema, 2005]. Depending on the type of crystal, very different functional consequences exist. For example, semicrystalline arrays of C2S2 supercomplexes, as found in mutants, are packed so tightly that PQ cannot diffuse between the protein complexes leading to a very severe restriction of electron transport by over-reduced and damaged PSII [de Bianchi *et al.*, 2008; Morosinotto *et al.*, 2006]. In contrast, C2S2M2 crystals that were found in *Arabidopsis* wild type plants [Dekker and Boekema, 2005; Tietz *et al.*, 2015] show facilitated PQ diffusion and improved photoprotection [Tietz *et al.*, 2015] which will be discussed in the next section.

2.2. Potential advantages and disadvantages of semi-crystalline protein arrays

The abundance of the semicrystalline state in grana thylakoids increases under abiotic stresses [Dekker and Boekema, 2005; Tietz *et al.*, 2015]. This indicates that the ordered state has probably a high physiological meaning. However, the functional consequences of semicryalline arrays in higher plant thylakoid membranes is still not clear. Mutants with higher abundance of C2S2-semicrystalline arrays show impaired excitonic connectivity, photochemical quantum yield of PSII, and photoprotective high energy quenching [Kovács *et al.*, 2006; Goral *et al.*, 2012]. In contrast, mutants with C2S2M2-type arrays do not show these negative impacts of semi-crystalline arrays on light harvesting and regulation [Tietz *et al.*, 2015]. C2S2M2-type arrays is the dominant type in wild type plants [Dekker and Boekema, 2005]. It is likely that protein arrays impact lateral membrane diffusion and the efficiency of protein reorganizations. For example, small lipophilic molecules (plastoquinone and xanthophylls) involved in the diffusion-dependent electron transport and photoprotective high energy-dependent quenching (qE) have to travel through highly crowded membrane regions. qE is a photoprotective mechanism that dissipates harvested light energy in the PSII antenna system safely into heat under light stress [Li *et al.*, 2009; Ruban *et al.*, 2012; Jahns and Holzwarth, 2012]. Although the exact molecular mechanism is still under debate, there is agreement that three factors are required for the full activation of qE. An indispensable factor for establishing qE is a proton gradient (ΔpH) across the thylakoid membrane. The magnitude of qE can be increased by the presence of the PsbS protein [Li *et al.*, 2000] and by the accumulation of the xanthophyll zeaxanthin [Demmig *et al.*, 1987]. The latter is converted from violaxanthin by the so-called xanthophyll cycle [Demmig *et al.*, 1987; Jahns *et al.*, 2009] that is catalyzed by the enzyme violaxanthin-deepoxidase (VDE). The 40 kDa VDE [Schubert *et al.*, 2002] is localized in the lumen of the thylakoids and is activated by dimerization triggered by acidification of the lumen in the light [Jahns *et al.*, 2009]. Furthermore, the repair of the large PSII supercomplexes requires protein transfer between stacked and unstacked thylakoid membrane regions. As detailed below, the switching of the supramolecular organization into more ordered states has different effects on diffusion-dependent electron transport, photoprotection, or protein repair processes, thus fine-tuning different functions of photosynthetic energy conversion [Tietz *et al.*, 2015].

2.2.1. *PSII repair*

Recovery of damaged PSII is realized by a multi-step repair cycle where the repair components are mainly localized in unstacked thylakoid domains [Mattoo and Edelman, 1987; Puthiyaveetil *et al.*, 2014; Järvi *et al.*, 2015]. In higher plants PSII is damaged in stacked grana. Therefore, the complex must travel over several hundred nanometers to reach the repair machinery in unstacked regions. Thus, under conditions of higher demand of PSII repair (*e.g.* high light stress) a brisk lateral traffic of PSII between stacked grana thylakoids and unstacked regions is required to keep a high photosynthetic performance. However, it is known that under non-stressed conditions, the mobility of PSII proteins in stacked grana is very low [Kirchhoff *et al.*, 2004, 2008; Goral *et al.*, 2010], raising the question how damaged PSII gets mobilized from core grana to reach stroma lamellae. Some observations indicate that under high light (HL) stress the mobility of grana-hosted pigment protein complexes (PSII and LHCII) increases [Kirchhoff, 2016]. By using fluorescence recovery after photobleaching (FRAP), an increased amount of protein exchange between different grana discs [Goral *et al.*, 2010] and also the mobility in isolated grana membranes [Herbstová *et al.*, 2012] was measured under HL. The higher mobility of PSII under HL will support efficient protein repair. In context of semicryalline arrays, the question arises whether protein ordering in grana interferes with mobilization of damaged PSII? It is reasonable that damaged PSII complexes localized in semi-crystalline arrays cannot easily escape out of these arrays, *i.e.* they have a highly restricted mobility. Therefore, one can hypothesize that under HL conditions highly ordered semi-crystalline protein arrays hinder PSII repair. Indeed, impairment of the PSII repair cycle was observed for a mutant with a constitutively high abundance of C2S2M2-types semicrystalline arrays [Tietz *et al.*, 2015]. This impaired repair was not caused by differences in repair enzymes, indicating that the mesoscopic organization in protein arrays is responsible for retarded recovery of damaged PSII. This observation is in line with reports that the amount of PSII arrays in wild type plants decreases under high light [Kouřil *et al.*, 2013], *i.e.* protein arrays are less abundant if efficient PSII repair is required. These results indicate that semi-crystalline protein arrays are disadvantageous for the PSII repair cycle because the required mobilization of damaged PSII from stacked grana is hampered.

2.2.2. *Diffusion in the lipid matrix*

Lateral diffusion of PQ and xanthophylls through a lipid matrix is a random walk process that can be described by Einstein's diffusion equation [Einstein, 1906]. For a two-dimensional diffusion process (as in biomembranes) the equation reads:

$$\langle x^2 \rangle = 4 \cdot D \cdot t$$

where $\langle x^2 \rangle$ is mean square displanement of a tracer (*e.g.* PQ) [cm], D is diffusion coefficient [$cm^2 s^{-1}$] and t is time [s].

However, this equation is only strictly valid for diluted membranes, *i.e.* without diffusion obstacles. Real biomembranes, in particular photosynthetic membranes, are the opposite of diluted, they are heavily crowded by integral membrane proteins. In fact, macromolecular crowding severely affects the PQ diffusion and it was postulated that fast PQ diffusion is restricted to small (roughly a few ten nm sized) temporarily formed lipid microdomains [Lavergne *et al.*, 1992; Kirchhoff *et al.*, 2000]. Since active PSII is highly concentrated in stacked grana [Andersson and Anderson, 1980; Albertsson, 2001], the sublocalization of cyt *bf* complex becomes critical for efficient PQ-dependent electron transport because it will determine the diffusion distance that PQ has to travel to connect PSII with cyt *bf* complexes. Screening the literature, it turns out that for the cyt *bf* complex all different scenarios for its sublocalization were reported [Kirchhoff *et al.*, 2017]. We hypothesized a mechanism that can explain this discrepancy in the literature that is based on small dynamic changes in the stroma gap between adjacent grana membranes [Kirchhoff *et al.*, 2017]. Bottom line is that under certain conditions cyt *bf* complexes could be depleted in stacked grana and accumulate in unstacked membrane regions causing a problem for PQ to connect PSII and cyt *bf* complexes by diffusion.

Can semicrystalline arrays in grana help solve the lateral diffusion problem for small lipophilic molecules? As mentioned above, clear evidence exists that C2S2-type semicrystalline arrays lead to an over-reduction of PSII caused by problems of PQ to get access to PSII in the arrays [de Bianchi *et al.*, 2008; Morosinotto *et al.*, 2006]. Thus, this type of protein array is clearly a disadvantage for diffusion of lipid-like molecules. In contrast, FRAP data of fluorescence-labeled fatty acids clearly show accelerated diffusion in C2S2M2-types arrays as found in a *fatty acid desaturase 5* (*fad 5*) mutant [Tietz *et al.*, 2015]. Consequently, PQ-diffusion-dependent electron transport steps as well as xanthophyll-dependent high energy quenching are accelerated in *fad 5* relative to wildtype plants. It was speculated that the reason for facilitated PQ and xanthophyll diffusion in the *fad 5* mutant is formation of a lipidic diffusion channel between PSII rows in the array [Tietz *et al.*, 2015]. This channel could function as a diffusion highway. Thus, the mesoscopic reorganization into a C2S2M2-type semicrystalline protein state in stacked grana could be a mechanism to facilitate lateral diffusion in a challenging crowded membrane environment.

2.3. *Factors controlling mesoscopic protein arrangements*

Two potential candidates for controlling mesoscopic dynamics are the reversible phosphorylation of grana-hosted protein complexes (mainly LHCII and PSII) and alterations in lipid bilayer compositions [Kirchhoff, 2008a; Tikkanen *et al.*, 2008; Pesaresi *et al.*, 2011; Herbstová *et al.*, 2012; Tietz *et al.*, 2015]. However, our knowledge of how lipids and protein phosphorylation modulate the architecture of the protein ensemble in grana is fragmentary to non-existing. Also, the protein PsbS was discussed to be involved in supramolecular protein organization based on the observation that knock-out or overexpression of this protein leads to higher or lower abundance of C2S2M2-type arrays [Kereïche *et al.*, 2010], indicating that PsbS prevents semicyralline state formation. However, these changes are moderate (maximal 9% change in protein array abundance from PsbS overexpressor to knock-out).

2.3.1. *Lipids and fatty acids*

The abundance of the non-bilayer forming lipid monogalactosyldiacylglycerol (MGDG) and the level of fatty acid desaturation in thylakoid lipids are both parameters that could be central for determining the mesoscopic protein landscape in grana thylakoids [Kirchhoff, 2008a; Tietz, 2015]. Both control the lateral membrane pressure profile (LMPP) in thylakoid membranes [Van den Brinck-van der Laan *et al.*, 2004; Anishkin *et al.*, 2014]. The LMPP gives the physical pressure value as a function of the position along the lipid bilayer normal (vertical position). The LMPP shows characteristic positive and negative pressure values that sum up to zero for a relaxed bilayer. However, local pressure values in LMPP can easily reach several MPa [Orsi and Essex, 2013]. For certain membrane proteins (*e.g.* mechanosensitive channels), it was shown that changes in LMPP induced by alterations of the lipid bilayer composition control the protein conformation and function [Battle *et al.*, 2015]. For thylakoid membranes, changes in phase state of MGDG (conversion from bilayer to non-bilayer HII phase) and fatty acid composition [Garab *et al.*, 2016] could trigger significant changes in LMPP [Orsi and Essex, 2013]. Thus, it is possible that alterations in physiochemical properties of the lipid matrix trigger conformational changes of grana hosted proteins leading to alterations in protein–protein interactions and mesoscopic rearrangements. At this point, this link between lipid composition and supramolecular protein organization is highly speculative and requires experimental verification. However, unraveling the interrelationship between lipid/fatty acid composition and protein ordering in grana membranes will likely give novel insights into the role of lipid matrix dynamics for light-harvesting and electron transport mediated by mesoscopic changes.

2.3.2. *Protein phosphorylation*

It is well-known that reversible protein phosphorylation facilitates repair of photodamaged PSII [Tikkanen *et al.*, 2008; Herbstová *et al.*, 2012; Kirchhoff, 2013b] and causes adjustments of photosystem antenna sizes for both photosystems by state transitions [Rochaix, 2007, 2014]. State transitions equilibrate the relative absorption between the two photosystems [Allen, 1992; Rochaix, 2014; Pribil *et al.*, 2014]. Under conditions when PSII is preferentially excited, a STN7 kinase [Bonardi *et al.*, 2005; Vainonen *et al.*, 2005] phosphorylates LHCII, causing its separation from PSII and docking to PSI. Dephosphorylation of LHCII is catalyzed by Tap38 (PPH1) phosphatase [Pribil *et al.*, 2010; Shapiguzov *et al.*, 2010] leading to back-migration of LHCII from PSI to PSII (state I). It has also been observed that LHCII trimers in C2S2M2 supercomplexes can be phosphorylated but does not lead to unbinding from PSII [Crepin and Caffarri, 2015]. In contrast to STN7 kinase, a second thylakoid membrane-bound kinase, STN8, mainly catalyzes PSII subunits D1, D2, CP43, and PsbH [Tikkanen *et al.*, 2008; Pesaresi *et al.*, 2011]. Dephosphorylation of these subunits is facilitated by a PBCP phosphatase [Samol *et al.*, 2012]. Protein phosphorylation catalyzed by STN7 and STN8 kinases is regulated by redox signals and strictly controlled by the environment and metabolism [Allen, 1992, Rochaix, 2014; Pribil *et al.*, 2014]. Whether phosphatases are regulated as well is unknown. Besides the well-documented role of LHCII and PSII phosphorylation/dephosphorylation in state transition and PSII repair, it is not well-understood how protein phosphorylation impacts the mesoscopic arrangements of grana-hosted proteins. Adding phosphate groups at the surface of photosynthetic membrane proteins significantly increase the negative surface charge density [Puthiyaveetil *et al.*, 2017] that likely affect the lateral interaction between them and therefore the mesoscopic protein organization [Barber, 1982]. We recently estimated that attractive and repulsive forces within the C2S2M2 supercomplex have similar magnitudes [Puthiyaveetil *et al.*, 2017]. If this is also the case for interactions between free LHCIIs and PSII supercomplexes, then the mesoscopic system is in a kind of metastable state, *i.e.* supramolecular rearrangements could be induced by small changes in electrostatic repulsion triggered by protein phosphorylation. Until today, systematic studies on the relationship between mesoscopic protein organization and reversible protein phosphorylation are missing.

2.4. *The elusive mesoscopic level*

Unraveling the dynamic structure-function relationships of protein ensembles in grana thylakoid membranes is one of the prime tasks in photosynthesis research. In contrast to studying the overall thylakoid membrane architecture by ultrastructural techniques and the structure of individual isolated protein complexes by *i.e.* crystallographic techniques, the characterization of mesoscopic features is challenging because of the lack of adequate methods. Freeze-fracture (or freeze-etching) electron microscopy or atomic force microscopy (AFM) provides details of protein arrangements in grana but has several drawbacks [Ort and Yocum, 1996; Kirchhoff *et al.*, 2004, 2007; Sznee *et al.*, 2011; Nevo *et al.*, 2012; Johnson *et al.*, 2014]. Very recently, cryo-scanning electron microscopy (cryo-SEM) on intact freeze-fractured leaf material [Charuvi *et al.*, 2015, 2016] has been developed; this is currently the state-of-the-art technique to study protein organization in grana under near native conditions. To add, cryo-electron tomography is also very promising [Engel *et al.*, 2015]; however, this technique is currently restricted to samples with a thickness of a few 100 nanometers, *i.e.* it cannot be applied for intact leaves. For a holistic view on supramolecular protein ensembles in photosynthetic membranes, computer simulations offer huge potentials. Previous computer models have already revealed interesting insights in emergent mesoscopic behaviors [Tremmel *et al.*, 2003; Schneider and Geissler, 2013]. Elaborating computer models fed by experimental data to describe the mesoscopic level could be promising for finding critical parameters that control and regulate supramolecular protein organization in grana thylakoids.

3. Thylakoid Membrane Dynamics

3.1. *Types of thylakoid lumen swelling and shrinkage*

There are good indications that in higher plants the overall grana structure is not static but highly dynamic, changing its shape in response to different environmental conditions [Kyle *et al.*, 1983; Staehelin, 1986; Chuartzman *et al.*, 2008; Kirchhoff *et al.*, 2011; Herbstová *et al.*, 2012; Puthiyaveetil *et al.*, 2014a; Yoshioka-Nishimura *et al.*, 2014]. Recently, it was found by analyzing TEM images of samples prepared by high-pressure freezing followed by freeze-substituion (HPF-FS) and also by cryo-EM that the transition from dark to moderate light intensities leads to an increased repeat distance in grana, *i.e.* the vertical distance between one stroma gap to the adjacent one increases [Kirchhoff *et al.*, 2011]. Detailed image analysis and modelling revealed that this increase in grana repeat is mainly caused by significant swelling (~100% increase) of the narrow thylakoid lumen space [Kirchhoff *et al.*, 2011]. Complementing the ultrastructural data with functional measurements show that the swelling of the lumen allows higher mobility of the lumen hosted 12 kDa electron carrier plastocyanin (PC) that facilitates electron shuttling between cyt *bf* complexes and PSI [Kirchhoff *et al.*, 2011]. This finding demonstrated that light-induced changes in the overall membrane architecture (swelling of the thylakoid lumen) control essential functions of photosynthetic electron transport (diffusion-dependent electron transport). Since the thylakoid lumen contains many other proteins involved in regulation, degradation, and processing processes (proteomic studies predict about 80 different proteins [Kieselbach and Schröder, 2003]), it can be expected that functions other than electron transport are affected by dynamic swelling and shrinkage of the lumen as well. Similar increase of granum thickness (or repeat distance) was observed during transition from state 1 to state 2 [Chuartzman *et al.*, 2008]. In addition to the vertical changes, state transitions also change lateral grana dimensions with shrinkage at state 2 compared with state 1 [Kyle *et al.*, 1983; Chuartzman *et al.*, 2008].

In contrast to moderate light intensities, architectural thylakoid changes are different under HL. HL treatment also leads to swelling of the thylakoid lumen but this seems restricted to the grana margins only (Puthiyaveetil *et al.*, 2014a; Yoshioka-Nishimura *et al.*, 2014; Tsabari *et al.*, 2015). This preferential swelling in the grana periphery leads to a bending of the overall grana structure that is typically very flat. The swelling and bending in the grana margins causes a partial destacking of appressed grana in this region, *i.e.* to a partial conversion of stacked to destacked membranes [Herbstová *et al.*, 2012; Puthiyaveetil *et al.* 2014a]. It was hypothesized that the partial destacking of grana allows better access of the bulky FtsH protease to damaged PSII. The FtsH protease is a key enzyme involved in PSII repair cycle catalyzing the degradation of damaged D1 subunit of PSII [Yoshioka-Nishimura and Yamamoto, 2014]. In strictly stacked thylakoid membranes (*e.g.* dark state), the stromal gap of about 3.5 nm is too narrow to allow access of the bulky FtsH (stroma protrusion about 6.5 nm, [Suno *et al.*, 2006]) to stacked regions [Kirchhoff, 2014a]. Therefore, the partial destacking in the grana periphery can help the protease accesses its substrate [Puthiyaveetil *et al.*, 2014a; Kirchhoff, 2014a]. Also, for other proteins involved in PSII repair (CtpA peptidases, Deg proteases, STN8 kinase, PBCP phosphatase, ribosomes), it was discussed that steric constraints of grana thylakoids control accessibility to stacked regions [Kirchhoff, 2014a]. These examples highlight that changing architectural features of the thylakoid membrane can be a central strategy for regulating and maintaining photosynthetic performance in response to a challenging environment. However, the mechanisms how architectural dynamics of the entire thylakoid membrane system are controlled are largely unknown.

3.2. *Factors controlling thylakoid membrane dynamics*

3.2.1. *Ion transporters*

From recent studies, it emerges that three groups of proteins are involved in controlling and regulating changes in thylakoid membrane architecture: ion transporters, protein phosphorylation/dephosphorylation, and CURT1 proteins. The first group is thylakoid-localized ion transporter/channels (review Finazzi *et al.* [2015]). A likely role of ion transporter for alterations in thylakoid structure is that light-induced ion flux into the thylakoid lumen leads to an increased osmotic potential in this small compartment which causes passive water influx and swelling [Murakami and Packer, 1970; Kirchhoff *et al.*, 2011; Pfeil *et al.*, 2014; Pribil *et al.*, 2014]. Among the group of ion transporter/channels in thylakoid membranes, potassium antiporter/channels and chloride channels are the most promising candidates for controlling the osmotic potential in the thylakoid lumen since these two ions are most abundant in chloroplasts. In the past few years, the two-pore K^+ uniporter (TPK3, [Carraretto *et al.*, 2013]) and the K^+-efflux antiporter 3 (KEA3, [Kunz *et al.*, 2014; Armbruster *et al.*, 2014]) were discovered. The molecular identity of thylakoid chloride channels is less clear. Early on, their existence was confirmed by patch clamp studies [Schönknecht *et al.*, 1988; Enz *et al.*, 1993] and recently two putative chloride channels were identified in the thylakoid membranes. The first is the voltage-dependent Cl^- channel 1 (VCCN1), and the second candidate is a member of the chloride channel (ClC) family named ClCe [Marmagne *et al.*, 2007; Herdean, 2015]. Very recently, VCCN1 was confirmed as a bestrophin-like thylakoid membrane voltage-gated Cl^- channel [Herdean *et al.*, 2016]. All of these ion transporters have been shown to play important roles in regulating the proton motive force (pmf) and its components ($\Delta\Psi$ or ΔpH) in response to changing light intensities. Ion transporters probably regulate the osmotic potential in the thylakoid lumen by controlling ion fluxes across the thylakoid membrane driven by pmf, thereby affecting swelling and shrinking of the thylakoid lumen. The influx of potassium controlled by the K^+/H^+ antiporter KEA3 and chloride controlled by ClCe and VCCN1 channels are expected to regulate osmotic swelling of the lumen. However, the fact that several ion transporters exist in thylakoid membranes (further may be discovered) and they seem to be regulated, puts a certain level of complexity to ion fluxes across thylakoid membranes and osmotic driven swelling/shrinkage processes.

3.2.2. *Protein kinases and phosphatases*

Another group of proteins involved in structural alterations of the thylakoid membrane system is kinases and phosphatases mentioned earlier. The STN8 kinase is mainly involved in regulating the turnover of photodamaged PSII under HL, whereas STN7 is involved in state transition under low light (see above). However, these functions do not seem to be strictly separated since an overlap in protein targets for both kinases was reported [Pesaresi *et al.*, 2011]. In the context of architectural thylakoid dynamics, it is important that SNT7/8 kinases-dependent phosphorylation shapes the thylakoid membrane architecture. In the loss-of-function, double mutant *stn7stn8*, grana are wider [Fristedt *et al.*, 2009] and dynamic shrinking of the grana diameter observed under HL is missing [Herbstová *et al.*, 2012]. In combination with the TAP38 and PBCP phosphatases, the reversible phosphorylation of thylakoid proteins by these kinases and phosphatases is likely involved in changes of the thylakoid structure under different light conditions. An interesting facet of the thylakoid kinase/phosphatase systems is that both systems work at different light intensities. The STN7/TAP38 system is mainly operative under low/moderate light, whereas the STN8/PBCP system controls PSII core protein phosphorylation mainly at HL intensities. It has to be elucidated whether the different light responses of the thylakoid architectural changes (general swelling of lumen under moderate light *versus* preferential swelling in grana margins under high light) is just a coincidence to the light response of STN kinases or a causal link exists.

3.2.3. *Membrane curvature proteins*

The curvature thylakoid 1 (CURT1) protein family was discovered to be essential for the strong curvature of grana margins [Armbruster *et al.*, 2013]. The amount of the CURT1 proteins can change the lateral diameter of grana as well as the height of grana [Armbruster *et al.*, 2013]. The preferential unstacking of margins under high light depends on the phosphorylation of thylakoid membrane proteins [Herbstová *et al.*, 2012; Puthiyaveetil *et al.*, 2014a]. An interesting connection to the CURT1 protein family is that these proteins contain phosphorylation sites [Armbruster *et al.*, 2013]. While the kinases and phosphatases for CURT1 proteins have not been identified so far, the reversible phosphorylation of CURT1 proteins could be involved in preferential destacking in grana margins under HL. All four CURT1 members (CURT1a, 1b, 1c, 1d) have two transmembrane helices with both the C- and N-terminus facing to the stroma. The N-terminus carries a phosphorylation site [Hansson and Vener, 2003; Armbruster *et al.*, 2013] that may represent an STN kinase target. Thus, it is possible that reversible CURT1 phosphorylation, as well as the quantity of CURT1, could control dynamic changes of the overall grana architecture induced by HL.

3.3. *Controversies of observed thylakoid dynamics*

While recent electron microscopic studies showed the increased grana stack repeat distance and lumen width upon illumination [Kirchhoff *et al.*, 2011; Tsabari *et al.*, 2015], other works showed opposite observations [Murakami and Packer, 1970; Pfeiffer and Krupinska, 2005]. Pioneering work done by Murakami and Packer showed that isolated spinach thylakoids shrink after illumination [Murakami and Packer, 1970]. The shrinkage upon illumination was also observed in barley leaves using HPF-FS sample preparation [Pfeiffer and Krupinska, 2005] but the numbers derived for grana repeat in dark adapted samples in this study are in conflict with cryo-EM tomographic data from spinach and pea [Daum *et al.*, 2010]. The latter numbers in turn are in very good agreement with [Kirchhoff *et al.*, 2011] collected with *Arabidopsis*. Besides TEM studies, small-angle neutron scattering was used to study thylakoid dynamics. Neutron scattering detects periodic structures like grana that are seen as diffraction peaks. However, it turns out that interpretation of neutron scattering data is not as straightforward as that of comparative TEM studies on high-pressure frozen cyanobacteria cells [Liberton *et al.*, 2013]. Neutron scattering studies on isolated thylakoid membranes from higher plants indicate shrinkage of grana thylakoids in the light [Nagy *et al.*, 2014; Ünnep *et al.*, 2017], that is in contrast to cyanobacteria where a light-induced swelling was observed [Liberton *et al.*, 2013]. Furthermore, grana repeat distances derived from neutron scattering analysis on isolated thylakoid membranes [Ünnep *et al.*, 2017] tend to be significantly higher than reported from TEM data [Daum *et al.*, 2010; Kirchhoff *et al.*, 2011]. The situation might also get more complicated since light also induces membrane undulations [Ünnep *et al.*, 2017] that makes determination of repeat distances from neutron scattering more difficult. This short survey illustrates that the simple question whether illumination leads to swelling or shrinkage of thylakoids is not easy to answer. There may not be a one-for-all answer to this question. First, different plants were used for thylakoid dynamics studies. In the case of barley chloroplasts, the diurnal behavior of stacking and unstacking of thylakoids is unique [Pfeiffer and Krupinska, 2005] among other studies. Second, environmental factors of chloroplasts, such as pH, ionic strength, and temperature affect the thylakoid structure. Also, swelling/shrinkage characteristics of the lumen are sensitive to CO_2 and O_2 concentrations [Tsabari *et al.*, 2015]. Many studies (including neutron scattering) were done on isolated thylakoids or isolated chloroplasts [Murakami and Packer, 1970; Ünnep *et al.*, 2017], which inevitably lose the *in vivo* environment for chloroplast. In addition, the methods used to determine thylakoid structure by electron microscopy are also critical. While most studies used conventional fixation for TEM, HPF-FS is generally considered better at preserving native structure of thylakoids. Yet, systematic studies of thylakoid dynamics using non-disturbing methods are needed to gain comprehensive understanding of the controversial observations. Without consensus on thylakoid swelling or shrinkage upon light illumination, the thylakoid membrane structure change is affected by the proton gradient across the thylakoid membrane [Kirchhoff *et al.*, 2011; Murakami and Packer, 1970; Pfeiffer and Krupinska, 2005; Ünnep *et al.*, 2017].

Acknowledgements

This work was supported by grants from the National Science Foundation (MCB-1616982), the US Department of Energy (DE-SC 0017160), and USDA National Institute of Food and Agriculture Hatch projects No. 1005351 and No. 0119.

References

Albertsson, P.-A. (2001). A quantitative model of the domain structure of the photosynthetic membrane, *Trends Plant Sci.*, 6, 349–354.

Allen, J.F. (1992). Protein phosphorylation in regulation of photosynthesis, *Biochim. Biophys. Acta*, 1098, 275–335.

Andersson, B. and Anderson, J.M. (1980). Lateral heterogeneity in the distribution of chlorophyll-protein complexes of the thylakoid membranes of spinach, *Biochim. Biophys. Acta*, 593, 427–440.

Anishkin, A., Loukin, S.H., Teng, J. and Kung C. (2014). Feeling the hidden mechanical forces in lipid bilayer is an original sense, *Proc. Natl. Acad. Sci.* USA, 111, 7898–7905.

Armbruster, U., Carrillo, L.R., Venema, K., Pavlovic, L., Schmidtmann, E., Kornfeld, A., Jahns, P., Berry, J.A., Kramer, D.M. and Jonikas, M.C. (2014). Ion antiport accelerates photosynthetic acclimation in fluctuating light environments, *Nat. Commun.*, 13, p. 5439.

Armbruster, U., Labs, M., Pribil, M., Viola, S., Xu, W., Scharfenberg, M., Hertle, A.P., Rojahn, U., Jensen, P.E., Rappaport, F., Joliot, P., Dörmann, P., Wanner, G. and Leister, D. (2013). Arabidopsis CURVATURE THYLAKOID1 proteins modify thylakoid architecture by inducing membrane curvature, *Plant Cell*, 25, 2661–2678.

Barber, J. (1982). Influence of surface charges on thylakoid structure and function, *Annu. Rev. Plant Physiol.*, 33, 261–295.

Battle, A.R., Ridone, P., Bavi, N., Nakayama, Y., Nikolaev, Y.A. and Martinac, B. (2015). Lipid-protein interactions: lessons learned from stress, *Biochim. Biophys. Acta*, 1848, 1744–1756.

Bennett, D.I.G., Amarnath, K. and Fleming, G.R. (2013). A structure-based model of energy transfer reveals the principles of light harvesting in photosystem II supercomplexes, *J. Am. Chem. Soc.*, 135, 9164–9173.

Betterle, N., Ballottari, M., Zorzan, S., de Bianchi, S., Cazzaniga, S., Dall'Osto, L., Morosinotto, T. and Bassi, R. (2009). Light-induced dissociation of an antenna hetero-oligomer is needed for non-photochemical quenching induction, *J. Biol. Chem.*, 284, 15255–15266.

Bonardi, V., Pesaresi, P., Becker, T., Schleiff, E., Wagner, R., Pfannschmidt, T., Jahns, P. and Leister, D. (2005). Photosystem II core phosphorylation and photosynthetic acclimation require two different protein kinases, *Nature*, 437, 1179–1182.

Caffarri, S., Broess, K., Croce, R. and van Amerongen, H. (2011). Excitation energy transfer and trapping in higher plant photosystem II complexes with different antenna sizes, *Biophys. J.*, 100, 2094–2103.

Caffarri, S., Kouřil, R., Kereiche, S., Boekema, E.J. and Croce, R. (2009). Functional architecture of higher plant photosystem II supercomplexes, *EMBO J.*, 28, 3052–3063.

Carraretto, L., Formentin, E., Teardo, E., Checchetto, V., Tomizioli, M., Morosinotto, T., Giacometti, G.M., Finazzi, G. and Szabó, I. (2013). A thylakoid-located two-pore K^+ channel controls photosynthetic light utilization in plants, *Science*, 342, 114–118.

Charuvi, D., Nevo, R., Kaplan-Ashiri, I., Shimoni, E. and Reich, Z. (2016). Studying the supramolecular organization of photosynthetic membranes within freeze-fractured leaf tissues by cryo-scanning electron microscopy, *J. Vis. Exp.*, doi:10.3791/54066.

Charuvi, D., Nevo, R., Shimoni, E., Naveh, L., Zia, A., Adam, Z., Farrant, J.M., Kirchhoff, H. and Reich, Z. (2015). Photoprotection conferred by changes in photosynthetic protein levels and organization during dehydration of a homoiochlorophyllous resurrection plant, *Plant Physiol.*, 167, 1554–1565.

Chow, W.S., Kim, E.-H., Horton, P. and Anderson, J.M. (2005). Granal stacking of thylakoid membranes in higher plant chloroplasts: the physicochemical forces at work and the functional consequences that ensue, *Photochem. Photobiol. Sci.*, 4, 1081–1090.

Chuartzman, S.G., Nevo, R., Shimoni, E., Charuvi, D., Kiss, V., Ohad, I., Brumfeld, V. and Reich, Z. (2008). Thylakoid membrane remodeling during state transitions in *Arabidopsis*, *Plant Cell*, 20, 1029–1039.

Crepin, A. and Caffarri, S. (2015). The specific localizations of phosphorylated Lhcb1 and Lhcb2 isoforms reveal the role of Lhcb2 in the formation of the PSI-LHCII supercomplex in *Arabidopsis* during state transitions, *Biochim. Biophys. Acta*, 1847, 1539–1548.

Danielsson, R., Suorsa, M., Paakkarinen, V., Albertsson, P.-A., Styring, S., Aro, E.-M. and Mamedov, F. (2006). Dimeric and monomeric organization of photosystem II. Distribution of five distinct complexes in the different domains of the thyalkoid membrane, *J. Biol. Chem.*, 281, 14241–14249.

Daum, B., Nicastro, D., Austin, J., II, McIntosh, R. and Kühlbrandt, W. (2010). Arrangement of photosystem II and ATP synthase in chloroplast membranes of spinach and pea, *Plant Cell*, 22, 1299–1312.

de Bianchi, S., Dall'Osto, L., Tognon, G., Morosinotto, T. and Bassi, R. (2008). Minor antenna proteins CP24 and CP26 affect the interactions between photosystem

II subunits and the electron transport rate in grana membranes of *Arabidopsis*, *Plant Cell*, 20, 1012–1028.

Dekker, J.P. and Boekema, E.J. (2005). Supramolecular organization of thylakoid membrane proteins in green plants, *Biochim. Biophys. Acta*, 1706, 12–39.

Demmig, B., Winter, K., Krüger, A. and Czygan, F.C. (1987). Photoinhibition and zeaxanthin formation in intact leaves: a possible role of the xanthophyll cycle in the dissipation of excess light energy, *Plant Physiol.*, 84, 218–224.

Einstein, A. (1906). Zur Theorie der Brownschen Bewegung, *Ann. d. Physik*, 19, 371–381.

Engel, B.D., Schaffer, M., Kuhn Cuellar, L., Villa, E., Plitzko, J.M. and Baumeister, W. (2015). Native architecture of the *Chlamydomonas* chloroplast revealed by *in situ* cryo-electron tomography, *Elife*, 4, doi:10.7554/eLife.04889.

Enz, C., Steinkamp, T. and Wagner, R. (1993). Ion channels in the thylakoid membrane (a patch-clamp study), *Biochim. Biophys. Acta*, 1143, 67–76.

Finazzi, G., Petroutsos, D., Tomizioli, M., Flori, S., Sautron, E., Villanova, V., Rolland, N. and Seigneurin-Berny, D. (2015). Ions channels/transporters and chloroplast regulation, *Cell Calcium*, 58, 86–97.

Fristedt, R., Willig, A., Granath, P., Crèvecoeur, M., Rochaix, J.-D. and Vener A.V. (2009). Phosphorylation of photosystem II controls functional macroscopic folding of photosynthetic membranes in *Arabidopsis*, *Plant Cell*, 21, 3950–3964.

Garab, G., Ughy, B. and Goss, R. (2016). Role of MGDG and non-bilayer lipid phases in the structure and dynamics of chloroplast thylakoid membranes, *Subcell. Biochem.*, 86, 127–157.

Goral, T.K., Johnson, M.P., Dufy, C.D.P., Brain, A.P.R., Ruban, A.V. and Mullineaux, C.W. (2012). Light-harvesting antenna composition controls the macrostructure and dynamics of thylakoid membranes in *Arabidopsis*, *Plant J.*, 69, 289–301.

Goral, T.K., Johnson, M.P., Kirchhoff, H., Ruban, A.V. and Mullineaux, C.W. (2010). Visualizing the diffusion of chlorophyll-proteins in higher plant thylakoid membranes: effects of photoinhibition and protein phosphorylation, *Plant J.*, 62, 948–959.

Hansson, M. and Vener, A.V. (2003). Identification of three previously unknown *in vivo* protein phosphorylation sites in thylakoid membranes of *Arabidopsis thaliana*, *Mol. Cell. Proteomics*, 2, 550–559.

Herbstová, M., Tietz, S., Kinzel, C., Turkina, M.V. and Kirchhoff, H. (2012). Architectural switch in plant photosynthetic membranes induced by light stress, *Proc. Natl. Acad. Sci.* USA 109, 20130–20135.

Herdean, A. (2015). Ion transport in chloroplasts with role in regulation of photosynthesis, Ph.D. thesis, Department of Biological and Environmental Sciences, University of Gothenburg, Sweden.

Herdean, A., Teardo, E., Nilsson, A.K., Pfeil, B.E., Johansson, O.N., Ünnep, R., Nagy, G., Zsiros, O., Dana, S., Solymosi, K., Garab, G., Szabó, I., Spetea, C. and Lundin, B. (2016). A voltage-dependent chloride channel fine-tunes photosynthesis in plants, *Nat. Commun.*, 7, p. 11654. doi:10.1038/ncomms11654.

Jahns, P. and Holzwarth, A.R. (2012). The role of the xanthophyll cycle and of lutein in photoprotection of photosystem II, *Biochim. Biophys. Acta*, 1817, 182–193.

Jahns, P., Latowski, D. and Strzalka, K. (2009) Mechanism and regulation of the violax-anthin cycle: the role of antenna proteins and membrane lipids, *Biochim. Biophys. Acta*, 1787, 3–14.

Järvi, S., Suorsa, M. and Aro, E.M. (2015). Photosystem II repair in plant chloroplasts—regulation, assisting proteins and shared components with photosystem II biogenesis, *Biochim. Biophys. Acta*, 1847, 900–909.

Johnson, M.P., Goral, T.K., Duffy, C.D., Brain, A.P., Mullineaux, C.W. and Ruban, A.V. (2011). Photoprotective energy dissipation involves the reorganization of photosystem II light-harvesting complexes in the grana membranes of spinach chloroplasts, *Plant Cell*, 23, 1468–1479.

Johnson, M.P., Vasilev, C., Olsen, J.D. and Hunter, C.N. (2014). Nanodomains of cyto-chrome b_6f and photosystem II complexes in spinach grana thylakoid membranes, *Plant Cell*, 26, 3051–3061.

Kereïche, S., Kiss, A.Z., Kouřil, R., Boekema, E.J. and Horton, P. (2010). The PsbS protein controls the macro-organisation of photosystem II complexes in the grana membranes of higher plant chloroplasts, *FEBS Lett.*, 584, 759–764.

Kieselbach, T. and Schröder, W.P. (2003). The proteome of the chloroplast lumen of higher plants. *Photosynth. Res.*, 78, 249–264.

Kirchhoff, H. (2008a). Molecular crowding and order in photosynthetic membranes, *Trends Plant Sci.*, 13, 201–207.

Kirchhoff, H. (2008b). Significance of protein crowding, order and mobility for photosyn-thetic functions, *Biochem. Soc. Trans.*, 36, 967–970.

Kirchhoff, H. (2013a). Diffusion of molecules and macromolecules in thylakoid mem-branes, *Biochim. Biophys. Acta.*, 1837, 495–502.

Kirchhoff, H. (2013b). Architectural switches in plant thylakoid membranes, *Photosynth. Res.*, 116, 481–487.

Kirchhoff, H. (2014a). Structural changes of the thylakoid membrane network induced by high-light stress in plant chloroplasts, *Philos. Trans. R. Soc. Lond. B Biol. Sci.*, 369, p. 20130225.

Kirchhoff, H. (2014b). Dynamic architecture of photosynthetic membranes. In *Advances in Plant Biology: Plastid Biology,* Wollman, F.-A. and Theg, S., eds. Chap. 5 (Springer Press), pp. 129–145.

Kirchhoff, H., Borinski, M., Lenhert, S., Chi, L. and Büchel, C. (2004). Transversal and lateral exciton energy transfer in grana thylakoids of spinach, *Biochemistry*, 43, 14508–14516.

Kirchhoff, H., Haase, W., Wegner, S., Daniellsson, R., Ackermann, R. and Albertsson, P.-A. (2007). Low-light induced array formation of photosystem II in higher plant chloroplasts, *Biochemistry*, 46, 11169–11176.

Kirchhoff, H., Haferkamp, S., Allen, J.F., Epstein, D. and Mullineaux, C.W. (2008). Significance of macromolecular crowding for protein diffusion in thylakoid mem-branes of chloroplasts, *Plant Physiol.*, 146, 1571–1578.

Kirchhoff, H., Hall, C., Wood, M., Herbstová, M., Tsabari, O., Nevo, R., Charuvi, D., Shimoni, E. and Reich, Z. (2011). Dynamic control of protein diffusion within the granal thylakoid lumen, *Proc. Natl. Acad. Sci.* USA, 108, 20248–20253.

Kirchhoff, H., Horstmann, S. and Weis, E. (2000). Control of the photosynthetic electron transport by PQ diffusion microdomains in thylakoids of higher plants, *Biochim. Biophys. Acta*, 1459, 148–168.

Kirchhoff, H., Lenhert, S., Büchel, C., Chi, L. and Nield, J. (2008). Probing the organization of photosystem II in photosynthetic membranes by atomic force microscopy, *Biochemisrty*, 47, 431–440.

Kirchhoff, H., Li, M. and Puthiyaveetil, S. (2017). Sublocalization of cytochrome b_6f complexes in photosynthetic membranes, *Trends Plant Sci.*, 22, 574–583. doi:10.1016/j.tplants.2017.04.004.

Kirchhoff, H., Tremmel, I., Haase, W. and Kubitscheck, U. (2004). Supramolecular photosystem II organization in grana thylakoid membranes: Evidence for a structured arrangement, *Biochemistry*, 43, 9204–9213.

Kouřil, R., Wientjes, E., Bultema, J.B., Croce, R. and Boekema, E.J. (2013). High-light vs. low-light: effect of light acclimation on photosystem II composition and organization in *Arabidopsis thaliana, Biochim. Biophys. Acta*, 1827(3), 411–419.

Kovács, L., Damkjær, J., Kereïche, S., Ilioaia, C., Ruban, A.V., Boekema, E.J., Jansson, S. and Horton, P. (2006). Lack of the light-harvesting complex CP24 affects the structure and function of the grana membranes of higher plant chloroplasts, *Plant Cell*, 18(11), 3106–3120.

Kunz, H.H., Gierth, M., Herdean, A., Satoh-Cruz, M., Kramer, D.M., Spetea, C. and Schroeder, J.I. (2014). Plastidial transporters KEA1, -2, and -3 are essential for chloroplast osmoregulation, integrity, and pH regulation in *Arabidopsis, Proc. Natl. Acad. Sci.* USA, 111, 7480–7485.

Kyle, D.J., Staehelin, L.A. and Arntzen, C.J. (1983). Lateral mobility of the light-harvesting complex in chloroplast membranes controls excitation energy distribution in higher plants, *Arch. Biochem. Biophys.*, 222, 527–541.

Lavergne, J., Bouchaud, J.-P. and Joliot, P. (1992). Plastochinone compartimentation in chloroplasts. II. Theoretical aspects, *Biochim. Biophys. Acta*, 1101, 13–22.

Lavergne, J. and Trissl, H.W. (1995). Theory of fluorescence induction in photosystem II: derivation of analytical expressions in a model including exciton-radical-pair equilibrium and restricted energy transfer between photosynthetic units, *Biophys. J.*, 68, 2474–2492.

Li, X.-P., Björkman, O., Shih, C., Grossman, A.R., Rosenquist, M., Jansson, S. and Niyogi, K.K. (2000). A pigment-binding protein essential for regulation of photosynthetic light harvesting, *Nature*, 403, 391–395.

Li, Z., Wakao, S., Fischer, B.B. and Niyogi, K.K. (2009). Sensing and responding to excess light, *Annu. Rev. Plant Biol.*, 60, 239–260.

Liberton, M., Page, L.E., O'Dell, W.B., O'Neill, H., Mamontov, E., Urban, V.S. and Pakrasi, H.B. (2013). Organization and flexibility of cyanobacterial thylakoid membranes examined by neutron scattering, *J. Biol. Chem.*, 288, 3632–3640.

Marmagne, A., Marion Vinauger-Douard, M., Monachello, D., Falcon de Longevialle, A., Charon, C., Allot, M., Rappaport, F., Wollman, F.-A., Barbier-Brygoo, H. and Ephritikhine, G. (2007). Two members of the Arabidopsis CLC (chloride channel) family, AtCLCe and AtCLCf, are associated with thylakoid and Golgi membranes, respectively, *J. Exp. Bot.*, 58, 3385–3393.

Mattoo, A.K. and Edelman, M. (1987). Intramembrane translocation and posttranslational palmitoylation of the chloroplast 32-kDa herbicide-binding protein, *Proc. Natl. Acad. Sci. USA*, 84, 1497–1501.

Melis, A. (1999). Photosystem-II damage and repair cycle in chloroplasts: what modulates the rate of photodamage *in vivo*? *Trends Plant Sci.*, 4, 130–135.

Morosinotto, T., Bassi, R., Frigerio, S., Finazzi, G., Morris, E. and Barber, J. (2006). Biochemical and structural analyses of a higher plant photosystem II supercomplex of a photosystem I-less mutant of barley. Consequences of a chronic over-reduction of the plastoquinone pool, *FEBS J.*, 273, 4616–4630.

Mullineaux, C.W. (2005). Function and evolution of grana, *Trends Plant Sci.*, 10, 521–525.

Murakami, S. and Packer, L. (1970). Protonation and chloroplast membrane structure, *J. Cell Biol.*, 47, 332–351.

Nagy, G., Unnep, R., Zsiros, O., Tokutsu, R., Takizawa, K., Porcar, L., Moyet, L., Petroutsos, D., Garab, G., Finazzi, G. and Minagawa, J. (2014). Chloroplast remodeling during state transitions in *Chlamydomonas reinhardtii* as revealed by noninvasive techniques *in vivo*, *Proc. Natl. Acad. Sci. USA*, 111, 5042–5047.

Nelson, N. and Ben-Shem, A. (2004). The complex architecture of oxygenic photosynthesis, *Nat. Rev. Mol. Cell Biol.*, 5, 971–982.

Nevo, R., Charuvi, D., Tsabari, O. and Reich, Z. (2012). Composition, architecture and dynamics of the photosynthetic apparatus in higher plants, *Plant J.*, 70, 157–176.

Ohad, I., Kyle, D.J. and Arntzen, C.J. (1984). Membrane protein damage and repair: removal and replacement of inactivated 32-kilodalton polypeptide in chloroplast membranes, *J. Cell Biol.*, 99, 481–485.

Orsi, M. and Essex, J.W. (2013). Physical properties of mixed bilayers containing lamellar and nonlamellar lipids: insights from coarse-grain molecular dynamics simulations, *Faraday Discuss.*, 161, 249–272.

Pesaresi, P., Pribil, M., Wunder, T. and Leister, D. (2011). Dynamics of reversible protein phosphorylation in thylakoids of flowering plants: the roles of STN7, STN8 and TAP38, *Biochim. Biophys. Acta*, 1807, 887–896.

Pfeiffer, S. and Krupinska, K. (2005). New insights in thylakoid membrane organization, *Plant Cell Physiol.*, 46(9), 1443–1451.

Pfeil, B.E., Schoefs, B. and Spetea, C. (2014). Function and evolution of channels and transporters in photosynthetic membranes, *Cell. Mol. Life Sci.*, 71, 979–998.

Pribil, M., Labs, M. and Leister, D. (2014). Structure and dynamics of thylakoids in land plants, *J. Exp. Bot.*, 65, 1955–1972.

Pribil, M., Pesaresi, P., Hertle, A., Barbato, R. and Leister, D. (2010). Role of plastid protein phosphatase TAP38 in LHCII dephosphorylation and thylakoid electron flow, *PLOS Biol.*, 8, p. e1000288.

Puthiyaveetil, S., Tsabari, O., Lowry, T., Lenhert, S., Lewis, R.R., Reich, Z. and Kirchhoff H. (2014a). Compartmentalization of the protein repair machinery in photosynthetic membranes, *Proc. Natl. Acad. Sci. USA*, 111, 15839–15844.

Puthiyaveetil, S., van Oort, B. and Kirchhoff, H. (2017). Surface charge dynamics in photosynthetic membranes and the structural consequences, *Nat. Plants*, 3, p. 17020.

Puthiyaveetil, S., Woodiwiss, T., Knoerdel, R., Zia, A., Wood, M., Hoehner, R. and Kirchhoff, H. (2014b). Significance of the photosystem II core phosphatase PBCP for plant viability and protein repair in thylakoid membranes, *Plant Cell Physiol.*, 55, 1245–1254.

Rochaix, J.D. (2007). Role of thylakoid protein kinases in photosynthetic acclimation, *FEBS Lett.*, 581, 2768–2775.

Rochaix, J.D. (2014). Regulation and dynamics of the light-harvesting system, *Annu. Rev. Plant Biol.*, 65, 287–309.

Ruban, A.V., Johnson, M.P. and Duffy, C.D. (2012). The photoprotective molecular switch in the photosystem II antenna, *Biochim. Biophys. Acta*, 1817, 167–81.

Samol, I., Shapiguzov, A., Ingelsson, B., Fucile, G., Crèvecoeur, M., Vener, A.V., Rochaix, J.D. and Goldschmidt-Clermont M. (2012). Identification of a photosystem II phosphatase involved in light acclimation in Arabidopsis, *Plant Cell*, 24, 2596–2609.

Schneider, A.R. and Geissler, P.L. (2013). Coexistence of fluid and crystalline phases of proteins in photosynthetic membranes, *Biophys. J.*, 105(5), 1161–1170.

Schönknecht, G., Hedrich, R., Junge, W. and Raschke, K. (1988). A voltage-dependent chloride channel in the photosynthetic membrane of a higher plant, *Nature*, 336, 589–592.

Schubert, M., Petersson, U.A., Haase, B.J., Funk, C., Schröder, W.P. and Kieselbach, T. (2002). Proteome map of the chloroplast lumen of Arabidopsis thaliana, *J. Biol. Chem.*, 277, 8354–8365.

Shapiguzov, A., Ingelsson, B., Samol, I., Andres, C., Kessler, F., Rochaix, J.D., Vener, A.V. and Goldschmidt-Clermont, M. (2010). The PPH1 phosphatase is specifically involved in LHCII dephosphorylation and state transitions in *Arabidopsis, Proc. Natl. Acad. Sci. USA*, 107, 4782–4787.

Staehelin, L.A. (1986). Photosynthesis III: photosynthetic membranes and light-harvesting systems. In *Chloroplast Structure and Supramolecular Organization of Photosynthetic Membranes*, Staehelin, L.A. and Arntzen, C.J., eds. Chap. 1 (Berlin: Springer-Verlag), pp. 1–84.

Staehelin, L.A. and van der Staay, G.W.M. (1996). Oxygenic photosynthesis: the light reactions. In *Structure, Composition, Functional Organization and Dynamic Properties of Thylakoid Membranes*, Ort, D.A. and Yocum, C.F., eds. Chap. 2 (Netherlands: Kluwer Academic Publishers), pp. 11–30.

Suno, R., Niwa, H., Tsuchiya, D., Zhang, X., Yoshida, M. and Morikawa, K. (2006). Structure of the whole cytosolic region of ATP-dependent protease FtsH, *Mol. Cell*, 22, 575–585.

Sznee, K., Dekker, J.P., Dame, R.T., van Roon, H., Wuite, G.J. and Frese, R.N. (2011). Jumping mode atomic force microscopy on grana membranes from spinach, *J. Biol. Chem.*, 286, 39164–39171.

Tietz, S., Puthiyaveetil, S., Enlow, H.M., Yarbrough, R., Wood, M., Semchonok, D.A., Lowry, T., Li, Z., Jahns, P., Boekema, E.J., Lenhert, S., Niyogi, K.K. and Kirchhoff, H. (2015). Functional implications of photosystem II crystal formation in photosynthetic membranes, *J. Biol. Chem.*, 290, 14091–14106.

Tikkanen, M., Nurmi, M., Kangasjarvi, S. and Aro, E.M. (2008). Core protein phosphorylation facilitates the repair of photodamaged photosystem II at high light, *Biochim. Biophys. Acta*, 1777, 1432–1437.

Tremmel, I., Kirchhoff, H., Weis, E. and Farquhar, G.D. (2003). Dependence of the plastoquinone diffusion coefficient on the shape, size, density of integral thylakoid proteins, *Biochim. Biophys. Acta*, 1607, 97–109.

Tsabari, O., Nevo, R., Meir, S., Carrillo, L.R., Kramer, D.M. and Reich, Z. (2015). Differential effects of ambient or diminished CO_2 and O_2 levels on thylakoid membrane structure in light-stressed plants, *Plant J.*, 81, 884–894.

Ünnep, R., Zsiros, O., Hörcsik, Z., Markó, M., Jajoo, A., Kohlbrecher, J., Garab, G. and Nagy, G. (2017). Low-pH induced reversible reorganizations of chloroplast thylakoid membranes — As revealed by small-angle neutron scattering, *Biochim. Biophys. Acta*, 1858(5), 360–365.

Vainonen, J.P., Hansson, M., Vener, A.V. (2005). STN8 protein kinase in Arabidopsis thaliana is specific in phosphorylation of photosystem II core proteins, *J. Biol. Chem.*, 280, 33679–33686.

van den Brink-van der Laan, E., Killian, J.A. and de Kruijff, B. (2004). Nonbilayer lipids affect peripheral and integral membrane proteins *via* changes in the lateral pressure profile, *Biochim. Biophys. Acta*, 1666, 275–288.

Yoshioka-Nishimura, M., Nanba, D., Takaki, T., Ohba, C., Tsumura, N., Morita, N., Sakamoto, H., Murata, K. and Yamamoto, Y. (2014). Quality control of photosystem II: direct imaging of the changes in the thylakoid structure and distribution of FtsH proteases in spinach chloroplasts under light stress, *Plant Cell Physiol.*, 55, 1255–1265.

Yoshioka-Nishimura, M. and Yamamoto, Y. (2014). Quality control of Photosystem II: the molecular basis for the action of FtsH protease and the dynamics of the thylakoid membranes, *J. Photochem. Photobiol. B*, 137, 100–106.

Chapter 11

Oxygenic Photosynthesis — Light Reactions within the Frame of Thylakoid Architecture and Evolution

Sari Järvi, Marjaana Rantala and Eva-Mari Aro*

Molecular Plant Biology, Department of Biochemistry, University of Turku, FI–20520 Turku, Finland
evaaro@utu.fi

Linear electron transfer chain in the thylakoid membrane of oxygenic photosynthetic organisms is rather similar from cyanobacteria to higher plants. On the contrary, the light harvesting systems and various regulation mechanisms of energy distribution and electron transfer routes show distinct evolution. Development of chlorophyll *b*-containing light harvesting systems in plants and complex regulatory networks of energy and electron transfer reactions led to the development of distinct lateral heterogeneity of the thylakoid membrane. Light-induced dynamics in lateral heterogeneity of higher plant thylakoid membrane allows fluent photosynthetic electron transfer and equal light harvesting capacity as well as efficient photoprotection of both photosystems in response to changes in the light environment. On the contrary, the fluency of electron flow in cyanobacteria thylakoid membrane is largely dependent upon a broad range of electron valves that have gradually disappeared during evolution of plant chloroplasts. After a great breakthrough in demonstration of the lateral heterogeneity of the thylakoid membrane in higher plant chloroplasts in 1980, our knowledge on light-induced dynamics of such a heterogeneity has slowly evolved in parallel with the development of isolation and characterization methods of thylakoid subdomains.

1. Introduction

Thylakoid membrane-embedded high molecular mass pigment-protein complexes photosystem I (PSI) and photosystem II (PSII) together with their light harvesting antenna systems collect the light energy and use it to drive the photosynthetic electron transfer reactions. In 1960, Hill and Bendall published their revolutionary concept of the Z-scheme demonstrating the serial trans-thylakoid electron transfer reactions, which constitute the linear electron transfer (LET) chain [Hill and Bendall, 1960]. Thereafter, our knowledge on the electron transfer reactions and their regulation has increased remarkably. The electrons originally derived from water to replace the electron hole in PSII can be directed to various routes in order to balance the function of the two photosystems in relation to each other and to eventually provide electrons to the carbon fixing and other reductive reactions. The diversity of regulation mechanisms of the photosynthetic light reactions show distinct evolution among oxygenic photosynthetic organisms, from cyanobacteria to various algal species and continuing in the land from mosses to lycophytes and gymnosperms, and eventually to the angiosperms.

2. Thylakoid Membrane Heterogeneity — From Cyanobacteria to Higher Plants

LET chain in the thylakoid membrane is composed of similar components in all oxygenic organisms; the water splitting PSII, the plastoquinone (PQ) pool, cytochrome (Cyt) b_6f complex, plastocyanin and PSI (Figure 1). Conversely, the light harvesting systems and regulation of light reactions show distinct changes during evolution of photosynthetic organisms. Cyanobacteria, the progenitors of plastids, typically use the soluble phycobilisomes (PBS) as light harvesting antennae of PSII, lack the thylakoid membrane stacking and show only modest lateral heterogeneity of the protein complexes [Sherman *et al.*, 1994]. Further, cyanobacteria mainly rely upon alternative electron transfer pathways and electron valves to protect the photosynthetic apparatus against damage upon abrupt, short-term alterations in the light environment. Among others, the flavodiiron (Flv) proteins are important players in the network of alternative electron transfer pathways of cyanobacteria. The photoprotective function of the *flv4-2 operon*, encoding the Flv2, Sll0218 and Flv4 proteins, is limited to β-cyanobacteria, whilst the Flv1 and Flv3 proteins function as an electron valve catalyzing the water–water cycle and protecting PSI under fluctuating light conditions in a much broader range of photosynthetic organisms [Allahverdiyeva *et al.*, 2013; Bersanini *et al.*, 2017; Ermakova *et al.*, 2014; Gerotto *et al.*, 2016] including also eukaryotes green algae, mosses, lycophytes and gymnosperms but, importantly, missing angiosperms.

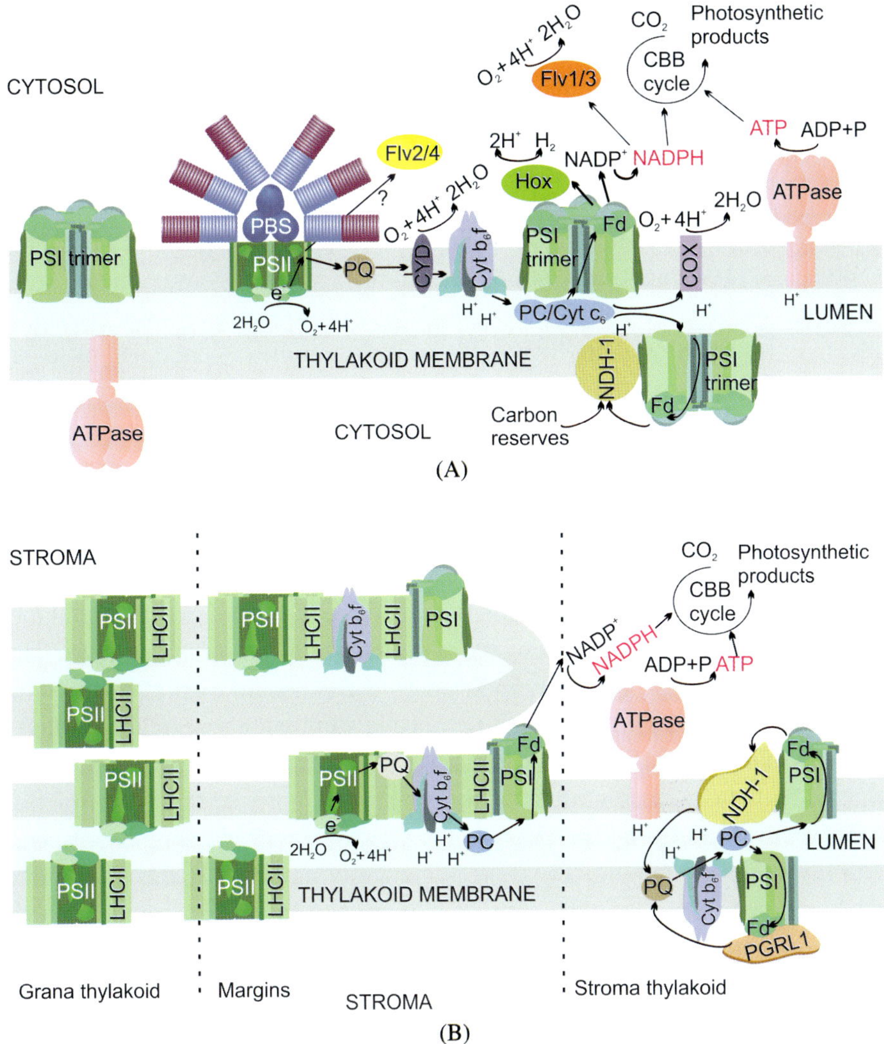

Figure 1. Schemes showing the thylakoid membrane organization, photosynthetic protein complexes, the light harvesting systems and electron transfer routes in cyanobacteria (A) and higher plants (B). While the connectivity between the photosystem PSII and PSI is minimal in cyanobacteria due to huge and soluble phycobilisome antenna (PBS), the PSII and PSI in the higher plant thylakoid membrane are in connection in the grana margin regions *via* the Chl *a/b* binding light-harvesting (LHC)II antenna. For details, see the text.

Evolution of the membrane-embedded chlorophyll (Chl) *a/b* binding light-harvesting (LHCII) antenna in green algae and land plants led to a replacement of the PBS antenna and consequently allowed the development of the lateral heterogeneity of the thylakoid membrane. The thylakoid membrane network in higher plant chloroplasts, and to a lesser extent also in green algae and mosses, is organized into appressed grana thylakoids and non-appressed stroma thylakoids. Specific thylakoid domains — the grana margins, grana end-membranes and stroma thylakoids — are exposed to the soluble chloroplast stroma where the carbon reduction reactions (CBB cycle) take place (Figure 1B). In 1980, Andersson and Anderson published their classical paper on the lateral heterogeneity of the thylakoid membrane in spinach chloroplasts and, by using mechanical thylakoid fractionation, demonstrated an extreme heterogeneity in the distribution of PSII and PSI in the thylakoid membrane [Andersson and Anderson, 1980]. Steric hindrance of the PSI complex, generated by a large domain protruding into the stroma, was demonstrated to prevent the location of the PSI complex in the tightly appressed grana thylakoids and PSI was shown to reside predominantly in the non-appressed thylakoid domains [Andersson and Anderson, 1980]. In turn, the most active form of PSII, the PSII-LHCII super-complexes are localized in appressed thylakoid membranes [Andersson and Anderson, 1980; Danielsson *et al.*, 2006].

Since 1980, our knowledge of the lateral heterogeneity of the thylakoid membrane has further increased. Discovery of a plethora of photosynthetic regulatory mechanisms, which are generally induced by the exposure of the photosynthetic organism to light and relaxed in darkness, have stimulated researchers to investigate their role in thylakoid heterogeneity. In-depth characterization of specific thylakoid regulatory protein mutants, together with improved thylakoid fractionation and analysis methods, has changed the view of the strict lateral heterogeneity of the thylakoid membrane. Today, it is generally agreed that the light-induced dynamics in the lateral heterogeneity of higher plant thylakoid membrane allows fluent photosynthetic electron transfer and equal light harvesting capacity as well as efficient photoprotection of both photosystems in response to changes in the light environment.

3. Key Regulatory Mechanisms of Thylakoid Electron Transfer Reactions

3.1. *No clear model organism for regulation of light reactions*

The concept of the light intensity- and quality-dependent dynamics in the lateral heterogeneity of the thylakoid membrane has developed concomitantly with accumulating evidence acquired from various mechanisms involved in regulation of photosynthetic light reactions, including the distribution of excitation energy to PSII and PSI and the various electron transfer routes in the thylakoid membrane.

In higher plants lacking the Flv proteins, a strict regulation of electron flow through the Cyt b_6f complex, which occupies a central position in the LET route (Figure 1), is of crucial importance [Joliot and Johnson, 2011; Suorsa *et al.*, 2012]. Likewise, the STN7 and STN8 kinase-dependent balancing of light energy distribution between PSII and PSI plays an important role in regulation of photosynthesis in response to changes in both the intensity and quality of light [Bellafiore *et al.*, 2005; Bonardi *et al.*, 2005; Tikkanen *et al.*, 2008b]. On the other hand, *Chlamydomonas reinhardtii* (Chlamydomonas) has demonstrated distinct variance from both the cyanobacteria and higher plant regulatory mechanisms of thylakoid electron transfer reactions, making green algae unsuitable as a model system for plant chloroplasts. As far as it comes to mechanisms adjusting the photosynthetic light reactions according to environmental cues, the green algae systems represent an intermediate form providing evidence for gradual evolution of the higher plant-type photosynthesis regulation in the thylakoid membrane, first in the water environment and continuing upon the movement of organisms from the water to the land. Upon evolution of the higher plant thylakoid network, a wide variety of regulation mechanisms has developed and many of those typical for lower oxygenic photosynthetic organisms have vanished, thus ruling out the possibility that any specific organism or group of organisms could serve as a representative model.

3.2. *Evolution of key regulatory mechanisms of thylakoid electron transfer reactions*

Our knowledge on thylakoid regulation mechanisms, and particularly on the molecular basis of such regulation, is currently only emerging. Generally, there appears to be three key issues to be considered when evaluating how to maximize the functionality of both photosystems and how to avoid unwanted damage of the photosynthetic apparatus upon changes in environmental conditions. These include (i) the maintenance of the balance between forward electron transfer and thermal dissipation of excess excitation energy, (ii) the maintenance of redox balance in the electron transfer chain and (iii) ensuring equal distribution of excitation energy to the two photosystems, PSII and PSI. Different evolutionary groups of oxygenic photosynthetic organisms have solved these regulation needs in distinct ways, which are shortly discussed below (i)–(iii).

(i) Excess excitation energy absorbed by the antenna systems, *i.e.* the energy that cannot be utilized for photochemistry, has to be safely dissipated by non-photochemial quenching (NPQ) mechanisms. In cyanobacteria, the quenching of excess excitation energy absorbed by the PBS antenna at high light intensities is mediated by a soluble carotenoid binding protein (orange carotenoid protein, OCP) that binds to the PBS antenna and triggers the dissipation of excess excitation energy [Wilson *et al.*, 2006]. In chloroplasts of plants and green algae, the thermal dissipation of excess excitation energy is to a large extent dependent on the protonation of the PsbS and LHCSR3 proteins, respectively, triggered by low lumenal pH upon illumination [Bonente *et al.*, 2011; Li *et al.*, 2000]. These thylakoid membrane proteins are likely responsible for conformational changes of the LHCII antenna proteins, which then alter the configuration of LHCII-bound pigments resulting in a safe energy dissipation as heat. In addition, violaxanthin de-epoxidase, a key enzyme of xanthophyll cycle is activated by low lumenal pH, and is required for full activation of thermal energy dissipation in chloroplasts [Niyogi *et al.*, 1997]. Despite a plethora of literature published during the past decades on various NPQ mechanisms, including the PSII photoinhibition-related energy quenching, it is likely that still uncharacterized quenching mechanisms remain to be discovered. Moreover, the exact molecular mechanisms and the full evolutionary diversity of the quenching mechanisms remains to be elucidated.

(ii) Alternative electron transfer pathways play a crucial role in relieving the excitation pressure in the electron transfer chain. In cyanobacteria, the functional roles of the *flv4-2* operon and the *flv1* and *flv3* genes were already mentioned above. In addition to the Flv proteins, the respiratory terminal oxidases Cyd and Cox in the thylakoid membrane of cyanobacteria are important in protection of PSII during abrupt changes in the light intensity [Ermakova *et al.*, 2016; Lea-Smith *et al.*, 2013]. Moreover, both NAD(P)H:quinone oxidoreductases, NDH-1 and NDH-2, are functional in cyanobacteria. Compared to cyanobacteria, green algae have less complex electron transfer pathways. For example, Cyd, Cox, the *flv4-2* operon proteins and the NDH-1 complex are completely missing from Chlamydomonas, which in turn has a Proton Gradient Regulation-Like 1 (PGRL1)-dependent PSI cyclic electron transfer (CET) [Dang *et al.*, 2014]. The Chlamydomonas genome also contains *flv1* and *flv3* homologues, the *flvA* and *flvB* genes that are highly expressed in the *pgrl1* mutant cells, providing evidence for the presence of a FLV-dependent photoprotection mechanism of PSI [Dang *et al.*, 2014]. Mosses were recently demonstrated to possess a FLV-dependent safety valve [Gerotto *et al.*, 2016], and the related genes also exist in the genomes of lycophytes and gymnosperms [Yamamoto *et al.*, 2016; Zhang *et al.*, 2009]. Flowering plants, instead, seem to completely lack the FLV-dependent electron transfer pathways and instead rely on a proton motive force-dependent down-regulation of LET, being strictly dependent on the Proton Gradient Regulation 5 (PGR5) protein [Suorsa *et al.*, 2012]. Such regulation of electron flow *via* the Cyt $b_6 f$ complex is crucial for protection of PSI against fluctuating light intensity-induced damage [Munekage *et al.*, 2002; Suorsa *et al.*, 2012; Tiwari *et al.*, 2016]. Further, higher plant chloroplasts contain a bacterial-like NDH-1 complex interacting with PSI and playing an important role in PSI-CET [Peltier *et al.*, 2016].

(iii) Mechanisms that allow the balanced distribution of excitation energy to PSII and PSI, including the "state transitions", play a central role in short-term light-acclimation processes of photosynthetic organisms. Since the demonstration of the concept of "state transitions" in 1969 [Bonaventura and Myers, 1969; Murata, 1969], our understanding on the complexity and physiological significance of these mechanisms has greatly advanced. The underlying molecular mechanism(s) of "state transitions" in cyanobacteria has remained fairly elusive. In Chlamydomonas and higher plants, reversible phosphorylation of the LHCII antenna plays an important role in "state transitions", yet several concomitant mechanisms are indispensable for balancing the distribution of excitation energy between PSII and PSI. The homologous serine/threonine-protein kinases responsible for phosphorylation of the LHCII proteins have been identified from Chlamydomonas (Stt7) [Depége *et al.*, 2003] and from *Arabidopsis thaliana* (Arabidopsis) (STN7) [Bellafiore *et al.*, 2005], the activity of the kinases being mainly regulated by changes in the redox state of the PQ pool. Binding of plastoquinol to the Qo site of the Cyt b_6f complex activates the STN7/Stt7 serine threonine kinases [Vener *et al.*, 1997]. Reduction of the lumenal Cys residues of STN7/Stt7 is likely to inhibit the kinase activity under high light intensity through the ferredoxin-thioredoxin system [Rintamäki *et al.*, 2000]. Dephosphorylation of LHCII proteins is dependent on the Thylakoid-Associated Phosphatase of 38 KDa/Protein Phosphatase 1 (TAP38/PPH1) [Pribil *et al.*, 2010; Shapiguzov *et al.*, 2010]. Importantly, not only the reversible phosphorylation of the LHCII proteins but also that of the PSII core proteins is essential for balancing the excitation energy between PSII and PSI [Mekala *et al.*, 2015; Rintamäki *et al.*, 1997]. A homologous kinase to STN7/Stt7, the STN8/Stl1, is responsible for phosphorylation of the PSII core proteins and the antagonist phosphatase, the Photosystem II Core Phosphatase (PBCP), has likewise been demonstrated [Bonardi *et al.*, 2005; Samol *et al.*, 2012].

Despite apparent similarity between the higher plant and green alga LHCII and PSII core protein phosphorylation, distinct differences also exist. It has been estimated that in Chlamydomonas up to 80% of the LHCII proteins are potentially phosphorylated giving rise not only to the control of energy balance between PSII and PSI, but also to a shift between the LET and PSI CET modes in order to balance the NADPH/ATP ratio according to environmental and cellular needs [Delosme *et al.*, 1996; Finazzi *et al.*, 1999, 2002]. PSI-CET specific megacomplex

PSI-LHCI-LHCII-Cyt$b_6$$f$-PGRL1 has been isolated from Chlamydomonas cells grown under PSII-favoring light, which induces the phosphorylation of both the LHCII and PSII core proteins [Iwai *et al.*, 2010]. Yet, contradictory results exist indicating that the PSI-CET in Chlamydomonas might be independent of LHCII phosphorylation and "state transitions" [Takahashi *et al.*, 2013]. In angiosperms like Arabidopsis, the STN7 kinase-mediated phosphorylation of LHCII leads to enhanced affinity of LHCII to PSI and consequently to an improved energy transfer from LHCII to PSI, which fully depends on the presence of the LHCII docking proteins PsaH and PsaL in PSI [Lunde *et al.*, 2000]. Intriguingly, phosphorylation of the LHCII proteins also regulates the formation of a labile PSI-PSII-LHCII megacomplex in non-appressed thylakoid membranes, boosting the energy distribution between PSII and PSI [Suorsa *et al.*, 2015], although direct interaction between PSI and PSII has also been documented [Yokono *et al.*, 2015]. Lack of LHCII phosphorylation or absence of LHCII docking proteins in PSI prevent the stable PSI-PSII-LHCII megacomplex formation in Arabidopsis [Leoni *et al.*, 2013; Suorsa *et al.*, 2015]. As mentioned above, in addition to LHCII phosphorylation, the balancing of excitation energy distribution between PSII and PSI in Arabidopsis has been demonstrated to also depend on the PSII core protein phosphorylation [Mekala *et al.*, 2015]. PSII core protein phosphorylation has a distinct role in regulation of thylakoid stacking (for details, see Section 4) and influences the physiological interactions between PSII, LHCII and PSI. Importantly, under natural fluctuations of the light intensity, the opposite phosphorylation of the LHCII antenna proteins and the PSII core proteins with respect to light intensity together ensure the optimal excitation energy distribution between PSII and PSI [Mekala *et al.*, 2015].

4. Factors Regulating the Thylakoid Architecture in Oxygenic Photosynthetic Organisms

The degree of thylakoid stacking is strongly dependent on growth light condition, yet also the light-induced short-term changes affect the thylakoid ultrastructure. Indeed, high growth light intensity reduces thylakoid stacking, while low growth light intensity enhances thylakoid stacking [Anderson and Aro, 1994]. Intriguingly, independent of growth light intensity, the phosphorylation of the LHCII and PSII core proteins are maintained at a similar, moderate level [Mekala *et al.*, 2015]. This is because the chloroplast acquires a redox balance when environmental conditions remain constant. Such redox-balance within the chloroplast provides an advantage to re-adjust the phosphorylation of LHCII and PSII core proteins whenever the light conditions suddenly change [Mekala *et al.*, 2015]. Strong phosphorylation of the PSII core proteins upon sudden increase in light intensity [Mekala *et al.*, 2015; Rintamäki *et al.*, 1997] has been suggested to increase the repulsion between thylakoid layers which, together with the osmotic swelling of thylakoid lumen caused by increased chloride and calcium influxes (counterbalanced by magnesium exflux), are likely to assist the destacking and lateral shrinkage of the grana stacks [Herbstová *et al.*, 2012; Kirchhoff, 2013]. In addition to the function of the proton and ion channels and the reversible thylakoid protein phosphorylation, also other post-translational protein modifications likely regulate the dynamic light-intensity-dependent alterations in thylakoid architecture.

So far, only a few proteins specifically determining the thylakoid architecture in higher plant chloroplasts have been identified. These include the oxygen evolving complex protein PsbP and the trimeric LHCII antenna proteins, which likely regulate the lumen diameter and stabilize the appressed structures of the thylakoid membrane, respectively [Wan *et al.*, 2014; Yi *et al.*, 2009]. Further, the lack of grana margin localized Curvature Thylakoid1 (CURT1) and Reduced Induction of Non-Photochemical Quencing (RIQ) 1 and 2 proteins have been shown to decrease and enhance the granal stacking, respectively [Armbruster *et al.*, 2013; Yokoyama *et al.*, 2016]. Intriguingly, neither green algae nor cyanobacterial genomes contain orthologs of the *RIQ* genes [Yokoyama *et al.*, 2016]. RIQ proteins, which regulate the granal structures likely through reorganization of the LHCII antenna, have been localized exclusively to the appressed thylakoid domains, providing evidence for co-evolution of the LHCII antenna, thylakoid stacking and RIQ proteins. In turn, the role of CURT1 is evolutionarily conserved and knocking down of the CURT1 homolog in a cyanobacterium *Synechocystis* sp. PCC6803 affects the thylakoid architecture by disrupting the formation of thylakoid biogenesis centers *e.g.* the functional sites of the PSII biogenesis [Heinz *et al.*, 2016].

5. Revealing the Dynamics of Thylakoid Architecture in Higher Plant Chloroplasts

5.1. *Methods to study thylakoid heterogeneity — Dark acclimated plants*

For detailed characterization of thylakoid heterogeneity, the thylakoid membrane has traditionally been fractionated into different subdomains by mechanical or chemical fractionation. Whenever the thylakoid membrane architecture is investigated, specific attention should be put on the fact that the thylakoid architecture, for instance the degree of grana stacking, is highly dependent upon various factors such as the concentrations of ions and the pH, and thus the fractionation protocol used should be carefully optimized. The strict lateral heterogeneity of the thylakoid membrane, published by Andersson and Anderson in 1980, was based on mechanical fractionation of thylakoids by Yeda press from dark-acclimated spinach leaves, followed by two-phase partitioning and differential centrifugation steps. Such a thylakoid fractionation protocol results in separation of the grana core, grana thylakoid membrane, margins, stroma thylakoid membrane and stroma core (known as Y100) fractions [Andersson and Anderson, 1980; Danielsson *et al.*, 2006]. The success of fractionation is generally estimated by the Chl *a/b* ratios of the obtained subfractions based on the fact that Chl *a* is bound to both photosystems, but Chl *b* is mostly bound to the LHCII antenna. In grana and stroma thylakoid membrane subfractions isolated by mechanical fractionation from spinach, the Chl *a/b* ratios are close to 2 and 6, respectively, indicating the enrichment of components of PSII-LHCII and PSI while in intact spinach thylakoids the Chl *a/b* ratio is around 3 [Tikkanen *et al.*, 2008b]. In the margin subfraction, the Chl *a/b* ratio is similar to that of the entire thylakoids, indicating the presence of both PSI and PSII complexes in the subdomain. Noteworthy, successful separation of the thylakoid membrane subdomains has been demonstrated only after acclimation of plants to darkness or low light [Tikkanen *et al.*, 2008b], while even short high light illumination leads to substantial loss of the yield of specific thylakoid membrane subfractions (for details, see Section 5.2).

Chemical approach for thylakoid membrane fractionation is based on the use of detergents. Mild, non-ionic detergents, such as dodecyl maltoside (DM), digitonin and Triton X-100 have uncharged, hydrophilic headgroups and have been used successfully for thylakoid fractionation, particularly to isolate native and enzymatically still active protein complexes from the lipid bilayer. Further, the design of a combination of detergents and differential centrifugation steps allows the separation and isolation of distinct thylakoid subfractions. β-DM solubilizes the entire thylakoid membrane network (both the appressed and non-appressed membrane domains) but concomitantly destroys the labile hydrophopic interactions between the protein complexes (for example, the interaction of LHCII antenna with PSI). Digitonin, on the contrary, preserves the weak protein–protein interactions, but has an access only to the non-appressed PSI-rich membrane domains, while the tightly appressed thylakoid domains are left insolubilized. Triton X-100 in high-salt buffer, in turn, can be used to isolate intact grana subfractions (BBY) by breaking-down the unstacked thylakoid regions.

It should be noted that, contrary to the mechanical fractionation of the thylakoid membrane, even mild detergents remove most of the lipids from the thylakoid membrane bilayer and cause a break-down of the labile protein–protein interactions thereby destroying the *in vivo* network of the photosynthetic protein complexes. The detergent treatments, especially β-DM and to a lesser extent digitonin, always result in a detachment of a fraction of LHCII trimers from the thylakoid membrane integrity. This is clearly shown by the 77 K fluorescence emission spectra demonstrating the uncoupled LHCII trimer as a strong fluorescence emission peak at 680 nm [Grieco *et al.*, 2015]. In sharp contrast to thylakoid fractionation with detergents, the mechanical fractionation of the thylakoid membrane never reveals the 680 nm band in 77 K fluorescence emission spectra of any of the obtained fractions [Tikkanen *et al.*, 2008b] indicating that the release of LHCII from thylakoid integrity is an artefact due to the use of detergents.

For reasons mentioned above, the detergent-free methods for isolation of membrane protein complexes have recently drawn a lot of interest. A styrene-maleic acid (SMA) copolymer treatment of membranes has been shown to be capable of isolating membrane fractions in the form of nanodiscs that contain both the membrane lipids and the protein complexes. SMA has been successfully used to isolate the PSI complexes together with up to five LHCII trimers from the appressed regions of spinach thylakoid membrane [Bell *et al.*, 2015]. The PSI-LHCII complexes enriched within these membranes are highly susceptible to detergents, and addition of 1% β-DM leads to uncoupling of the LHCII antenna from PSI, indicated by an appearance of a large fluorescence emission peak at 680 nm at 77 K.

Despite partial release of LHCII antenna from the thylakoid integrity, the detergent-based thylakoid fractionation is still a valid method and the success of fractionation is, as in the case of mechanical fractionation, generally followed by measuring the Chl *a/b* ratios of obtained thylakoid fractions. Typically, three distinct fractions are isolated by detergent treatments and differential centrifugations *i.e.* the grana, margin and stroma thylakoid membrane subfractions with Chl *a/b* ratios for Arabidopsis typically being 2.5, 3.9 and 6.2 and for pumpkin 2.3, 3.3 and 9.3, respectively, when thylakoids are isolated from dark-acclimated leaves (Table 1).

Table 1. Typically obtained Chl *a/b* ratios of thylakoid subfractions isolated from differentially light-treated Arabidopsis and pumpkin leaves. Arabidopsis and pumpkin thylakoids were isolated from dark-acclimated leaves (dark) or from leaves exposed to low light (LL) or high light (HL) for 2 hours. Prior to Chl *a/b* determinations, thylakoids were fractionated into grana, margin and stroma thylakoid membranes using digitonin as described in Wunder *et al.* [2013] (Arabidopsis) or digitonin together with Triton X-100 as described in Rokka *et al.* [2000] (pumpkin). $n = 3$. Noteworthy, the yield of different subfractions is highly dependent on the pre-illumination light intensity of plants.

Arabidopsis	Dark	LL	HL	Pumpkin	Dark	LL	HL
Intact thylakoid	3.1	3.1	3.0	Intact thylakoid	3.7	3.9	3.7
Grana	2.5	2.6	2.4	Grana	2.3	2.5	2.3
Margin	3.9	3.7	3.3	Margin	3.3	3.5	3.5
Stroma thylakoid	6.2	5.0	5.5	Stroma thylakoid	9.3	6.7	6.8

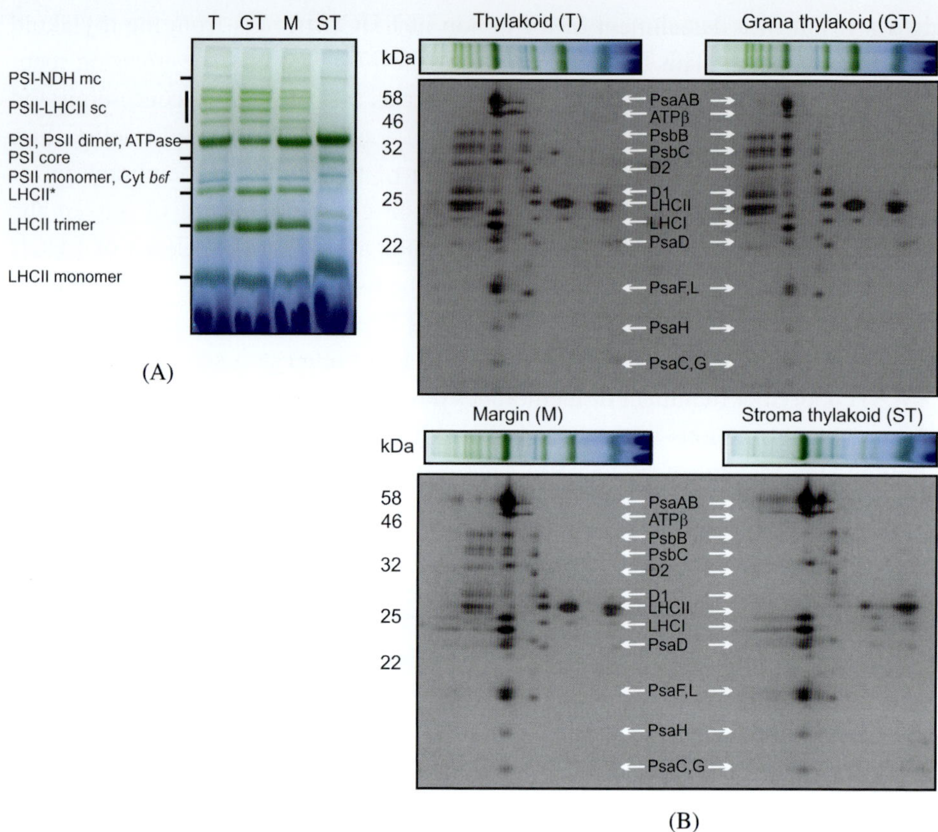

Figure 2. Large pore blue native (lpBN)-PAGE analysis of thylakoid subfractions solubilized with β-DM. Thylakoids were isolated from Arabidopsis leaves harvested in the end of the diurnal dark period. Intact thylakoid membrane (T) was fractionated into grana thylakoid (GT), margin (M) and stroma thylakoid (ST) subfractions with digitonin and differential centrifugation as described in Wunder *et al.* [2013]. Intact thylakoids and the thylakoid subfractions were solubilized with 1% β-DM before separation on lpBN-PAGE. (A) lpBN-PAGE. (B) Two-dimensional lpBN/SDS-PAGE stained with silver nitrate. Identification of the thylakoid membrane proteins is based on Aro *et al.* [2005].

A typical pattern of thylakoid protein complexes obtained from dark-acclimated Arabidopsis leaves (T) as well as from the digitonin-fractionated thylakoids yielding the appressed grana thylakoid (GT), margins (M) and non-appressed stroma thylakoid (ST) subfractions is demonstrated in Figure 2. Prior to the separation of the protein complexes by a large pore blue native (lpBN)-PAGE (Figure 2A), both the intact thylakoids and all thylakoid subfractions were solubilized with 1% β-DM. The subunits of the thylakoid protein complexes, separated by denaturing SDS-PAGE gel in the second dimension (Figure 2B), have been identified using protein specific antibodies and/or mass spectrometry [Aro *et al.*, 2005].

Intact thylakoid membrane isolated from dark-acclimated leaves and treated with 1% β-DM (Figure 2, lane T) reveals the following major photosynthetic protein complexes: the PSI-NDH-1 megacomplex, several PSII-LHCII supercomplexes with varying molecular mass, the co-migrating PSI complex, PSII dimer and ATP synthase, the PSI core complex, the co-migrating PSII monomer and Cyt b_6f complexes, the LHC* (also known as M-LHCII), the LHCII trimer and the LHCII monomer. The comparison of the protein complexes of intact thylakoids (T) and thylakoid subfractions (GT, M, ST) reveals thylakoid domain specific differences. In the grana thylakoid subfraction (Figure 2, lane GT) the amount of the PSII-LHCII supercomplexes is significantly higher while the amount of PSI and ATP synthase is significantly lower compared to the total thylakoids (T). The most prominent protein complexes in stroma thylakoid subfraction (Figure 2, lane ST), in turn, are the PSI-NDH-1 megacomplex, the PSI complex, ATP synthase, the PSI core complex (lacking the LHCI), PSII monomer, LHCII trimer and LHCII monomer, being in line with the classical thylakoid heterogeneity model of Andersson and Anderson [Andersson and Anderson, 1980]. The protein complex pattern of the margin subfraction (Figure 2, lane M) represents an intermediate of the grana and stroma thylakoids. Highly similar pattern of thylakoid protein complexes was obtained when thylakoids isolated from dark-acclimated leaves were fractionated mechanically prior to β-DM solubilization and BN/SDS-PAGE separation of the protein complexes [Danielsson *et al.*, 2006]

In order to gain information of more natural *in vivo* distribution of various protein complexes in distinct thylakoid subdomains, it is preferable to avoid extensive fractionation procedures and make use of the properties of mild detergents. 1% digitonin in 25BTH20G buffer solubilizes exclusively the stroma-exposed membrane domains during a five-minute incubation period and a centrifugation step to remove the detergent-inaccessible (insolubilized) thylakoid membrane fractions [Järvi *et al.*, 2011]. Figure 3 demonstrates the difference in the protein complexes solubilized from the thylakoid membrane by 1% β-DM and 1% digitonin. The lpBN/SDS-PAGE gel analysis reveals the presence of the subunits of all different protein complexes in 1% β-DM solubilized thylakoid membranes whereas the 1% digitonin solubilized complexes comprise mainly the subunits of PSI (*e.g.* PsaA/B), ATP synthase (*e.g.* ATPβ) and LHCII. Figure 3 also demonstrates that the tightly appressed thylakoid domain, which is left insolubilized by digitonin treatment, is mainly composed of PSII (*e.g.* PsbB) and LHCII, typical to the grana subdomain of the thylakoid membrane.

Taking together, Figures 2 and 3 clearly demonstrate that the thylakoid membrane solubilization with digitonin, which is capable of maintaining the weak protein–protein interactions, gives a unique pattern of protein complexes that differs from those obtained from the grana, margin or stroma thylakoid subfractions solubilized with β-DM prior to the native gel analysis. The key difference is that

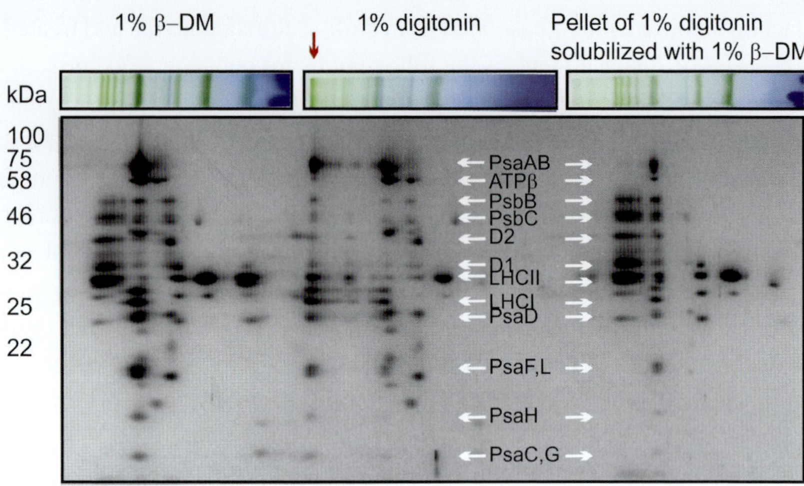

Figure 3. Large pore blue native (lpBN)/SDS-PAGE analysis of Arabidopsis thylakoid membrane protein complexes solubilized with β-DM and digitonin. Intact thylakoid membrane, isolated from dark-acclimated Arabidopsis leaves, was solubilized with 1% β-DM (solubilizing the entire thylakoid membrane) or 1% digitonin (solubilizing the non-appressed membranes) prior to lpBN-PAGE. The pellet from 1% digitonin sample was further solubilized with 1% β-DM. lpBN-PAGE lanes were subsequently submitted to separation of the individual subunits of the protein complexes in the second dimension by SDS-PAGE followed by silver staining. Identification of protein subunits is based on [Aro [*et al.*, 2005]. White arrows denote the major subunits of the photosynthetic protein complexes, red arrow indicates the PSI-PSII-LHCII-megacomplex.

the labile high molecular mass megacomplexes, of which the largest one represents the LHCII phosphorylation-dependent PSI-PSII-LHCII megacomplex (marked with red arrow in Figure 3), are enriched in the digitonin-solubilized thylakoid subfraction.

5.2. *Thylakoid fractionation from light-acclimated plants demonstrates light-dependent, reduced stringency of lateral heterogeneity*

It is important to improve the fractionation methods for light-acclimated plants in order to get better insights into the flexibility of the lateral heterogeneity and of the interactions between the thylakoid protein complexes upon changes in environmental conditions. Environmental changes, especially the light conditions, up- and down-regulate tens, even hundreds, of different regulatory mechanisms that, in concert, are responsible for fluent function of the photosynthetic light reactions (discussed in Section 4). It is particularly important to start understanding the relationships between the thylakoid architecture, protein complex interactions and the thylakoid regulatory mechanisms, the latter being the driving force for all changes in the thylakoid structure.

As already mentioned above (Section 5.1), contrary to the relatively easy fractionation of the thylakoid membrane from dark-acclimated leaves, by both the mechanical and detergent-based fractionation mechanisms, the fractionation of the thylakoid membrane from light-acclimated leaves always produces more randomized thylakoid fractions, and results in different yields of the subfractions, as compared to the fractionation of dark-acclimated leaves. This is due to light-induced regulatory mechanisms and subsequent re-organization occurring in the thylakoid architecture. Such regulation mechanisms induce the movement of thylakoid protein complexes laterally in the membrane, mixing the clear segregation of PSII, LHCII and PSI, typically present in dark-acclimated thylakoids. Enrichment of the "state transition"-related PSI-LHCII complex [Pesaresi *et al.*, 2009] is one example manifesting the low light-induced reorganization of the thylakoid membrane protein complexes (Figure 4A). Another low light-induced enhancement in protein-protein interactions is the accumulation of the PSI-PSII-LHCII megacomplex in grana margins as a result of differential thylakoid protein phosphorylation and accelerated need for maximal light harvesting to both photosystems [Suorsa *et al.*, 2015] (Figure 4A). Similarly, the light intensity-dependent PSII photodamage-repair cycle, evidenced by monomerization of the PSII-LHCII supercomplexes with increasing light intensity, is a typical example of the high light-induced consequences on thylakoid membrane architecture [Tikkanen *et al.*, 2008a] (Figure 4B).

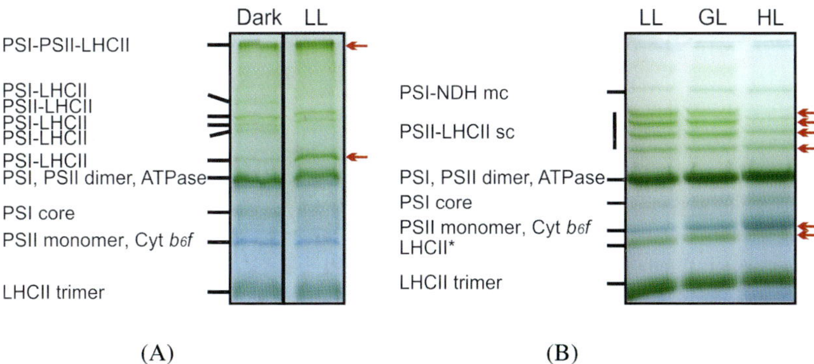

Figure 4. Large pore blue native (lpBN)-PAGE analysis of Arabidopsis thylakoid membrane protein complexes. Arabidopsis leaves were harvested in the end of the diurnal dark period and after 3 hours of exposure to low light (LL), growth light (GL) or high light (HL), and then immediately subjected to thylakoid isolation. Prior to gel electrophoresis, the thylakoid membrane was solubilized with (A) 1% digitonin and (B) 1% β-DM. Red arrows denote the light-induced dynamics at the level of thylakoid protein complexes.

Light-induced partial destacking of the grana and randomization of the protein complexes become evident also by monitoring the changes in the Chl *a/b* ratios of the margin and stroma thylakoid subfractions isolated from light-acclimated leaves and compared to those obtained from dark-acclimated leaves (Table 1). Spinach thylakoids isolated from dark-acclimated leaves and fractionated with the aqueous two-phase system typically give the Chl *a/b* ratios 2.2, 2.9 and 6.3 for the grana, margin and stroma thylakoid subfractions, respectively, while for low light-acclimated leaves the corresponding values were 2.3, 3.3 and 5.4 [Tikkanen *et al.*, 2008b]. Repeating lack of success in strict mechanical fractionation of the thylakoid membrane from either the growth- or high light-exposed spinach leaves, with respect to both the Chl *a/b* ratios and the yields of the subfractions, has to be interpreted as a result from light-induced modifications in the thylakoid membrane (unpublished results from Aro's lab). Similar problems are faced in detergent-based fractionation of the thylakoid membrane from light-acclimated leaves, *i.e.* the reorganizations of thylakoid protein complexes in light lead to a loss of strict thylakoid heterogeneity, as evidenced by light intensity-dependent changes in the yield of various subfractions and as a decrease in the Chl *a/b* ratio of the margin and stroma thylakoid subfractions isolated from Arabidopsis and pumpkin leaves (Table 1).

5.3. *Significance of thylakoid plasticity for photosynthetic light reactions and beyond*

Light-induced dynamic changes in thylakoid architecture are intimately linked to a high number of regulatory processes involved in the adjustment of the photosynthetic light reactions and carbon reduction reactions to rapid changes in environmental conditions, particularly to varying light intensities [Chow *et al.*, 2005]. Recently, there has been great advances in the high-resolution imaging techniques. However, as long as the imaging approaches of thylakoid membrane protein complexes, such as high-resolution cryo-electron tomography, do not provide sufficient resolution of the organization of the thylakoid pigment protein complexes, it is important to improve the fractionation methods for light-acclimated plants. It is highly conceivable that the development of the lateral heterogeneity of the thylakoid membrane is important not only for the performance of the photosynthetic processes in changing environments, but also reflects the wider signaling function of the photosynthetic apparatus in controlling the growth, development and stress acclimation of multicellular land plants [Gollan *et al.*, 2015; Kangasjärvi *et al.*, 2014; Piippo *et al.*, 2006]. To that end, it is highly likely that the mechanisms involved in regulation of thylakoid plasticity will remain in the focus of plant science also during the forthcoming years.

Acknowledgments

This publication is dedicated to the memory of our beloved mentor, Dr. Jan Anderson, and reflects the work done in Aro's laboratory as inspired by her dedicated photosynthesis research. We apologize to any authors whose work may not have been cited due to the length restrictions. We thank Virpi Paakkarinen for excellent technical assistance and the Academy of Finland for providing long-term financial support to Aro's lab (currently the projects 307335 and 303757).

References

Allahverdiyeva, Y., Mustila, H., Ermakova, M., Bersanini, L., Richaud, P., Ajlani, G., Battchikova, N., Cournac, L. and Aro, E.M. (2013). Flavodiiron proteins Flv1 and Flv3 enable cyanobacterial growth and photosynthesis under fluctuating light, *Proc. Natl. Acad. Sci. USA*, 110(10), 4111–4116.

Anderson, J.M. and Aro, E.M. (1994). Grana stacking and protection of photosystem II in thylakoid membranes of higher plant leaves under sustained high irradiance: an hypothesis, *Photosynth. Res.*, 41(2), 315–326.

Andersson, B. and Anderson, J.M. (1980). Lateral heterogeneity in the distribution of chlorophyll-protein complexes of the thylakoid membranes of spinach chloroplasts, *Biochim. Biophys. Acta*, 593(2), 427–440.

Armbruster, U., Labs, M., Pribil, M., Viola, S., Xu, W., Scharfenberg, M., Hertle, A.P., Rojahn, U., Jensen, P.E., Rappaport, F., Joliot, P., Dörmann, P., Wanner, G. and Leister, D. (2013). Arabidopsis CURVATURE THYLAKOID1 proteins modify thylakoid architecture by inducing membrane curvature, *Plant Cell*, 25(7), 2661–2678.

Aro, E.M., Suorsa, M., Rokka, A., Allahverdiyeva, Y., Paakkarinen, V., Saleem, A., Battchikova, N. and Rintamäki, E. (2005). Dynamics of photosystem II: a proteomic approach to thylakoid protein complexes, *J. Exp. Bot.*, 56(411), 347–356.

Bell, A.J., Frankel, L.K. and Bricker, T.M. (2015). High yield non-detergent isolation of photosystem I-light-harvesting chlorophyll II membranes from spinach thylakoids: implications for the organization of the PS I antennae in higher plants, *J. Biol. Chem.*, 290(30), 18429–18437.

Bellafiore, S., Barneche, F., Peltier, G. and Rochaix, J.D. (2005). State transitions and light adaptation require chloroplast thylakoid protein kinase STN7, *Nature*, 433(7028), 892–895.

Bersanini, L., Allahverdiyeva, Y., Battchikova, N., Heinz, S., Lespinasse, M., Ruohisto, E., Mustila, H., Nickelsen, J., Vass, I. and Aro, E.M. (2017). Dissecting the photoprotective mechanism encoded by the *flv4-2* operon: a distinct contribution of Sll0218 in photosystem II stabilization, *Plant Cell Environ.*, 40(3), 378–389.

Bonardi, V., Pesaresi, P., Becker, T., Schleiff, E., Wagner, R., Pfannschmidt, T., Jahns, P. and Leister, D. (2005). Photosystem II core phosphorylation and photosynthetic acclimation require two different protein kinases, *Nature*, 437(7062), 1179–1182.

Bonaventura, C. and Myers, J. (1969). Fluorescence and oxygen evolution from *Chlorella pyrenoidosa, Biochim. Biophys. Acta,* 189(3), 366–383.

Bonente, G., Ballottari, M., Truong, T.B., Morosinotto, T., Ahn, T.K., Fleming, G.R., Niyogi, K.K. and Bassi, R. (2011). Analysis of LhcSR3, a protein essential for feedback de-excitation in the green alga *Chlamydomonas reinhardtii, PLOS Biol.,* 9(1), p. e1000577.

Chow, W.S., Kim, E., Horton, P. and Anderson, J.M. (2005). Granal stacking of thylakoid membranes in higher plant chloroplasts: the physicochemical forces at work and the functional consequences that ensue, *Photochem. Photobiol. Sci.,* 4(12), 1081–1090.

Dang, K.V., Plet, J., Tolleter, D., Jokel, M., Cuiné, S., Carrier, P., Auroy, P., Richaud, P., Johnson, X., Alric, J., Allahverdiyeva, Y. and Peltier, G. (2014). Combined increases in mitochondrial cooperation and oxygen photoreduction compensate for deficiency in cyclic electron flow in *Chlamydomonas reinhardtii, Plant Cell,* 26(7), 3036–3050.

Danielsson, R., Suorsa, M., Paakkarinen, V., Albertsson, P.A., Styring, S., Aro, E.M. and Mamedov, F. (2006). Dimeric and monomeric organization of photosystem II. Distribution of five distinct complexes in the different domains of the thylakoid membrane, *J. Biol. Chem.,* 281(20), 14241–14249.

Delosme, R., Olive, J. and Wollman, F. (1996). Changes in light energy distribution upon state transitions: an *in vivo* photoacoustic study of the wild type and photosynthesis mutants from *Chlamydomonas reinhardtii, Biochim. Biophys. Acta,* 1273(2), 150–158.

Depége, N., Bellafiore, S. and Rochaix, J.D. (2003). Role of chloroplast protein kinase Stt7 in LHCII phosphorylation and state transition in *Chlamydomonas. Science,* 299(5612), 1572–1575.

Ermakova, M., Battchikova, N., Richaud, P., Leino, H., Kosourov, S., Isojärvi, J., Peltier, G., Flores, E., Cournac, L., Allahverdiyeva, Y. and Aro, E.M. (2014). Heterocyst-specific flavodiiron protein Flv3B enables oxic diazotrophic growth of the filamentous cyanobacterium *Anabaena* sp. PCC 7120, *Proc. Natl. Acad. Sci. USA,* 111(30), 11205–11210.

Ermakova, M., Huokko, T., Richaud, P., Bersanini, L., Howe, C.J., Lea-Smith, D.J., Peltier, G. and Allahverdiyeva, Y. (2016). Distinguishing the roles of thylakoid respiratory terminal oxidases in the cyanobacterium *Synechocystis* sp. PCC 6803, *Plant Physiol.,* 171(2), 1307–1319.

Finazzi, G., Furia, A., Barbagallo, R.P. and Forti, G. (1999). State transitions, cyclic and linear electron transport and photophosphorylation in *Chlamydomonas reinhardtii, Biochim. Biophys. Acta,* 1413(3), 117–129.

Finazzi, G., Rappaport, F., Furia, A., Fleischmann, M., Rochaix, J.D., Zito, F. and Forti, G. (2002). Involvement of state transitions in the switch between linear and cyclic electron flow in *Chlamydomonas reinhardtii, EMBO Rep.,* 3(3), 280–285.

Gerotto, C., Alboresi, A., Meneghesso, A., Jokel, M., Suorsa, M., Aro, E.M. and Morosinotto, T. (2016). Flavodiiron proteins act as safety valve for electrons in *Physcomitrella patens, Proc. Natl. Acad. Sci. USA,* 113(43), 12322–12327.

Gollan, P.J., Tikkanen, M. and Aro, E.M. (2015). Photosynthetic light reactions: integral to chloroplast retrograde signalling, *Curr. Opin. Plant Biol.,* 27, 180–191.

Grieco, M., Suorsa, M., Jajoo, A., Tikkanen, M. and Aro, E. (2015). Light-harvesting II antenna trimers connect energetically the entire photosynthetic machinery — including both photosystems II and I, *Biochim. Biophys. Acta*, 1847(6–7), 607–619.

Heinz, S., Rast, A., Shao, L., Gutu, A., Gügel, I.L., Heyno, E., Labs, M., Rengstl, B., Viola, S., Nowaczyk, M.M., Leister, D. and Nickelsen, J. (2016). Thylakoid membrane architecture in *Synechocystis* depends on CurT, a homolog of the granal CURVATURE THYLAKOID1 proteins, *Plant Cell*, 28, 2238–2260.

Herbstová, M., Tietz, S., Kinzel, C., Turkina, M.V. and Kirchhoff, H. (2012). Architectural switch in plant photosynthetic membranes induced by light stress, *Proc. Natl. Acad. Sci. USA*, 109(49), 20130–20135.

Hill, R. and Bendall, F. (1960). Function of the two cytochrome components in chloroplasts: a working hypothesis, *Nature*, 186, 136–137.

Iwai, M., Takizawa, K., Tokutsu, R., Okamuro, A., Takahashi, Y. and Minagawa, J. (2010). Isolation of the elusive supercomplex that drives cyclic electron flow in photosynthesis, *Nature*, 464(7292), 1210–1213.

Järvi, S., Suorsa, M., Paakkarinen, V. and Aro, E.M. (2011). Optimized native gel systems for separation of thylakoid protein complexes: novel super- and mega-complexes, *Biochem. J.*, 439(2), 207–214.

Joliot, P. and Johnson, G.N. (2011). Regulation of cyclic and linear electron flow in higher plants, *Proc. Natl. Acad. Sci. USA*, 108(32), 13317–13322.

Kangasjärvi, S., Tikkanen, M., Durian, G. and Aro, E.M. (2014). Photosynthetic light reactions — an adjustable hub in basic production and plant immunity signaling, *Plant Physiol. Biochem.*, 81, 128–134.

Kirchhoff, H. (2013). Architectural switches in plant thylakoid membranes, *Photosynth. Res.*, 116(2–3), 481–487.

Lea-Smith, D.J., Ross, N., Zori, M., Bendall, D.S., Dennis, J.S., Scott, S.A., Smith, A.G. and Howe, C.J. (2013). Thylakoid terminal oxidases are essential for the cyanobacterium *Synechocystis* sp. PCC 6803 to survive rapidly changing light intensities, *Plant Physiol.*, 162(1), 484–495.

Leoni, C., Pietrzykowska, M., Kiss, A.Z., Suorsa, M., Ceci, L.R., Aro, E.M. and Jansson, S. (2013). Very rapid phosphorylation kinetics suggest a unique role for Lhcb2 during state transitions in *Arabidopsis*, *Plant J.*, 76(2), 236–246.

Li, X.P., Björkman, O., Shih, C., Grossman, A.R., Rosenquist, M., Jansson, S. and Niyogi, K.K. (2000). A pigment-binding protein essential for regulation of photosynthetic light harvesting, *Nature*, 403(6768), 391–395.

Lunde, C., Jensen, P.E., Haldrup, A., Knoetzel, J. and Scheller, H.V. (2000). The PSI-H subunit of photosystem I is essential for state transitions in plant photosynthesis, *Nature*, 408(6812), 613–615.

Mekala, N.R., Suorsa, M., Rantala, M., Aro, E.M. and Tikkanen, M. (2015). Plants actively avoid state transitions upon changes in light intensity: role of light-harvesting complex II protein dephosphorylation in high light, *Plant Physiol.*, 168(2), 721–734.

Munekage, Y., Hojo, M., Meurer, J., Endo, T., Tasaka, M. and Shikanai, T. (2002). PGR5 is involved in cyclic electron flow around photosystem I and is essential for photoprotection in *Arabidopsis*, *Cell*, 110(3), 361–371.

Murata, N. (1969). Control of excitation transfer in photosynthesis I. Light-induced change of chlorophyll a fluorescence in *Porphyridium cruentum, Biochim. Biophys. Acta*, 172(2), 242–251.

Niyogi, K.K., Björkman, O. and Grossman, A.R. (1997). The roles of specific xanthophylls in photoprotection, *Proc. Natl. Acad. Sci. USA*, 94(25), 14162–14167.

Peltier, G., Aro, E.M. and Shikanai, T. (2016). NDH-1 and NDH-2 plastoquinone reductases in oxygenic photosynthesis, *Annu. Rev. Plant. Biol.*, 67, 55–80.

Pesaresi, P., Hertle, A., Pribil, M., Kleine, T., Wagner, R., Strissel, H., Ihnatowicz, A., Bonardi, V., Scharfenberg, M., Schneider, A., Pfannschmidt, T. and Leister, D. (2009). Arabidopsis STN7 kinase provides a link between short- and long-term photosynthetic acclimation, *Plant Cell*, 21(8), 2402–2423.

Piippo, M., Allahverdiyeva, Y., Paakkarinen, V., Suoranta, U.M., Battchikova, N. and Aro, E.M. (2006). Chloroplast-mediated regulation of nuclear genes in *Arabidopsis thaliana* in the absence of light stress, *Physiol. Genomics*, 25(1), 142–152.

Pribil, M., Pesaresi, P., Hertle, A., Barbato, R. and Leister, D. (2010). Role of plastid protein phosphatase TAP38 in LHCII dephosphorylation and thylakoid electron flow, *PLOS Biol.*, 8(1), p. e1000288.

Rintamäki, E., Martinsuo, P., Pursiheimo, S. and Aro, E.M. (2000). Cooperative regulation of light-harvesting complex II phosphorylation *via* the plastoquinol and ferredoxin-thioredoxin system in chloroplasts, *Proc. Natl. Acad. Sci. USA*, 97(21), 11644–11649.

Rintamäki, E., Salonen, M., Suoranta, U.M., Carlberg, I., Andersson, B. and Aro, E.M. (1997). Phosphorylation of light-harvesting complex II and photosystem II core proteins shows different irradiance-dependent regulation *in vivo* — application of phosphothreonine antibodies to analysis of thylakoid phosphoproteins, *J. Biol. Chem.*, 272(48), 30476–30482.

Rokka, A., Aro, E.M., Herrmann, R.G., Andersson, B. and Vener, A.V. (2000). Dephosphorylation of photosystem II reaction center proteins in plant photosynthetic membranes as an immediate response to abrupt elevation of temperature, *Plant Physiol.*, 123(4), 1525–1536.

Samol, I., Shapiguzov, A., Ingelsson, B., Fucile, G., Crevécoeur, M., Vener, A.V., Rochaix, J.D. and Goldschmidt-Clermont, M. (2012). Identification of a photosystem II phosphatase involved in light acclimation in *Arabidopsis, Plant Cell*, 24(6), 2596–2609.

Shapiguzov, A., Ingelsson, B., Samol, I., Andres, C., Kessler, F., Rochaix, J.D., Vener, A.V. and Goldschmidt-Clermont, M. (2010). The PPH1 phosphatase is specifically involved in LHCII dephosphorylation and state transitions in *Arabidopsis, Proc. Natl. Acad. Sci. USA,* 107(10), 4782–4787.

Sherman, D.M., Troyan, T.A. and Sherman, L.A. (1994). Localization of membrane proteins in the cyanobacterium Synechococcus sp. PCC7942 (radial asymmetry in the photosynthetic complexes), *Plant Physiol.*, 106(1), 251–262.

Suorsa, M., Järvi, S., Grieco, M., Nurmi, M., Pietrzykowska, M., Rantala, M., Kangasjärvi, S., Paakkarinen, V., Tikkanen, M., Jansson, S. and Aro, E.M. (2012). PROTON GRADIENT REGULATION5 is essential for proper acclimation of *Arabidopsis* photosystem I to naturally and artificially fluctuating light conditions, *Plant Cell,* 24(7), 2934–2948.

Suorsa, M., Rantala, M., Mamedov, F., Lespinasse, M., Trotta, A., Grieco, M., Vuorio, E., Tikkanen, M., Järvi, S. and Aro, E.M. (2015). Light acclimation involves dynamic re-organization of the pigment-protein megacomplexes in non-appressed thylakoid domains, *Plant J.,* 84(2), 360–373.

Takahashi, H., Clowez, S., Wollman, F.A., Vallon, O. and Rappaport, F. (2013). Cyclic electron flow is redox-controlled but independent of state transition, *Nat. Commun.,* 4, p. 1954.

Tikkanen, M., Nurmi, M., Kangasjärvi, S. and Aro, E.M. (2008a). Core protein phosphorylation facilitates the repair of photodamaged photosystem II at high light, *Biochim. Biophys. Acta,* 1777(11), 1432–1437.

Tikkanen, M., Nurmi, M., Suorsa, M., Danielsson, R., Mamedov, F., Styring, S. and Aro, E.M. (2008b). Phosphorylation-dependent regulation of excitation energy distribution between the two photosystems in higher plants, *Biochim. Biophys. Acta,* 1777(5), 425–432.

Tiwari, A., Mamedov, F., Grieco, M., Suorsa, M., Jajoo, A., Styring, S., Tikkanen, M. and Aro, E.M. (2016). Photodamage of iron-sulphur clusters in photosystem I induces non-photochemical energy dissipation, *Nat. Plants,* 2, p. 16035.

Vener, A.V., van Kan, P.J., Rich, P.R., Ohad, I. and Andersson, B. (1997). Plastoquinol at the quinol oxidation site of reduced cytochrome *bf* mediates signal transduction between light and protein phosphorylation: thylakoid protein kinase deactivation by a single-turnover flash, *Proc. Natl. Acad. Sci. USA,* 94(4), 1585–1590.

Wan, T., Li, M., Zhao, X., Zhang, J., Liu, Z. and Chang, W. (2014). Crystal structure of a multilayer packed major light-harvesting complex: implications for grana stacking in higher plants, *Mol. Plant,* 7(5), 916–919.

Wilson, A., Ajlani, G., Verbavatz, J.M., Vass, I., Kerfeld, C.A. and Kirilovsky, D. (2006). A soluble carotenoid protein involved in phycobilisome-related energy dissipation in cyanobacteria, *Plant Cell,* 18(4), 992–1007.

Wunder, T., Xu, W., Liu, Q., Wanner, G., Leister, D. and Pribil, M. (2013). The major thylakoid protein kinases STN7 and STN8 revisited: effects of altered STN8 levels and regulatory specificities of the STN kinases, *Front. Plant Sci.,* 4, p. 417.

Yamamoto, H., Takahashi, S., Badger, M.R. and Shikanai, T. (2016). Artificial remodelling of alternative electron flow by flavodiiron proteins in *Arabidopsis, Nat. Plants,* 2, p. 16012.

Yi, X., Hargett, S.R., Frankel, L.K. and Bricker, T.M. (2009). The PsbP protein, but not the PsbQ protein, is required for normal thylakoid architecture in *Arabidopsis thaliana, FEBS Lett.,* 583(12), 2142–2147.

Yokono, M., Takabayashi, A., Akimoto, S. and Tanaka, A. (2015). A megacomplex composed of both photosystem reaction centres in higher plants, *Nat. Commun.,* 6, p. 6675.

Yokoyama, R., Yamamoto, H., Kondo, M., Takeda, S., Ifuku, K., Fukao, Y., Kamei, Y., Nishimura, M. and Shikanai, T. (2016). Grana-localized proteins, RIQ1 and RIQ2, affect the organization of light-harvesting complex II and grana stacking in *Arabidopsis, Plant Cell,* 28(9), 2261–2275.

Zhang, P., Allahverdiyeva, Y., Eisenhut, M. and Aro, E.M. (2009). Flavodiiron proteins in oxygenic photosynthetic organisms: photoprotection of photosystem II by Flv2 and Flv4 in *Synechocystis* sp. PCC 6803, *PLOS ONE,* 4(4), p. e5331.

Chapter 12

Estimation of the Cyclic Electron Flux around Photosystem I in Leaf Discs

Jiancun Kou[*,†], Duncan Fitzpatrick[†], Da-Yong Fan[†],
Shunichi Takahashi[†], Riichi Oguchi[†] and Wah Soon Chow[†,‡]

*College of Animal Science and Technology,
Northwest A&F University,
Yangling 712100, China
†Research School of Biology,
The Australian National University,
Acton, ACT 2601, Australia
‡fred.chow@anu.edu.au

Quantification of the cyclic electron flux (*CEF*) around photosystem I (PSI) in leaves has been hampered by the absence of net product formation or net substrate consumption. We estimate *CEF* as the difference between the total electron flux through PSI and the linear electron flux through both photosystems, both fluxes measured in the same leaf tissue in identical conditions. This method can be applied to leaves of plants such as spinach grown in high light and in which the Mehler reaction and charge recombination are minimal. However, leaves of plants such as *Arabidopsis* grown in low light but tested in high light may have a large electron flux component that cycles within the PSI reaction centre by charge separation followed by charge recombination. In such leaves, *CEF* can be estimated as the antimycin A-sensitive component of the total electron flux through PSI. However, if *CEF* is significantly mediated by the nicotinamide adenine dinucleotide dehydrogenase-like complex (NDH), inhibition with a combination of antimycin A and an inhibitor of NDH would be necessary. The method has the potential to be fine-tuned to give a closer estimation of the actual *CEF*.

1. Introduction

ATP synthesis driven by cyclic electron flow around PSI was discovered more than 60 years ago [Arnon *et al.*, 1955], but quantification of the *CEF* is difficult even today. This problem arises because there is neither net formation of a product nor net consumption of a substrate that can be measured. Yet quantification of *CEF* is imperative, given the roles that *CEF* plays in (1) enhancing the supply of ATP, the energy carrier which often limits photosynthesis [Shen, 1990] and (2) photoprotection of the photosynthetic apparatus [Munekage *et al.*, 2002; Takahashi *et al.*, 2009; Huang *et al.*, 2011; Kukuczka *et al.*, 2014].

Traditionally, an estimate of *CEF* is made by first establishing a steady state during continuous illumination. Then on cessation of illumination the initial rate of re-oxidation of a component such as ferredoxin in anaerobic conditions and in the presence of 3-(3,4-dichlorophenyl)-1,1-dimethylurea (DCMU), [Cleland and Bendall, 1992] or of re-reduction of the oxidized form of the primary electron donor in PSI (P700$^+$) [*e.g.* Maxwell and Biggins, 1976] is assumed to equal the electron flux through the component *during* steady-state illumination. This initial redox rate includes any linear electron flux through the measured redox component, and can be equated with *CEF* only in the absence of a linear electron flux. This means that the post-illumination method is frequently used with far-red actinic light which only minimally excites photosystem II (PSII) to initiate the linear electron flux, or in the presence of the PSII inhibitor DCMU. Such a restriction does not allow evaluation of *CEF* under conditions that are physiologically relevant. A further limitation is that the initial post-illumination electron flux underestimates the actual electron flux during illumination if, for example, a proton diffusion potential develops after cessation of illumination, thereby impairing the docking of plastocyanin (Pc) with the donor side of PSI [Fan *et al.*, 2016]. The underestimation can be serious, by a factor of about 3, making the initial post-illumination electron flux through PSI impossibly smaller than the linear electron flux [Fan *et al.*, 2016].

Our recent approach has been to estimate *CEF* by the difference ($\Delta Flux$) between the total electron flux through PSI (*ETR1*) determined spectroscopically, and the linear electron flux through both photosystems given by the gross oxygen evolution rate × 4 (LEF_{O2}), measured simultaneously or consecutively, in 1% CO_2. *ETR1* is measured as $Y(I) \times I \times 0.85 \times f_I$, where $Y(I)$ is the photochemical yield of PSI averaged over all PSI reaction centres, I the irradiance, 0.85 the leaf absorptance which is relatively insensitive to chlorophyll contents > 400 μmol m^{-2} [Evans and Poorter, 2001], and f_I the fraction of the absorbed light that is partitioned to PSI. $Y(I)$ is measured under steady-state actinic illumination from the P700$^+$ signal by superimposing a strong far-red light pulse on the actinic light prior to the application of a saturating pulse [Kou *et al.*, 2013]. The calculation further assumes that the quantum yield of charge separation in open PSI reaction traps is 1.0. However, to calculate *ETR1* as $Y(I) \times I \times 0.85 \times f_I$, the fraction f_I first needs to be determined. Thus we set out to determine the partitioning of absorbed light between the two photosystems.

2. Partitioning of Absorbed Light Between the Two Photosystems

For the determination of f_I, we choose conditions in which *CEF* is inhibited by infiltration of leaf segments with 200 μM antimycin A and in which the irradiance is not so excessive that charge recombination, which arises after electrons have been transported through PSI, starts to contaminate the measurement of *ETR1*. In such conditions, we assume that $ETR1 = LEF_{O2} = Y(I) \times I \times 0.85 \times f_I$, from which f_I can be obtained. Figure 1A shows that, in glasshouse-grown spinach, *ETR1* is close to 0.5 for irradiances up to half of full sunlight. At higher irradiances,

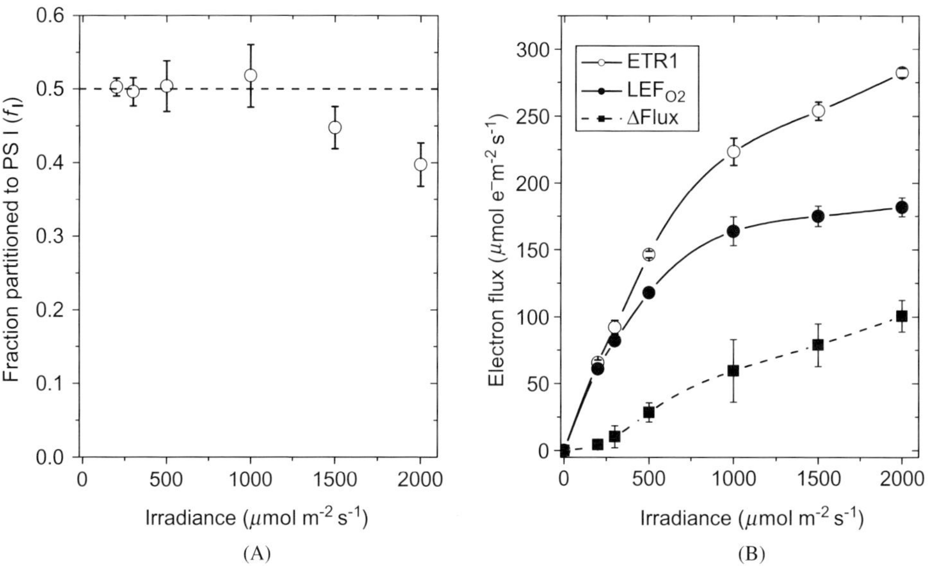

Figure 1. (A) The fraction of absorbed light partitioned to PS I (f_I). Spinach leaf discs were infiltrated with 200 μM antimycin A to inhibit PGR5-mediated *CEF*, allowing the approximation that $LEF_{O2} \approx Y(I) \times I \times 0.85 \times f_I$, an approximation that does not seem to hold at high irradiance. (B) The total electron flux through PSI (*ETR1*), the linear electron flux (*LEF$_{O2}$*) measured with membrane inlet mass spectrometry, as gross O_2 evolution rate $\times$ 4, and their difference ($\Delta Flux$) in spinach leaf discs, plotted as a function of irradiance of white light from a halogen lamp. Plants were grown in a glasshouse. Replotted from Fan *et al.* [2016].

$Y(I) \times I \times 0.85 \times f_I$ no longer equals LEF_{O2}. As charge recombination in PSI sets in, the local electron cycle adds to $ETR1$, thereby increasing it without affecting $ETR2$; the increased value of $ETR1$ compared to LEF_{O2} means that f_I falls when we equate $Y(I) \times I \times 0.85 \times f_I$ with LEF_{O2}.

For the value of f_I in steady-state photosynthesis, we assume 0.5 throughout the irradiance range up to 2,000 μmol photons m^{-2} s^{-1} for glasshouse-grown spinach.

In *Arabidopsis* grown in low light (100 μmol photons m^{-2} s^{-1}), the value of f_I seems to be smaller, about 0.4 [Kou *et al.*, 2015]. It seems that f_I is not always 0.5, making it necessary to determine the value for the leaves under study.

A way of thinking about the partitioning of absorbed light between the two photosystems is as follows. Of the absorbed irradiance, the fraction partitioned to PSI is f_I. The irradiance of the light partitioned to PSI, ($I \times 0.85 \times f_I$), consists of three components: the fraction $Y(I)$ which drives photochemical conversion, the fraction $Y(ND)$ which is dissipated non-photochemically due to limitation on the donor side (because P700 in some PSI complexes is already oxidized) and the fraction $Y(NA)$ which is dissipated non-photochemically due to limitation on the acceptor side (because electron acceptors in some PSI complexes are already reduced). The measurement is described by Klughammer and Schreiber [2008a]. The only modification that we have adopted is that we introduce a pulse of strong far-red light immediately prior to the saturating pulse; without it, $Y(I)$ may be underestimated by about 18% [Siebke *et al.*, 1997].

The fraction of absorbed light partitioned to PSII is $(1 - f_I)$. The irradiance of the light partitioned to PSII, $[I \times 0.85 \times (1 - f_I)]$, also consists of three components: the fraction ϕ_{PSII} which drives photochemical conversion in PSII, the fraction ϕ_{NPQ} which dissipates the energy non-photochemically in a light-dependent manner, and the fraction $\phi_{f,D}$ which dissipates the energy non-photochemically in a relatively light-independent manner and also in the form of chlorophyll fluorescence [Hendrickson *et al.*, 2004; Kramer *et al.*, 2004; Klughammer and Schreiber, 2008b].

3. Cyclic Electron Flux in Spinach Leaves

With the value of f_I determined, one is in a position to calculate $ETR1$ from $Y(I) \times I \times 0.85 \times f_I$. Figure 1B shows $ETR1$ in spinach leaf segments increasing with irradiance (open circles). It also depicts LEF_{O2} increasing with irradiance and then saturating at high irradiance (solid circles). The difference between them ($\Delta Flux$, solid squares) is very low at irradiance < 300 μmol m^{-2} s^{-1}, increasing steadily at higher irradiance. The measurements of LEF_{O2} in Figure 1B were made in CO_2-enriched air, using membrane-inlet mass spectrometry to obtain the gross rate of oxygen evolution through stable isotope discrimination of oxygen producing and consuming reactions [Fan *et al.*, 2016]. Earlier measurements of LEF_{O2}, made with a gas-phase O_2 electrode, yielded similar results [Kou *et al.*, 2013]. $\Delta Flux$ is mostly *CEF* in spinach leaves, since it is almost completely inhibited by antimycin A, accompanied by a decrease in ϕ_{NPQ} as the contribution to the transmembrane pH gradient from *CEF* is abolished [Kou *et al.*, 2013].

At a chosen irradiance (980 μmol m^{-2} s^{-1}), $\Delta Flux$ increases with temperature, and with the loss of leaf water content [Kou *et al.*, 2013].

4. Cyclic Electron Flux in *Arabidopsis* Leaves

Arabidopsis plants were grown in relatively low light (100 μmol m^{-2} s^{-1}), and tested at irradiances up to 11 fold higher.

4.1. *CEF in wild type Arabidopsis*

Figure 2A shows the increase in *ETR1* with irradiance (open circles), rapid at first and then more gradual later. LEF_{O2} was near saturation at an irradiance of 200–300 μmol m^{-2} s^{-1}, about 2–3 times the growth irradiance. $\Delta Flux$ was relatively significant at growth or near growth irradiance, and continued to increase at higher irradiance.

Antimycin A had no significant effect on LEF_{O2} in *Arabidopsis* (Figure 2B), as is also the case in spinach [Kou *et al.*, 2013]. However, it abolished the $\Delta Flux$ at low irradiance < 200 μmol m^{-2} s^{-1}. $\Delta Flux$ increased steadily above 300 μmol m^{-2} s^{-1}. This residual $\Delta Flux$ is likely to be unrelated to *CEF* mediated by the Proton Gradient Regulation 5 protein, PGR5, which is the main pathway of

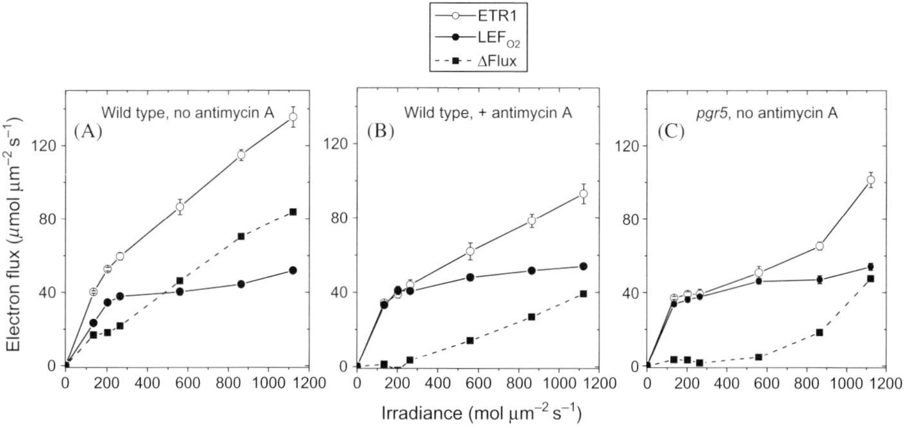

Figure 2. *ETR1*, LEF_{O2} and $\Delta Flux$, as a function of irradiance of white light from a halogen lamp given to *Arabidopsis* leaf discs of the wild type in the absence of antimycin A (A), wild type in the presence of antimycin A (B), and the *pgr5* mutant in the absence of antimycin A (C). Plants were grown in low light. Replotted from Kou *et al.* [2015].

cyclic electron flow in C3 leaves [Munekage *et al.*, 2002] and which should have been inhibited by antimycin A. Instead, it is probably due to a combination of (1) charge recombination, (2) *CEF* mediated by the nicotinamide adenine dinucleotide dehydrogenase-like complex (NDH) and, to a small extent, (3) the Mehler reaction. The Mehler reaction is pseudo-cyclic electron flow which adds to *ETR1*, but which, in the complete water–water cycle, does not affect the measurement of oxygen evolution in a simple gas-phase O_2 electrode [Kou *et al.*, 2013]; therefore, $\Delta Flux$ includes any Mehler reaction-mediated electron transport that is present. On the other hand, when using membrane inlet mass spectrometry to measure oxygen fluxes, it is possible to determine the magnitude of the Mehler reaction [Ruuska *et al.*, 2000] and subtract it from *ETR1*; the $\Delta Flux$ so obtained will then not include any Mehler reaction.

4.2. *CEF in the pgr5 mutant of Arabidopsis*

The cyclic electron flux mediated by PGR5 is the main pathway of cyclic electron flow in C3 leaves [Munekage *et al.*, 2002]. In the *pgr5* mutant of *Arabidopsis*, there is little or no $\Delta Flux$ at irradiance < 400 μmol m^{-2} s^{-1} (Figure 2C). This suggests that the PGR5-mediated pathway is practically lost in the *pgr5* mutant, as expected from the absence of PGR5. At an irradiance that is several times the growth irradiance, $\Delta Flux$ appeared and it increased with irradiance, similar to the case of the wild type treated with antimycin A.

4.3. *CEF in the ndh mutant of Arabidopsis*

In leaves there is a minor pathway of cyclic electron flow mediated by NDH. In the *ndh* mutant untreated with antimycin A, the PGR5-dependent pathway should still operate, yet $\Delta Flux$ appears quite small below 200 μmol m^{-2} s^{-1} (Figure 3A). It is only at high irradiance that $\Delta Flux$ markedly increased.

In the presence of antimycin A, the small $\Delta Flux$ is further decreased to practically zero at low irradiance <150 μmol m^{-2} s^{-1}. At higher irradiance, $\Delta Flux$ increased, but not by as much as in the absence of antimycin A (Figure 3B).

4.4. *The antimycin A-sensitive component of ETR1 in Arabidopsis*

ETR1 includes any charge-recombination and Mehler reaction that may occur, which is the reason for the existence of a residual electron flux when *CEF* is inhibited by antimycin A, even in the *ndh* mutant. Assuming that (1) the electron flux

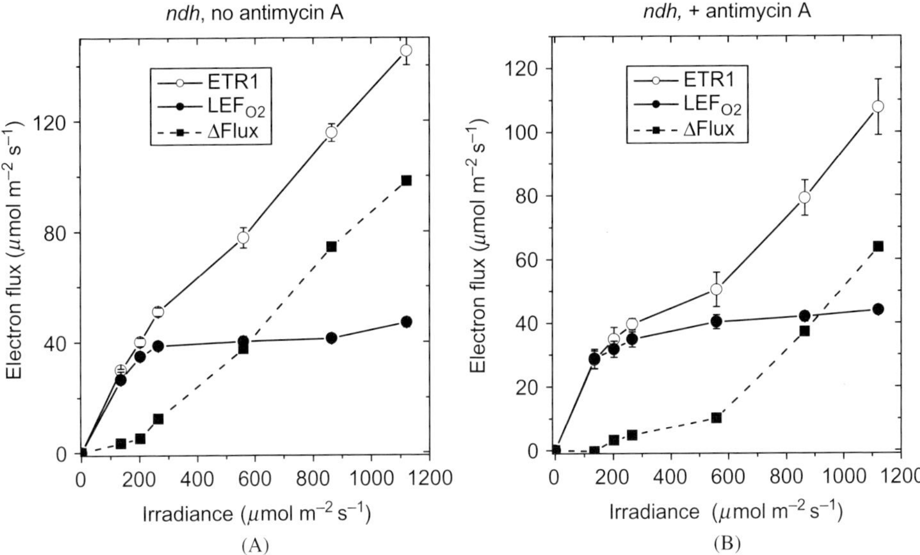

Figure 3. *ETR1*, *LEF$_{O2}$* and *ΔFlux*, as a function of irradiance of white light from a halogen lamp given to *Arabidopsis* leaf discs of the *ndh* mutant in the absence of antimycin A (A) or in the presence of antimycin A (B). Replotted from Kou *et al.* [2015].

component giving rise to charge recombination and (2) the Mehler reaction are both present equally in the absence and presence of antimycin A, then subtraction of the *ΔFlux* in the presence from that in the absence of antimycin A should give the antimycin A-sensitive part of *CEF*. Figure 4 shows that $\Delta Flux_{-AA} - \Delta Flux_{+AA}$ increases rapidly in the wild type with irradiance, saturating at about eight times growth irradiance. $\Delta Flux_{-AA} - \Delta Flux_{+AA}$ in the *ndh* mutant, after displaying a low value below 200 μmol photons m^{-2} s^{-1}, also increases rapidly with irradiance, saturating at a similar irradiance. The maximum $\Delta Flux_{-AA} - \Delta Flux_{+AA}$ is about 45 μmol e$^-$ m^{-2} s^{-1} in the wild type, and about 35 μmol e$^-$ m^{-2} s^{-1} in the *ndh* mutant; these electron fluxes are similar in magnitude to the *LEF$_{O2}$* at light saturation. Moreover, $\Delta Flux_{-AA} - \Delta Flux_{+AA}$ is likely to be an *underestimate* of the antimycin A-sensitive component of *CEF*: since the rate of charge recombination included in $\Delta Flux_{+AA}$ is expected to be greater than that included in $\Delta Flux_{-AA}$, the difference between these two terms is smaller than would be the case if charge recombination included in both terms were identical.

As noted above, at an irradiance below 200 μmol m^{-2} s^{-1}, $\Delta Flux_{-AA} - \Delta Flux_{+AA}$ of the *ndh* mutant is only a small fraction of that in the wild type. It seems that the presence of NDH in the wild type aids the PGR5-mediated pathway of cyclic electron flow. Indeed, it has been suggested that the presence of NDH may promote the antimycin-sensitive *CEF* in the wild type, perhaps through redox regulation [Shikanai, 2014; Martin *et al.*, 2015].

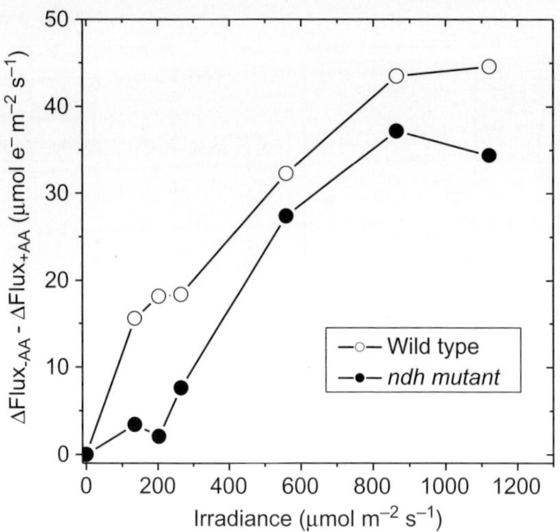

Figure 4. The difference between Δ*Flux* in the absence of antimycin A and that in the presence of antimycin A, plotted as a function of irradiance, in wild type and the *ndh* mutant of *Arabidopsis*, approximately representing the antimycin A-sensitive component of *CEF*.

5. Concluding Remarks

Our current method of estimating *ETR1* makes use of (a) the P700⁺ signal to determine the total electron flux through PSI and (b) the rate of gross O_2 evolution, measured in identical conditions. By difference, *CEF* is inferred. A critical criterion has been met, in that both signals originate from the whole leaf tissue [Kou *et al.*, 2013], so that the subtraction is valid. There are, however, a number of approximations made and, therefore, the results should be treated with caution.

First, the P700⁺ signal is not pure, even though a reference wavelength (870 nm) is used in conjunction with the measuring wavelength (810 nm) to lower the contamination of the signal by redox changes of plastocyanin (Pc). No doubt, the signal will be greatly improved by deconvolution [Klughammer and Schreiber, 2016].

Second, the greatest interference seems to come from charge recombination that occurs when low-light-grown plants are examined at an irradiance which is an order of magnitude above their growth irradiance. In that situation, the photosynthetic apparatus is unable to handle the rapid delivery of electrons to the acceptor side of PSI, which is forced to return the electrons to the donor side in a local circuit in PSI. A test of the seriousness of the interference from charge recombination can be gauged by antimycin A treatment to abolish the PGR5-dependent *CEF*, leaving mainly the electron flux associated with charge recombination. However, this approach of eliminating *CEF* by antimycin A treatment is not always convenient, and is certainly not adequate in systems in which NDH-mediated *CEF* may be a major component.

Third, the partitioning of absorbed light between the two photosystems needs to be determined so that the value of f_I can be used to calculate *ETR1*. So far, we have assumed that antimycin A treatment and measurement in relatively low light give a reasonable value. However, in systems where the NDH-mediated pathway is significant [*e.g.* C4 plants, Nakamura *et al.*, 2013], ideally both the PGR5- and NDH-pathways should be inhibited, thereby allowing *ETR1* to be equated with LEF_{O2} in the determination of f_I.

Fourth, current measurements of the gross O_2 evolution rate are conducted in CO_2-enriched air, thereby minimizing photorespiration or just keeping the supply of CO_2 non-limiting and invariant. A future challenge is to determine *CEF* in photorespiratory conditions, which prevail in nature.

Acknowledgements

This work was supported by the award of an Australian Research Council grant (DP 1093827) to WSC. The authors would like to dedicate this article to the fond memory of the late Jan M. Anderson, our esteemed supervisor, mentor and colleague, who enthusiastically supported young researchers.

References

Arnon, D.I., Whatley, F.R. and Allen, M.B. (1955). Vitamin K as a cofactor of photosynthetic phosphorylation, *Biochim. Biophys. Acta*, 16, 607–608.

Cleland, R.E. and Bendall, D.S (1992). Photosystem-I cyclic electron transport: measurement of ferredoxin-plastoquinone reductase activity, *Photosynth. Res.*, 34, 409–418.

Evans, J.R. and Poorter, H. (2001). Photosynthetic acclimation of plants to growth irradiance: the relative importance of specific leaf area and nitrogen partitioning in maximizing carbon gain, *Plant Cell Environ.*, 24, 755–767.

Fan, D.Y., Fitzpatrick, D., Oguchi, R., Ma, W., Kou, J. and Chow, W.S. (2016). Obstacles in the quantification of the cyclic electron flux around Photosystem I in leaves of C3 plants, *Photosynth. Res.*, 129, 239–251.

Hendrickson, L., Furbank, R.T. and Chow, W.S. (2004). A simple alternative approach to assessing the fate of absorbed light energy using chlorophyll fluorescence, *Photosynth. Res.*, 82, 73–81.

Huang, W., Zhang, S.B. and Cao, K.F. (2011). Cyclic electron flow plays an important role in photoprotection of tropical tress illuminated at temporal chilling temperature, *Plant Cell Physiol.*, 52, 297–305.

Klughammer, C. and Schreiber, U. (2008a). Saturation pulse method for assessment of energy conversion in PS I, *PAM Appl. Notes*, 1, 11–14, http://www.walz.com/.

Klughammer, C. and Schreiber, U. (2008b). Complementary PS II quantum yields calculated from simple fluorescence parameter measured by PAM fluorometry and the saturation pulse method, *PAM Appl. Notes*, 1, 27–35, http://www.walz.com/.

Klughammer, C. and Schreiber, U. (2016). Deconvolution of ferredoxin, plastocyanin, and P700 transmittance changes in intact leaves with a new type of kinetic LED array spectrophotometer, *Photosynth. Res.*, 128, 195–214.

Kou, J., Takahashi, S., Fan, D.-Y., Badger, M.R. and Chow, W.S. (2015). Partially dissecting the steady-state electron fluxes in Photosystem I in wild-type and *pgr5* and *ndh* mutants of *Arabidopsis*, *Front. Plant Sci.*, 6, p. 758.

Kou, J., Takahashi, S., Oguchi, R., Fan, D.-Y., Badger, M.R. and Chow, W.S. (2013). Estimation of the steady-state cyclic electron flux in white light, CO_2-enriched air and other varied conditions, *Funct. Plant Biol.*, 40, 1018–1028.

Kramer, D.M., Johnson, G., Kiirats, O. and Edwards, G.E. (2004). New fluorescence parameters for the determination of QA redox state and excitation energy fluxes, *Photosynth. Res.*, 79, 209–218.

Kukuczka, B., Magneschi, L., Petroutsos, D., Steinbeck, J., Bald, T., Powikrowska, M., Fufezan, C., Finazzi, G. and Hippler, M. (2014). Proton Gradient Regulation 5-Like 1-mediated cyclic electron flow is crucial for acclimation to anoxia and complementary to nonphotochemical quenching in stress adaptation, *Plant Physiol.*, 165, 1604–1617.

Martin, M., Noarbe, D.M., Serrot, P.H. and Sabater, B. (2015). The rise of the photosynthetic rate when light intensity increases is delayed in *ndh* gene-defective tobacco at high but not at low CO_2 concentrations, *Front. Plant Sci.*, 6, p. 34.

Maxwell, P.C. and Biggins, J. (1976). Role of cyclic electron transport in photosynthesis as measured by photoinduced turnover of P-700 *in vivo*, *Biochemistry*, 15, 3975–3981.

Munekage, Y., Hojo, M., Meurer, J., Endo, T. and Shikanai, T. (2002). PGR5 is involved in cyclic electron flow around photosystem I and is essential for photoprotection in Arabidopsis, *Cell*, 110, 361–371.

Nakamura, N., Iwano, M., Havaux, M., Yokota, A. and Munekage, Y.N. (2013). Promotion of cyclic electron transport around photosystem I during the evolution of NADP-malic enzyme-type C4 photosynthesis in the genus *Flavaria*, *New Phytol.*, 199, 832–842.

Ruuska, S.A., Badger, M.R., Andrews, J.T. and von Caemmerer, S. (2000). Photosynthetic electron sinks in transgenic tobacco with reduced amounts of Rubisco: little evidence for significant Mehler reaction, *J. Exp. Bot.*, 51, 357–368.

Shen, Y.-K. (1990). Some factors limiting photosynthesis in nature. In *Current Research in Photosynthesis*, Baltscheffsky, M., ed., Vol. IV (Dordrecht: Kluwer Academic Publishers), pp. 843–850.

Shikanai, T. (2014). Central role of cyclic electron transport around photosystem I in the regulation of photosynthesis, *Curr. Opin. Biotechnol.*, 26, 25–30.

Siebke, K., von Caemmerer, S., Badger, M.R. and Furbank, R.T. (1997). Expressing an *RbcS* antisense gene in transgenic *Flaveria bidentis* leads to an increased quantum requirement for CO_2 fixed in photosystems I and II, *Plant Physiol.*, 115, 1163–1174.

Takahashi, S., Milwood, S.E., Fan, D.-Y., Chow, W.S. and Badger, M.R. (2009). How does cyclic electron flow alleviate photoinhibition in *Arabidopsis*? *Plant Physiol.*, 149, 1560–1567.

Chapter 13

The Contribution of Electron Transfer after Photosystem I to Balancing Photosynthesis

Guy Hanke* and Renate Scheibe[†]

*Queen Mary University of London,
The Mile End Road E1 4NS, London, UK
g.hanke@qmul.ac.uk
[†]Plant Physiology, Faculty of Biology and Chemistry,
University of Osnabrück, 49069 Osnabrück, Germany
Renate.Scheibe@Biologie.Uni-Osnabrueck.DE

Light excitation at photosystem II and photosystem I drives assimilation and biosynthesis. This process is inherently dangerous, and a limitation in electron acceptors can result in damage by free radicals to photosystem II, and in extreme cases to photosystem I. This chapter deals with how regulation of electron transport events that occur *after* photosystem I can control the partitioning of these electrons into different metabolic pathways and sinks, and how this partitioning has the potential to influence protective quenching mechanisms. We focus on events in higher plants, and discuss the primary acceptor at photosystem I, ferredoxin, and how the diversity of ferredoxin iso-proteins can contribute to controlling electron flux into various pathways. In addition, we consider how electron flux into various alternative electron sinks and quenching pathways can be regulated to protect the plant, while maintaining competitive assimilation and biosynthesis. The integration of metabolic and reductive signals will be taken as a particular example.

1. Introduction

Professor Jan Anderson, to whom this volume is dedicated, was best known for her experiments proving the lateral heterogeneity of the thylakoid, and demonstrating that it contained two distinct, spatially separated photosystems [Anderson *et al.*, 1966; Boardman and Anderson, 1967]. Due to the action of photosystem II (PSII), the chloroplast is a highly oxygenic environment and therefore, in the absence of appropriate acceptors at the photosystems, electron donation to oxygen inevitably occurs, producing reactive oxygen species (ROS) [Allen and Hall, 1974; Durrant *et al.*, 1990]. These dangerous products can cause extensive damage to protein side chains within the photosystems [Aro *et al.*, 1993; Sonoike, 1995], and in consequence the subunits of PSII that are most exposed to such free radicals (D1 and D2) are rapidly turned over [Schuster *et al.*, 1988; Shipton and Barber, 1994].

PSII has a modular structure, which makes replacement of the reaction center from within the complex possible, but the photosystem I (PSI) complex is more compact [Caffarri *et al.*, 2014], precluding such a mechanism. Poor oxygen diffusion into the PSI reaction center may be the reason why, even in the triplet state, PSI reaction center chlorophylls do not seem to result in singlet oxygen [Setif *et al.*, 1981]. In the absence of an acceptor at PSI the majority of ROS formed are thought to originate instead at the acceptor side, where the Fe–S clusters are capable of donating electrons to oxygen, forming superoxide [Sonoike, 1995]. In contrast, PSII is considered to mainly form singlet oxygen [Durrant *et al.*, 1990], although superoxide is also produced with some hydroxyl radicals resulting from Fenton reactions [Pospisil *et al.*, 2004]. Singlet oxygen was not previously thought to be generated at PSI [Hideg and Vass, 1995], but it has recently been suggested that, in extreme circumstances, PSI may also be capable of generating singlet oxygen [Takagi *et al.*, 2017]. It is clear that, in the absence of acceptors at PSI, charge recombination occurs [Sétif and Brettel, 1990], resulting in photoinhibition and damage [Sonoike, 1996] that is repaired relatively slowly compared to PSII [Kudoh and Sonoike, 2002; Zivcak *et al.*, 2015]. There is a growing body of evidence that, presumably because of this difficulty in repairing PSI, photosynthetic organisms have developed a multitude of mechanisms to prevent excited electrons from being trapped in PSI [Suorsa *et al.*, 2012; Takagi *et al.*, 2017]. Many of these mechanisms have the same basic premise: acting as electron sinks or dumps [Mehler, 1951; Satoh, 1970; Scheibe and Stitt, 1988; Park *et al.*, 1996; Allahverdiyeva *et al.*, 2013], which can act to rapidly re-oxidise the reaction center chlorophylls by removing excess electrons, effectively preventing formation of any potential triplet states that would result in photoinhibition or even photodamage.

The principle sink for photosynthetic electrons is of course the reductive step of the Calvin–Benson–Bassham (CBB) cycle, where NADPH is consumed to power conversion of 1,3-bisphosphoglycerate to glyceraldehyde 3-phosphate in a light-regulated manner [Müller and Ziegler, 1969; Müller *et al.*, 1969; Wolosiuk and Buchanan, 1978; Baalmann *et al.*, 1994]. The CBB cycle is also highly regulated at various other places, to ensure that intermediates are not exhausted or accumulated to excess, and to ensure that enzyme activities coincide with the availability of substrates (CO_2 and inorganic phosphate (P_i), in addition to ATP and NADPH) [Fridlyand *et al.*, 1997]. Most importantly, individual intermediates fine-tune enzyme activities at specific steps of the cycle, by acting not only as substrates and products but also as effectors (positive or negative) of redox regulation (see discussion in Section 5 of this chapter for details). Thus the metabolic state of the chloroplast continuously adjusts the capacity of the CBB enzymes to respond to redox-derived regulation, which in turn has its origins in the photosynthetic electron transport chain [Scheibe, 1991]. Because light intensity can vary dramatically and frequently over the course of the day, the capacity for ATP and NADPH production from photosynthetic electron transport is highly variable, while regulatory responses associated with the CBB may be much slower [Kramer and Evans, 2011]. In order to prevent acceptor limitation at PSI on sudden increase in light intensity and during high-light periods, where electron transport activity exceeds CBB capacity, alternative electron sinks are required for protection [Niyogi, 2000; Wilhelm and Selmar, 2011; Scheibe and Dietz, 2012; Dietz and Hell, 2015].

Exhaustive research has been carried out on several pathways and membrane complexes capable of dissipating excess electrons. These include the reduction of O_2 to water, either *via* the plastid terminal oxidase [Cournac *et al.*, 2000; Josse *et al.*, 2000] or dedicated flavodiiron (FLV) proteins [Allahverdiyeva *et al.*, 2013], which, although absent from angiosperms, are present in some gymnosperms. Alternatively, the Mehler reaction can result in electron donation to oxygen, generating superoxide anion radicals, followed by their removal through dismutation to H_2O_2, and finally reduction to H_2O and O_2 *via* the Halliwell–Asada–Beck cycle (also known as the water–water cycle). Here, ascorbate and glutathione are used and sequentially regenerated using NADPH, in turn resulting in the regeneration of $NADP^+$, which provides an electron sink at the acceptor side of PSI. A second water–water cycle has been more recently described where electrons from NADPH are transferred by a plastid-specific NADPH:thioredoxin reductase, named NTRC [Cejudo *et al.*, 2012], to re-reduce the 2-Cys peroxiredoxin [Pulido *et al.*, 2010], which takes care of any H_2O_2 generated in pseudocyclic electron flow (Mehler reaction). Photorespiration can also serve as a pathway for dissipating electrons when the plant is under stress [Stuhlfauth *et al.*, 1990]. Here the inadvertent fixation of O_2 onto ribulose bisphosphate by RuBisCO generates

2-phosphoglycolate [Andrews *et al.*, 1973], removal of which requires both electrons and ATP, thereby dissipating the energy generated by photosynthetic electron transport [Gardeström and Wigge, 1988; Igamberdiev *et al.*, 2001; Voss *et al.*, 2013]. Moreover, several metabolic pathways can be co-opted as electron sinks in conditions of PSI acceptor limitation, including: nitrite reduction [Anderson and Done, 1977b; Baysdorfer and Robinson, 1985], ascorbate regeneration [Anderson *et al.*, 1983], itself necessary to remove ROS, and the conversion of oxaloacetate to malate [Scheibe and Beck, 1975; Scheibe, 2004]. This last reaction, conducted as part of the "malate valve", can result in controlled export of reductant from the chloroplast [Scheibe and Stitt, 1988], either to support nitrate reduction with NADH in the cytosol or for eventual consumption of electrons in the mitochondria [Krömer and Scheibe, 1996; Zhang *et al.*, 2012] generating ATP to support cytosolic sucrose synthesis [Krömer *et al.*, 1988]). Plant mitochondria possess a specific protein, namely the alternative oxidase (AOX), allowing direct reduction of O_2 to H_2O if no ATP is required and energization needs to be relaxed thermally. Interestingly, Jan Anderson also devoted significant research effort into the investigation of these alternative metabolic sinks for photosynthetic electrons and how they act in concert [Anderson and Done, 1977a, 1977b; Anderson, 1981; Jablonski and Anderson, 1981; Anderson *et al.*, 1983].

In addition to electron sinks or "dumps", the redox-state of PSI can of course be regulated at the step of reduction by plastocyanin (PC). In photosynthetic electron transport, water is oxidized and provides electrons to photo-excited PSII, with the eventual reduction of plastoquinone to plastoquinol. Plastoquinol oxidation pumps protons *via* the Q-cycle in concert with the cytochrome b_6f complex [Mitchell, 1975], eventually donating them to the luminal acceptor PC. In turn, photo-excited PSI transfers these electrons from PC to ferredoxin (Fd). As PSI is an oxidoreductase, protection of its reaction center by maintaining it in an oxidized state is clearly also dependent on the rate of arrival of electrons from electron carriers prior to PC. This is in turn controlled by several processes such as non-photochemical quenching (NPQ), state transitions, and photosynthetic control, which are all highly regulated [Rumberg and Siggel, 1969; Haehnel, 1976; Bennett *et al.*, 1980; Niyogi *et al.*, 1998; Pesaresi *et al.*, 2011; Colombo *et al.*, 2016; Sacharz *et al.*, 2017]. As most of these mechanisms, such as the reactions leading to NPQ, are dealt with extensively in other chapters, in this chapter we will focus on the influence of electron partitioning from PSI. Both NPQ (energy dissipation at PSII) and photosynthetic control (downregulation of plastoquinol oxidation at the cytochrome b_6f complex) are strongly regulated by the proton motive force [Haehnel, 1976; Shen *et al.*, 1996; Horton *et al.*, 2000; Colombo *et al.*, 2016], generated by the proton pumping reactions of the electron transport chain. Electron partitioning after PSI can strongly influence the rate of proton pumping if these electrons are returned to the plastoquinone pool in a cyclic electron flow. In this case, even low turnover at PSII can rapidly generate enough ΔpH to induce NPQ or photosynthetic control. Although several pathways of cyclic electron transport have been suggested, only two have been proven, with return of electrons to the plastoquinone pool either *via* the NDH pathway [Burrows *et al.*, 1998] or the antimycin A-sensitive ferredoxin:quinone reductase (FQR) PgrL1/Pgr5 pathway [Munekage *et al.*, 2002; DalCorso *et al.*, 2008].

Since electrons can also be drained from the cyclic pathways for use in metabolism, the fluxes into electron-consuming pathways, such as malate valve, starch biosynthesis *etc.*, need to be simultaneously adjusted to ensure both protection and optimal assimilation and biosynthesis. Huge research effort has resulted in the identification of many individual components in these pathways, which are well covered in other chapters, and we now have some understanding of their physiological importance. However, it stands to reason that dissipatory mechanisms must be highly regulated. If they are permanently active, then few electrons will be available for the reductive steps of the CBB cycle and assimilatory processes in general for biomass production. Partial information is available for some of these pathways regarding their up-regulation in conditions of severe acceptor limitation, but little is known about the equally important down-regulation during ambient conditions, when CO_2 assimilation should be at its maximum. In this chapter, we will describe how electron partitioning can contribute to these regulatory events and highlight important questions for future research. Although there are many similarities between cyanobacteria, algae and higher plants, we will focus principally on angiosperms and gymnosperms for the sake of brevity and clarity.

2. The Hierarchy of Electron Donation

In land plants, Fd is the physiological acceptor at PSI. The majority of electrons donated by photo-excited PSI to Fd are transferred to the enzyme Fd:NADP(H) oxidoreductase (FNR) [Shin *et al.*, 1963] resulting in formation of NADPH, which is primarily consumed in the CBB cycle. However, many other enzymes are also dependent on Fd as a direct electron donor. These have been well-described elsewhere [Hanke and Mulo, 2013], but will be mentioned here briefly as follows: the FQR pathway of cyclic electron flow, PgrL1 [Arnon and Chain, 1979; Hertle *et al.*, 2013]; the NDH pathway of cyclic electron flow [Yamamoto *et al.*, 2011]; enzymes of nitrogen assimilation, nitrite reductase [Vega and Kamin, 1977; Lancaster *et al.*, 1979] and Fd-GOGAT [Lea and Miflin, 1974; Suzuki and Knaff, 2005]; sulfur assimilation, sulfite reductase [Krueger and Siegel, 1982]; redox signaling, Fd:thioredoxin reductase [Dai *et al.*, 2007]; chlorophyll biosynthesis, chlorophyll *a* oxygenase [Tanaka *et al.*,

1998; Reinbothe *et al.*, 2006], and catabolism, pheophorbide *a* oxygenase and red chlorophyll reductase [Rodoni *et al.*, 1997]; phytochrome synthesis, haem oxygenase [Muramoto *et al.*, 2002] and phytochromobilin synthase [Kohchi *et al.*, 2001]; and fatty acid biosynthesis, acyl-ACP desaturases [Schmidt and Heinz, 1990]. Moreover, it is reported that Fd is also capable of directly reducing monodehydroascorbate to regenerate the redox buffer ascorbate [Miyake and Asada, 1992]. Finally, although it is reported that they contain a Rossmann fold, and are therefore proposed to use NAD(P)H as an electron donor [Vicente *et al.*, 2002], it remains a possibility that the FLV proteins also receive electrons directly from Fd, as they have been identified in a screen for Fd interaction partners [Hanke *et al.*, 2011]. It seems inevitable that, as we understand the function of more chloroplast redox enzymes, we will add to the known network of Fd-dependent enzymes.

Some of the enzymes described above have the capacity to act as heavy sinks for electrons, potentially competing with FNR, and therefore CO_2 assimilation. In some cases (for example the FLV proteins), this is their precise function — dissipating electrons to prevent their damaging build-up in the electron transport chain under high-light conditions. However, most chloroplasts will spend significant periods of time in lower light or shade conditions, where such competition would negatively impact assimilatory processes, growth and finally plant fitness [Kramer and Evans, 2011]. It is, therefore, critical that the plant is able to "ring-fence" sufficient reductant for the CBB cycle, by ensuring that electron donation from Fd to FNR takes precedence under these conditions. This is achieved through a hierarchy of donation to different enzymes [Fridlyand and Scheibe, 1999; Backhausen *et al.*, 2000], based on variable protein–protein interaction [Hase *et al.*, 2006], and the transient effects of small molecules acting as effectors for the thioredoxin-driven interconversion cycles of reduced and oxidized enzymes [Faske *et al.*, 1995] as will be discussed later (see Section 5).

The interaction of Fd with most electron acceptor enzymes (and indeed electron donors such as PSI), follows a well-established pattern [Hurley *et al.*, 1999; Kurisu *et al.*, 2001; Hurley *et al.*, 2002; Dai *et al.*, 2007]. Initially, a loose complex forms, based on electrostatic attraction between charged amino acid side chains on Fd (usually acidic) and the enzyme (usually basic). This ensures optimal orientation of the two active redox centers for electron transfer, and the side chains of some of these residues form salt bridges between Fd and the enzyme. Figure 1 shows the surface charge distribution on Fd (center) and a series of Fd-interaction partners, for which the structure is known. In each case, the interaction surface is displayed, with the redox center in the middle. It is clear from Figure 1 that in all cases the redox center is predominantly non-charged, and surrounded by a ring of

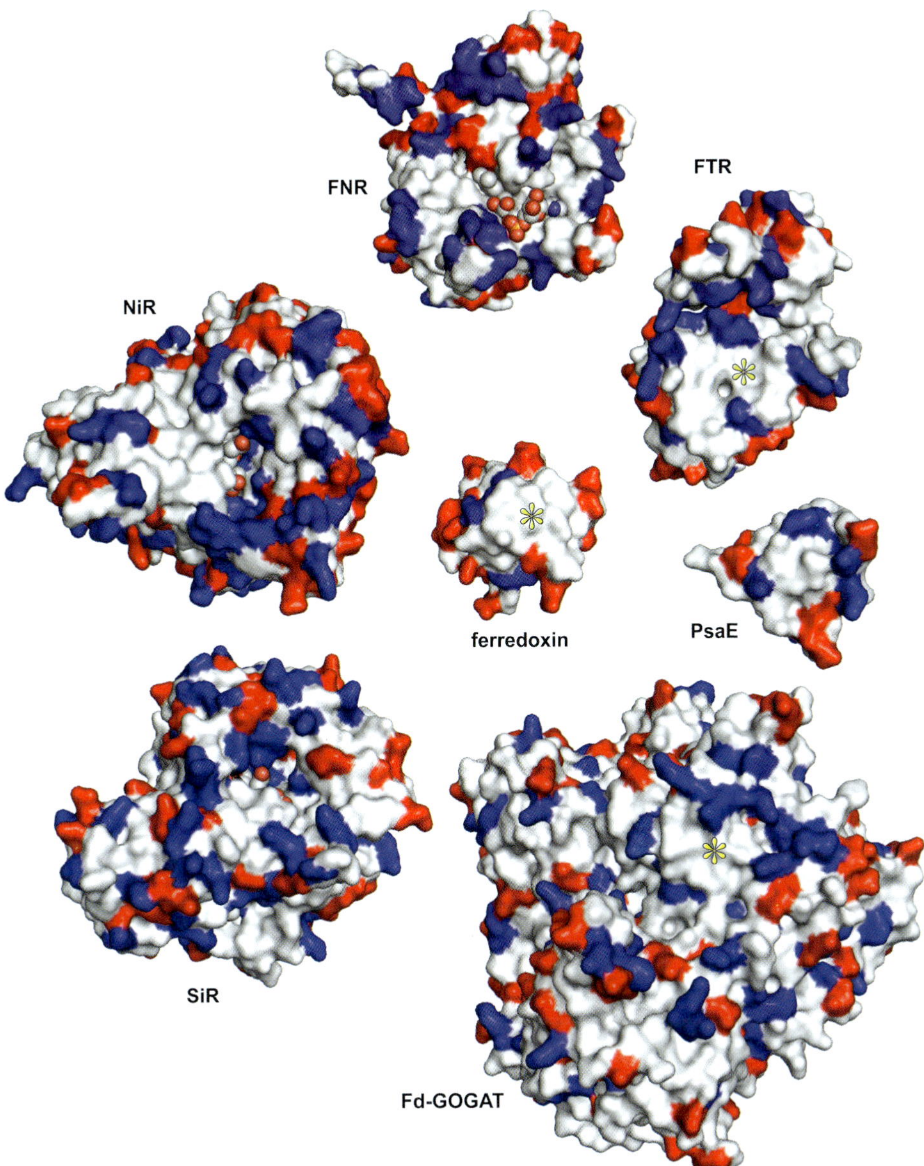

Figure 1. Comparison of charge distribution on ferredoxin and its interaction partners. Positive charge rendered in blue, negative charge rendered in red. Where no structural component of the redox center atoms is visible, location within the protein is marked by a yellow asterisk. Prior to electron transfer, optimal redox center orientation is ensured through electrostatic interactions between side chains on ferredoxin (mostly acidic) and enzymes (mostly basic). PDB files used were: ferredoxin, 3B2F; PSI subunit E (PsaE), 1GXI; ferredoxin:thioredoxin reductase (FTR), 2PVG; ferredoxin:NADP(H) oxidoreductase (FNR), 1GAW; nitrite reductase (NiR), 2AKJ; sulfite reductase (SiR), 5H92 and ferredoxin-dependent oxoglutarate aminotransferase (Fd-GOGAT), 1OFD.

charged residues that promote optimal orientation. It is equally obvious that there is no consensus "Fd-interaction" motif, and that there is significant variation in the pattern of charged residues between the enzymes. These differential interaction profiles are supported by very extensive site-directed mutagenesis studies on PSI [Setif *et al.*, 2002, 2010] and several well-known Fd-dependent enzymes, such as FNR [Aliverti *et al.*, 1994; Hurley *et al.*, 2002], nitrite reductase [Dose *et al.*, 1997; Hirasawa *et al.*, 2010], and sulfite reductase [Nakayama *et al.*, 2000; Kim *et al.*, 2016]. Moreover, chemical shift perturbation of Fd side chains following complex formation and co-crystalisation of Fd with several of these enzymes reveals that the side chains on Fd involved in salt-bridging also vary depending on the enzyme involved [Kurisu *et al.*, 2005; Saitoh *et al.*, 2006; Dai *et al.*, 2007; Sakakibara *et al.*, 2012; Shinohara *et al.*, 2017]. Following these initial interactions, catalytically competent Fd-enzyme complexes are formed, often associated with conformational changes [Lee *et al.*, 2011]. This is entropically favoured by the exclusion of water from the interface between the two proteins [Jelesarov and Bosshard, 1994; Kurisu *et al.*, 2001; Dai *et al.*, 2007; Shinohara *et al.*, 2017] and the resulting hydrophobic interactions [Martinez-Julvez *et al.*, 2009]. Dissociation is promoted by a drop in the affinity between Fd and the partner enzyme following electron transfer. This is well-characterized in the case of interaction with FNR, [Batie and Kamin, 1984a], and recently for the much more challenging system of Fd and PSI [Setif *et al.*, 2017]. In the case of FNR, it is known that negative cooperativity between Fd and NADP(H) binding plays a role in dissociation [Batie and Kamin, 1984b; Martinez-Julvez *et al.*, 2009].

Although a large surface area of interaction is often present, it is critical that the surface topology of the two proteins is not too complimentary. Fd must dissociate following electron transfer, and in cases where mutations have been introduced that further stabilize interaction and improve affinity, the catalytic cycle is inhibited [Thomsen-Zieger *et al.*, 2004]. A combination of charge interactions and the topology of the interface between the two proteins therefore exquisitely poises the interaction, ensuring that the enzyme occupies its correct position in the electron donation hierarchy, and that Fd effectively dissociates following electron transfer in order to allow the next turnover of the redox-reaction. More recently, chemical shift perturbation experiments have demonstrated that specific amino acids on Fd are involved in interaction with different enzymes [Xu *et al.*, 2006; Sakakibara *et al.*, 2012], and that there is co-evolution of the surfaces between the interaction partners, meaning they can become quite specialized, with poor inter-species activities [Sakakibara *et al.*, 2012].

Remarkably, it has been shown that redox enzymes normally located in the endoplasmic reticulum are capable of directly receiving electrons from Fd [Mellor *et al.*, 2016; Wlodarczyk *et al.*, 2016]. Under normal conditions these enzymes would be reduced by NADH, but show excellent activity with a Fd-based reduction system, both *in vivo* and *in vitro*. This promiscuity indicates that, in conditions with abundant light, and therefore ample electron supply, Fd is capable of freely donating electrons to many non-specific targets. However, these enzymes, lacking the fine-tuning of protein–protein interactions seen for the classical electron transfer partners, are therefore presumably at a disadvantage when the reductant is in short supply. As we seek to harness the power of photosynthesis to support various heterologous redox processes using synthetic biology, it will be critical to tune the interaction of novel enzymes to an appropriate position in the electron donation hierarchy. They should be high enough to maximize yield, while still allowing Fd dissociation and not draining resources from CO_2 assimilation inappropriately [Atkinson *et al.*, 2016; Mellor *et al.*, 2017].

3. Diversity Among Electron Carriers

Figure 2 compares the number of genes in four different higher plants for the soluble electron carriers of the photosynthetic electron transport chain ([2Fe-2S]-cluster-containing Fd and Cu-containing PC), showing their evolutionary relationship. It is immediately apparent that, while PC has remained relatively unduplicated, with approximately two genes per species, Fd has been duplicated multiple times, and that there exist several evolutionarily conserved Fd sub-types in higher plants. Many of these sub-types have homologues reaching back to algae [Terauchi *et al.*, 2009] and cyanobacteria [Cassier-Chauvat and Chauvat, 2014], emphasizing their unique, conserved functions. There are relatively more Fd iso-proteins in higher plants than in cyanobacteria and algae, although higher plants have completely lost an alternative electron acceptor at PSI, the flavoprotein flavodoxin, which replaces the [2Fe-2S]-cluster-containing Fd under conditions of iron limitation in some cyanobacteria and algae [Pierella Karlusich and Carrillo, 2017], providing further options for electron partitioning.

The diverse Fd proteins frequently vary in terms of their surface charge distribution, and the hydrophobic surface around the redox center, conferring unique relative affinity profiles for different redox enzymes (and therefore electron-donation hierarchy) [Onda *et al.*, 2000; Yonekura-Sakakibara *et al.*, 2000; Hanke *et al.*, 2004; Kurisu *et al.*, 2005; Voss *et al.*, 2011]. This provides a working model for electron distribution in chloroplasts, shown in Figure 2B, where the plant is capable of responding to altered redox demands over a period of hours by altering the relative expression of different Fds. Depending on the unique hierarchy of electron distribution for each Fd, the plant can therefore manipulate electron partitioning by altering their relative expression.

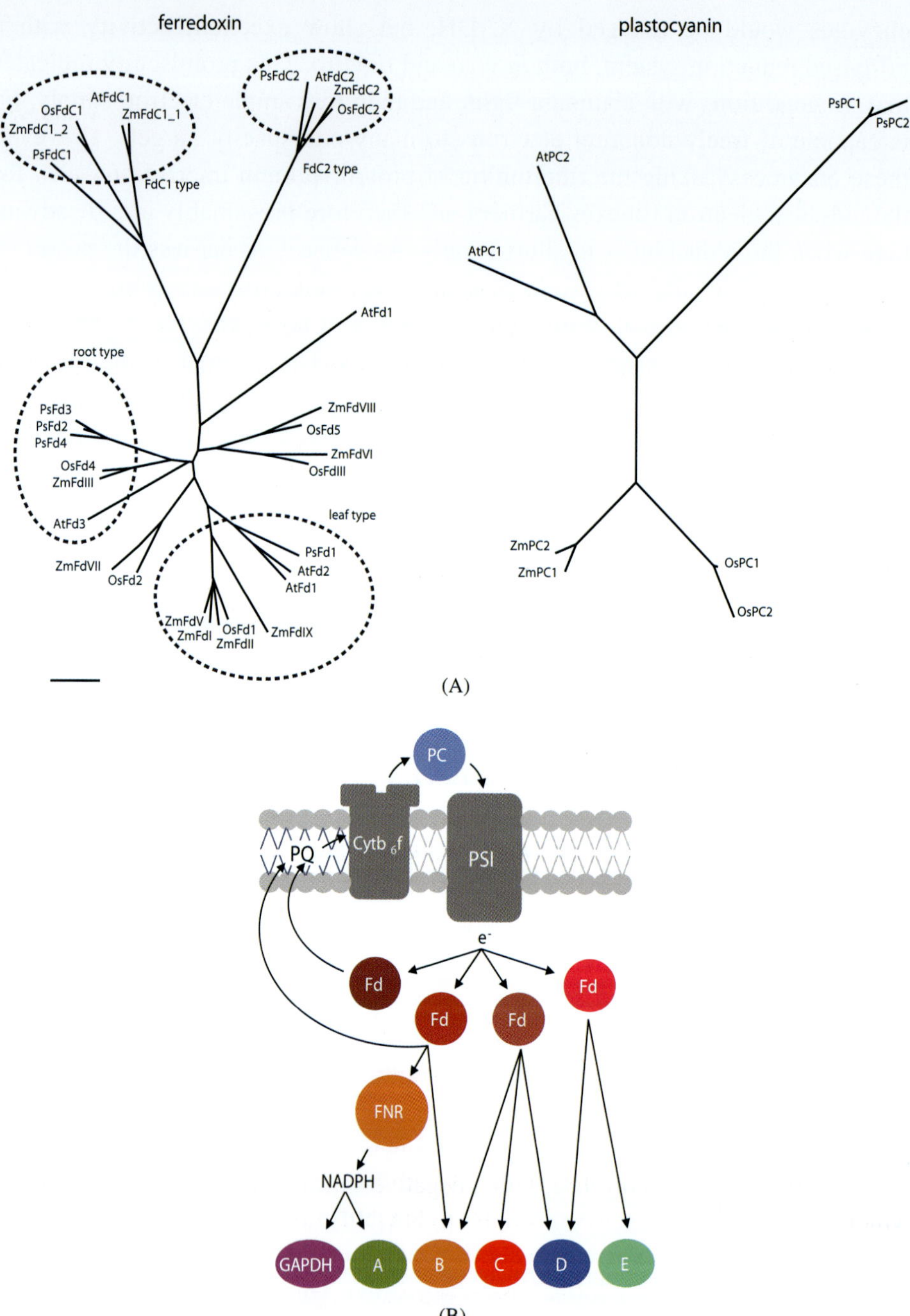

Figure 2.   Electron partitioning through ferredoxin diversity. (A) Phylogenetic tree comparing diversity of the two soluble electron carriers in photosynthetic electron transport: ferredoxin (Fd) and plastocyanin (PC) in four higher plants — *Pinus sylvestris* (Ps), *Arabidopsis thaliana* (At), *Zea mays* (Zm) and *Oryza sativa* (Os). Amino acid sequences minus the predicted transit peptide were aligned in ClustalOmega, and trees assembled in Phylodendron. Major conserved groups of Fd are indicated. (B) Cartoon showing differential electron partitioning by Fd iso-proteins, receiving electrons from

Figure 2. (*continued*) photosystem I (PSI) and partitioning them to CO_2 assimilation (the enzyme GAPDH in the CBB cycle) or multiple other enzymes (labelled A–E). Depending on relative affinity for these enzymes, electrons can be channeled into alternative pathways, including return to the plastoquinone (PQ) pool in several pathways of CEF (not shown), *via* the cytochrome b_6f (Cyt b_6f) and PC to PSI. The plant can therefore alter electron partitioning by changing the relative expression of Fd iso-proteins with variable hierarchies of affinity for electron acceptors.

Although the functional properties and physiological roles of most Fds in higher plants still remain to be solved more progress has been made for algal Fds [Terauchi *et al.*, 2009; Peden *et al.*, 2013; Yang *et al.*, 2015]. However, in many cases these genes have diversified significantly during evolution, and have no higher-plant homologues (for example, the well-characterized Fd 5 from *Chlamydomonas reinhardtii* [Yang *et al.*, 2015]). Despite this fact, significant progress has been made in understanding a few examples from higher plants, which will be highlighted below.

More abundant information is available in terms of translational responses, which give indirect clues about the specific physiological role of particular Fds. Assimilation of nitrogen is a highly reductant intensive process, with both NiR and Fd-GOGAT dependent on electrons from Fd. It is therefore logical that specific Fd genes are found to respond to nitrate availability [Matsumura *et al.*, 1997; Sakakibara, 2003]. These Fds are actually closely related to so-called "root-type" Fds, named for their abundance in non-photosynthetic tissues. It should be noted, however, that transcripts for these proteins have also been detected in the leaves of higher plants [Kimata and Hase, 1989; Hanke *et al.*, 2004]. Root-type Fds are marked by specific physico-chemical properties, which enable more efficient electron transfer in the "non-photosynthetic" direction, using NADPH to reduce Fd. These properties include high affinity for root-type FNR [Onda *et al.*, 2000], and a highly specialized mechanism of interaction with this enzyme [Shinohara *et al.*, 2017]. On interaction of photosynthetic, "leaf type" Fd with "leaf type" FNR, intra-molecular salt bridges in both molecules are broken in order to form new, intermolecular salt bridges between Fd and FNR [Kurisu *et al.*, 2001]. The associated structural changes correlate with a negative shift in the redox potential of Fd, which makes electron transport from Fd to NADPH more energetically favorable. If such a structural phenomenon also occurred in the root combination of Fd and FNR, it would negatively impact on electron transfer in the opposite direction (NADPH to Fd), and so the unique interaction mechanism of the heterotrophic proteins [Shinohara *et al.*, 2017] is optimized to avoid this. Finally, root-type Fds have a highly conserved, positively shifted redox potential, in the range of -340 mV rather than -420 mV for classical photosynthetic Fds [Onda *et al.*, 2000; Hanke *et al.*, 2004]. This is in the range for the -320 mV value for NADP(H), making Fd reduction more favorable in non-photosynthetic organelles. It has also been shown that a root-type Fd supply system is more efficient in supporting activity of sulfite reductase [Yonekura-Sakakibara *et al.*, 2000].

There is significant functional diversity within classical photosynthetic Fds, with at least two identified in most species examined so far [Kimata and Hase, 1989; Hanke *et al.*, 2004]. In many cases, divergent physiological roles have not been identified, but there is a body of evidence supporting the role of specific Fds in CEF. The best understood of these is maize ferredoxin II (ZmFdII), which is specifically localized to the bundle sheath cells of maize [Matsumura *et al.*, 1999; Majeran *et al.*, 2005]. Chloroplasts in these cells conduct predominantly CEF, and it was later shown that ZmFdII has a poor affinity for FNR, and preferentially drives CEF when used to replace the predominant Fd in the cyanobacterium *Plectonema boryanum* [Kimata-Ariga *et al.*, 2000]. Poor support for NADP$^+$ photoreduction by ZmFdII in the maize bundle sheath is the likely reason for the poor rates of glutathione regeneration in this tissue [Doulis *et al.*, 1997], as this process is NADPH-dependent.

There is also evidence that, even in C3 plants, specific Fds can preferentially drive CEF. As only a minority of electrons are thought to be partitioned into CEF under steady-state conditions, any CEF-specific Fd would be expected to be present at a relatively low abundance, but to increase in response to such CEF-demanding pressures as high light or drought stress. Such Fds have been identified in Arabidopsis [Hanke *et al.*, 2004; Lehtimaki *et al.*, 2010] and pea [Khristin and Akulova, 1976; Dutton *et al.*, 1980]. Interestingly, RNAi knock-down of the minor Fd isoform in Arabidopsis (Fd1) results in a phenotype consistent with some disturbance to CEF [Hanke and Hase, 2008]. In addition, transplastomic introduction of the major pea Fd isoform to tobacco had relatively little impact on photosynthesis [Yamamoto *et al.*, 2006], while introduction of the minor isoform resulted in enhanced CEF and a severe phenotype consistent with exaggerated photosynthetic control even under low light [Blanco *et al.*, 2013].

4. Regulation of Sink Demand Beyond the First Acceptor

Given the importance of protecting the photosynthetic machinery, it is logical that multiple protective mechanisms with some redundancy have evolved. Manipulation of electron flux to enhance proton gradient generation [Burrows *et al.*, 1998; Munekage *et al.*, 2002; DalCorso *et al.*, 2008; Livingston, Cruz, *et al.*, 2010] in order to downregulate the system [Rumberg and Siggel, 1969; Haehnel, 1976; Bennett *et al.*, 1980; Niyogi *et al.*, 1998; Pesaresi *et al.*, 2011; Colombo *et al.*, 2016; Sacharz *et al.*, 2017] or to redirect electrons into massive dissipation by alternative sinks [Stuhlfauth *et al.*, 1990; Scheibe, 2004; Allahverdiyeva *et al.*, 2013] all act protectively, but can prove a double-edged sword. These mechanisms will all drain electrons, and in many cases ATP, from bioassimilatory processes, leading to potential growth penalties. It is therefore critical that they are upregulated, and downregulated, appropriately. This must ensure protection when necessary, but minimal disruption to bioassimilation and biosynthesis. In the case of many of these systems, such information is completely lacking. For example, nothing is known about the regulation of the FLV proteins, which have the capacity to drain enormous amounts of electrons from photosynthesis, and little is understood about how the contributions of the water–water cycles and photorespiration to energy dissipation are controlled. Proton gradient generation by CEF is known to be regulated [Livingston, 2010], and there is evidence that this is through thiol regulation of the antimycin A-sensitive Pgr5/PgrL1 system [Strand *et al.*, 2016], and H_2O_2 regulation of the NDH complex [Strand *et al.*, 2015], but how this is integrated *in vivo* remains to be thoroughly investigated. Perhaps the only system into which we have deep insight is the integration of metabolic signals with thioredoxin regulation that controls the relative flux of electrons into the CBB cycle and the malate valve, discussed in detail below.

Regulation of several stromal target enzymes is critically dependent on transduction of redox signals from the photosynthetic electron transport chain by the signaling molecule thioredoxin (Trx). Trxs form a family of small, soluble ubiquitous proteins of about 100 amino acids. The protein exists in two different interconvertible redox states: the reduced Trx with two SH groups and the oxidized form in which the two regulatory thiol (–SH) groups form a disulfide (S–S) bridge [Buchanan and Balmer, 2005]. Electrons for the reduction of Trxs originate from the electron transport chain and are transferred from Fd by the Fd-Trx reductase (FTR). In addition, plastids possess an NADPH-dependent Trx reductase (NTRC) with a built-in Trx domain that allows electron flow from NADPH to target proteins [Cejudo *et al.*, 2012]. There are multiple targets of Trxs in plants [Montrichard *et al.*, 2009; Navrot *et al.*, 2011]. As for Fd (see Figure 1), the structural basis determining interaction of Trx with different targets is very diverse, originating from different evolutionary scenarios [Gütle *et al.*, 2017].

Reduced Trxs function as protein-disulfide oxido-reductases generating reduced target enzymes. Reoxidation is thought to occur spontaneously in the presence of oxygen due to the negative redox-potential of the regulatory cysteines. Regarding the large number of redoxins in plants, it is still an open question how any specificity might be explained. As can be taken from *in vitro* determinations of their redox-potentials, any distinct preferences of the electron flow from FTR to the various Trxs is difficult to imagine [Yoshida and Hisabori, 2017]. This is also the case for the electron transfer from the Trxs to the various targets when we only consider differences in their redox-potentials [Yoshida *et al.*, 2015]. However, there is also a high capacity for flexibility in this network, depending on changing conditions and requirements [Nikkanen and Rintamaki, 2014; Geigenberger *et al.*, 2017]. During short-term changes, as well as upon sustained stress, fluxes will be adjusted smoothly at the post-translational level. However, when required, the enzyme capacity can be altered by changes at the transcriptional level [Scheibe *et al.*, 2005; Li *et al.*, 2009]. In the latter case, emerging signals transmit the need for changed gene expression to the nucleus, whereby ROS, particularly H_2O_2, are frequently discussed as the origin of these signals [Petrov and Van Breusegem, 2012; Noctor and Foyer, 2016].

5. Flexible Adjustment of Target-Enzyme Activity According to Metabolic Demand

The light-modulation mechanism not only comprises an on/off switch for the target enzymes, but also has the capacity to integrate metabolic information from downstream reactions that consume electrons. When an enzyme is reduced by Trx in the light, it is reoxidized by oxygen at a constant rate, probably in a non-enzymatic reaction. The portion of the enzyme in the reduced state therefore varies with the changing redox-state of Trx. This means that a futile cycle of continuous reduction and reoxidation of the enzyme coordinates the availability of reducing equivalents from the electron transport chain with the actual activation state of the target enzymes in the CBB cycle and the other light/dark-modulated enzymes. Such a monocascade can be further fine-tuned in response to the various metabolic fluxes within individual pathways by allosteric regulation [Scheibe, 1991]. In this way specific effector metabolites influence the redox potential of the disulfide and hence the rate of reduction *versus* oxidation of the target enzymes. Ongoing changes in the steady-state concentration of active enzyme then allow for rapid adjustment of fluxes through various metabolic steps.

The metabolic factors integrated into light/dark modulation include the stromal proton concentration, which decreases in the light, and the Mg^{2+} concentration, which in turn increases. In addition, various intermediates of the CBB cycle as well as the concentration of P_i change rapidly upon dark/light transitions, as well as in response to metabolism during illumination. All these factors act in multiple ways at various sites to coordinate the fluxes through different metabolic pathways and to integrate energy supply from the light reactions with its consumption in assimilatory processes [Dietz *et al.*, 2002; Heineke and Scheibe, 2009]. When the CBB cycle is active as a result of the light-dependent activation of key enzymes, the phosphorylated intermediates of the cycle accumulate in the chloroplast and the content of P_i decreases. On the other hand, P_i is also required as the substrate for F-ATP synthase. To avoid inhibition of photophosphorylation by substrate limitation, a certain level of P_i has to be maintained in the stroma. This requires coordination of the activities of the key enzymes, which is achieved by the regulatory effect of intermediates, usually the substrates and products of the respective target enzyme [Faske *et al.*, 1995].

This concept of fine-tuning grew out of work on spinach chloroplast NADP-MDH whose reductive activation is inhibited by high NADP levels [Scheibe and Jacquot, 1983]. This effect ensures that only surplus NADPH is converted to malate for export through the malate valve in a controlled manner during photosynthesis [Backhausen *et al.*, 1994, 2000]. Fine-tuning also applies to other regulatory enzymes of chloroplasts, thus providing a common principle (redox cycling) to the control of flux through every metabolic step according to the demand [Scheibe, 1991].

6. Signaling Upon Redox-Imbalances for Long-Term Adaptation at the Transcriptional Level

As plants are sessile organisms, they cannot escape adverse conditions, and signaling processes are required to continuously adapt metabolism, in particular energy distribution, during abiotic and biotic stress, and during development on the long-term time scale. A complex network of signal transduction integrates information from multiple stimuli, and in most cases redox imbalance plays a role in the early steps to initiate the changes [Kocsy *et al.*, 2013]. Upon stress, *e.g.* due to sudden exposure to high light, plants can respond at different levels. They are capable of rapidly adjusting to changed conditions as described in the Sections 4 and 5 of this chapter. However, when the enzymes are running at near to maximal velocity, any further challenge, which would exceed their capacity to maintain homeostasis, must lead to an induction of gene expression resulting in long-term adaptation at the transcriptional level [Scheibe *et al.*, 2005; Scheibe and Dietz, 2012]. In order to achieve the latter, imbalances in the electron transport chains of chloroplasts or mitochondria (as central pathways of energy-converting reactions) must initiate signaling processes that are finally perceived in the nucleus, resulting in altered gene expression of organelle-localized proteins [Wojtera-Kwiczor *et al.*, 2013; Zachgo *et al.*, 2013; Gollan *et al.*, 2015; Guo *et al.*, 2016; Kmiecik *et al.*, 2016; Noctor and Foyer, 2016; Xu *et al.*, 2016; Ren *et al.*, 2017; Van Aken *et al.*, 2017]. Such changes in expression can then feed back into the abundance of the initial acceptors at PSI. As described in Sections 2 and 3 of this chapter, adjusting the relative expression levels of Fds with different hierarchies of donation can alter the partitioning of electrons, favoring protective or dissipatory sinks. A future research target of high importance will be to identify how electron flow into various alternative sinks, resulting in different quenching mechanisms, can be tightly regulated to ensure that adequate protection does not come at the cost of compromised productivity.

References

Aliverti, A., Corrado, M.E. and Zanetti, G. (1994). Involvement of lysine-88 of spinach ferredoxin-NADP$^+$ reductase in the interaction with ferredoxin, *FEBS Lett.*, 343(3), 247–250.

Allahverdiyeva, Y., Mustila, H., Ermakova, M., Bersanini, L., Richaud, P., Ajlani, G., Battchikova, N., Cournac, L. and Aro, E.M. (2013). Flavodiiron proteins Flv1 and Flv3 enable cyanobacterial growth and photosynthesis under fluctuating light, *Proc. Natl. Acad. Sci. USA*, 110(10), 4111–4116.

Allen, J.F. and Hall, D.O. (1974). The relationship of oxygen uptake to electron transport in photosystem I of isolated chloroplasts: the role of superoxide and ascorbate, *Biochem. Biophys. Res. Commun.*, 58(3), 579–585.

Anderson, J.M., Fork, D.C. and Amesz, J. (1966). P700 and cytochrome f in particles obtained by digitonin fragmentation of spinach chloroplasts, *Biochem. Biophys. Res. Commun.*, 23(6), 874–879.

Anderson, J.W., Ed. (1981). *Light-Energy-Dependent Processes Other Than CO$_2$ Assimilation. The Biochemistry of Plants.* (Academic Press, New York/London).

Anderson, J.W. and Done, J. (1977a). Polarographic study of ammonia assimilation by isolated chloroplasts, *Plant Physiol.*, 60(4), 504–508.

Anderson, J.W. and Done, J. (1977b). A polarographic study of glutamate synthase activity in isolated chloroplasts, *Plant Physiol.*, 60(3), 354–359.

Anderson, J.W., Foyer, C.H. and Walker, D.A. (1983). Light-dependent reduction of dehydroascorbate and uptake of exogenous ascorbate by spinach chloroplasts, *Planta*, 158(5), 442–450.

Andrews, T.J., Lorimer, G.H. and Tolbert, N.E. (1973). Ribulose diphosphate oxygenase. I. Synthesis of phosphoglycolate by fraction-1 protein of leaves, *Biochemistry*, 12(1), 11–18.

Arnon, D.I. and Chain, R.K. (1979). Regulatory electron transport pathways in cyclic photophosphorylation: reduction of C-550 and cytochrome b_6 by ferredoxin in the dark, *FEBS Lett.*, 102(1), 133–138.

Aro, E.M., McCaffery, S. and Anderson, J.M. (1993). Photoinhibition and D1 protein degradation in peas acclimated to different growth irradiances, *Plant Physiol.*, 103(3), 835–843.

Atkinson, J.T., Campbell, I., Bennett, G.N. and Silberg, J.J. (2016). Cellular assays for ferredoxins: a strategy for understanding electron flow through protein carriers that link metabolic pathways, *Biochemistry*, 55(51), 7047–7064.

Baalmann, E., Backhausen, J.E., Kitzmann, C. and Scheibe, R. (1994). Regulation of NADP-dependent glyceraldehyde 3-phosphate dehydrogenase activity in spinach chloroplasts, *Bot. Acta,* 107, 313–320.

Backhausen, J.E., Kitzmann, C., Horton, P. and Scheibe, R. (2000). Electron acceptors in isolated intact spinach chloroplasts act hierarchically to prevent over-reduction and competition for electrons, *Photosynth. Res.*, 64(1), 1–13.

Backhausen, J.E., Kitzmann, C. and Scheibe, R. (1994). Competition between electron acceptors in photosynthesis: regulation of the malate valve during CO_2 fixation and nitrite reduction, *Photosynth. Res.*, 42, 75–86.

Batie, C.J. and Kamin, H. (1984a). Electron transfer by ferredoxin:NADP$^+$ reductase. Rapid-reaction evidence for participation of a ternary complex, *J. Biol. Chem.*, 259(19), 11976–11985.

Batie, C.J. and Kamin, H. (1984b). Ferredoxin:NADP$^+$ oxidoreductase. Equilibria in binary and ternary complexes with NADP$^+$ and ferredoxin, *J. Biol. Chem.*, 259(14), 8832–8839.

Baysdorfer, C. and Robinson, J.M. (1985). Metabolic interactions between spinach leaf nitrite reductase and ferredoxin-NADP reductase: competition for reduced ferredoxin, *Plant Physiol.*, 77(2), 318–320.

Bennett, J., Steinback, K.E. and Arntzen, C.J. (1980). Chloroplast phosphoproteins: regulation of excitation energy transfer by phosphorylation of thylakoid membrane polypeptides, *Proc. Natl. Acad. Sci. USA*, 77(9), 5253–5257.

Blanco, N.E., Ceccoli, R.D., Via, M.V., Voss, I., Segretin, M.E., Bravo-Almonacid, F.F., Melzer, M., Hajirezaei, M.R., Scheibe, R. and Hanke, G.T. (2013). Expression of the minor isoform pea ferredoxin in tobacco alters photosynthetic electron partitioning and enhances cyclic electron flow, *Plant Physiol.*, 161(2), 866–879.

Boardman, N.K. and Anderson, J.M. (1967). Fractionation of the photochemical systems of photosynthesis. II. Cytochrome and carotenoid contents of particles isolated from spinach chloroplasts, *Biochim. Biophys. Acta*, 143(1), 187–203.

Buchanan, B.B. and Balmer, Y. (2005). Redox regulation: a broadening horizon, *Annu. Rev. Plant Biol.*, 56, 187–220.

Burrows, P.A., Sazanov, L.A., Svab, Z., Maliga, P. and Nixon, P.J. (1998). Identification of a functional respiratory complex in chloroplasts through analysis of tobacco mutants containing disrupted plastid ndh genes, *EMBO J.*, 17(4), 868–876.

Caffarri, S., Tibiletti, T., Jennings, R.C. and Santabarbara, S. (2014). A comparison between plant photosystem I and photosystem II architecture and functioning, *Curr. Protein Pept. Sci.*, 15(4), 296–331.

Cassier-Chauvat, C. and Chauvat, F. (2014). Function and regulation of ferredoxins in the cyanobacterium, Synechocystis PCC6803: recent advances, *Life (Basel)*, 4(4), 666–680.

Cejudo, F. J., Ferrandez, J., Cano, B., Puerto-Galan, L. and Guinea, M. (2012). The function of the NADPH thioredoxin reductase C-2-Cys peroxiredoxin system in plastid redox regulation and signaling, *FEBS Lett.*, 586, 2974–2980.

Colombo, M., Suorsa, M., Rossi, F., Ferrari, R., Tadini, L., Barbato, R. and Pesaresi, P. (2016). Photosynthesis control: an underrated short-term regulatory mechanism essential for plant viability, *Plant Signal Behav.*, 11(4), p. e1165382.

Cournac, L., Redding, K., Ravenel, J., Rumeau, D., Josse, E.M., Kuntz, M. and Peltier, G. (2000). Electron flow between photosystem II and oxygen in chloroplasts of photosystem I-deficient algae is mediated by a quinol oxidase involved in chlororespiration, *J. Biol. Chem.*, 275(23), 17256–17262.

Dai, S., Friemann, R., Glauser, D.A., Bourquin, F., Manieri, W., Schurmann, P. and Eklund, H. (2007). Structural snapshots along the reaction pathway of ferredoxin-thioredoxin reductase, *Nature*, 448(7149), 92–96.

DalCorso, G., Pesaresi, P., Masiero, S., Aseeva, E., Schunemann, D., Finazzi, G., Joliot, P., Barbato, R. and Leister, D. (2008). A complex containing PGRL1 and PGR5 is involved in the switch between linear and cyclic electron flow in Arabidopsis, *Cell*, 132(2), 273–285.

Dietz, K.-J., Link, G., Pistorius, E.K. and Scheibe, R. (2002). Redox regulation in oxigenic photosynthesis, *Prog. Bot.*, 63, 207–224.

Dose, M.M., Hirasawa, M., Kleis-SanFrancisco, S., Lew, E.L. and Knaff, D.B. (1997). The ferredoxin-binding site of ferredoxin: nitrite oxidoreductase. Differential chemical modification of the free enzyme and its complex with ferredoxin, *Plant Physiol.*, 114(3), 1047–1053.

Doulis, A.G., Debian, N., Kingston-Smith, A.H. and Foyer, C.H. (1997). Differential localization of antioxidants in maize leaves, *Plant Physiol.*, 114(3), 1031–1037.

Durrant, J.R., Giorgi, L.B., Barber, J., Klug, D.R. and Porter, G. (1990). Characterisation of triplet states in isolated Photosystem II reaction centres: oxygen quenching as a mechanism for photodamage, *Biochim. Biophys. Acta*, 1017(2), 167–175.

Dutton, J.E., Rogers, L.J., Haslett, B.G., Takruri, I.A.H., Gleaves, J.T. and Boulter, D. (1980). Comparative studies on the properties of two ferredoxins from *Pisum sativum* L., *J. Exp. Bot.*, 31(2), 379–391.

Faske, M., Holtgrefe, S., Ocheretina, O., Meister, M., Backhausen, J.E. and Scheibe, R. (1995). Redox equilibria between the regulatory thiols of light/dark-modulated enzymes and dithiothreitol: fine-tuning by metabolites, *Biochim. Biophys. Acta,* 1247, 135–142.

Fridlyand, L.E., Backhausen, J.E., Holtgrefe, S., Kitzmann, C. and Scheibe, R. (1997). Quantitative evaluation of the rate of 3-phosphoglycerate reduction in chloroplasts, *Plant Cell Physiol.*, 38, 1177–1186.

Fridlyand, L.E. and Scheibe, R. (1999). Controlled distribution of electrons between acceptors in chloroplasts: a theoretical consideration, *Biochim. Biophys. Acta*, 1413(1), 31–42.

Gardeström, P. and Wigge, B. (1988). Influence of photorespiration on ATP/ADP ratios in the chloroplasts, mitochondria, and cytosol, studied by rapid fractionation of barley (*Hordeum vulgare*) protoplasts, *Plant Physiol.*, 88(1), 69–76.

Geigenberger, P., Thormählen, I., Daloso, D.M. and Fernie, A.R. (2017). The unprecedented versatility of the plant thioredoxin system, *Trends Plant Sci.*, 22, 249–262.

Gollan P.J., Tikkanen M. and Aro E.M. (2015). Photosynthetic light reactions: integral to chloroplast retrograde signaling, *Curr Opin Plant Biol.*, 27, 180–191.

Guo, H., Feng, P., Chi, W., Sun, X., Xu, X., Li, Y., Ren, D., Lu, C., Rochaix, J.D., Leister, D. and Zhang L. (2016). Plastid-nucleus communication involves calcium-modulated MAPK signaling, *Nat. Commun.*, 7, p. 12173.

Gütle, D.D., Roret, T., Hecker, A., Reski, R. and Jacquot, J.-P. (2017). Dithiol disulphide exchange in redox regulation of chloroplast enzymes in response to evolutionary and structural constraints, *Plant Sci.*, 255, 1–11.

Haehnel, W. (1976). The ratio of the two light reactions and their coupling in chloroplasts, *Biochim. Biophys. Acta*, 423(3), 499–509.

Hanke, G.T. and Hase, T. (2008). Variable photosynthetic roles of two leaf-type ferredoxins in Arabidopsis, as revealed by RNA interference, *Photochem. Photobiol.*, 84(6), 1302–1309.

Hanke, G. and Mulo, P. (2013). Plant type ferredoxins and ferredoxin-dependent metabolism, *Plant Cell Environ.*, 36(6), 1071–1084.

Hanke, G.T., Holtgrefe, S., König, N., Strodtkötter, I., Voss, I. and Scheibe, R. (2009). Use of transgenic plants to uncover strategies for maintenance of redox homeostasis during photosynthesis, *Adv. Bot. Res.*, 52, 207–251.

Hanke, G.T., Kimata-Ariga, Y., Taniguchi, I. and Hase, T. (2004). A post genomic characterization of Arabidopsis ferredoxins, *Plant Physiol.*, 134(1), 255–264.

Hanke, G.T., Satomi, Y., Shinmura, K., Takao, T. and Hase, T. (2011). A screen for potential ferredoxin electron transfer partners uncovers new, redox dependent interactions, *Biochim. Biophys. Acta*, 1814(2), 366–374.

Hase, T., Schürmann, P. and Knaff, D.B. (2006). The interaction of ferredoxin with ferredoxin-dependent enzymes. In *Photosystem I*, Golbeck, J., ed. (Springer: Dordrecht, The Netherlands), pp. 477–498.

Hebbelmann, I., Selinski, J., Wehmeyer, C., Goss, T., Voss, I., Mulo, P., Kangasjärvi, S., Aro, E.M., Oelze, M.L., Dietz, K.J., Nunes-Nesi, A., Do, P.T., Fernie, A.R., Talla, S.K., Raghavendra, A.S., Linke, V. and Scheibe, R. (2012). Multiple strategies to prevent oxidative stress in Arabidopsis plants lacking the malate valve enzyme NADP-malate dehydrogenase, *J. Exp. Bot.*, 63, 1445–1459.

Heineke, D. and Scheibe, R. (2009). Photosynthesis: the Calvin cycle. In *Encyclopedia of Life Sciences.* (John Wiley & Sons, Ltd., Chichester).

Hertle, A.P., Blunder, T., Wunder, T., Pesaresi, P., Pribil, M., Armbruster, U. and Leister, D. (2013). PGRL1 is the elusive ferredoxin-plastoquinone reductase in photosynthetic cyclic electron flow, *Mol. Cell*, 49(3), 511–523.

Hideg, É.V. and Vass, I. (1995). Singlet oxygen is not produced in photosystem I under photoinhibitory conditions, *Photochem. Photobiol.*, 62(5), 949–952.

Hirasawa, M., Tripathy, J.N., Sommer, F., Somasundaram, R., Chung, J.S., Nestander, M., Kruthiventi, M., Zabet-Moghaddam, M., Johnson, M.K., Merchant, S.S., Allen, J.P. and Knaff, D.B. (2010). Enzymatic properties of the ferredoxin-dependent nitrite reductase from *Chlamydomonas reinhardtii*. Evidence for hydroxylamine as a late intermediate in ammonia production, *Photosynth. Res.*, 103(2), 67–77.

Holtgrefe, S., Bader, H.P., Horton, P., Scheibe, R., von Schaewen, A. and Backhausen, J.E. (2003). Decreased content of leaf ferredoxin changes electron distribution and limits photosynthesis in transgenic potato plants, *Plant Physiol.*, 133, 1768–1778.

Horton, P., Ruban, A.V. and Wentworth, M. (2000). Allosteric regulation of the light-harvesting system of photosystem II, *Philos. Trans. R. Soc. Lond. B. Biol. Sci.*, 355(1402), 1361–1370.

Hurley, J.K., Hazzard, J.T., Martinez-Julvez, M., Medina, M., Gomez-Moreno, C. and Tollin, G. (1999). Electrostatic forces involved in orienting Anabaena ferredoxin during binding to Anabaena ferredoxin:NADP$^+$ reductase: site-specific mutagenesis, transient kinetic measurements, and electrostatic surface potentials, *Protein Sci.*, 8(8), 1614–1622.

Hurley, J.K., Morales, R., Martinez-Julvez, M., Brodie, T.B., Medina, M., Gomez-Moreno, C. and Tollin, G. (2002). Structure-function relationships in Anabaena ferredoxin/ferredoxin:NADP⁺ reductase electron transfer: insights from site-directed mutagenesis, transient absorption spectroscopy and X-ray crystallography, *Biochim. Biophys. Acta*, 1554(1–2), 5–21.

Igamberdiev, A.U., Bykova, N.V., Lea, P.J. and Gardestrom, P. (2001). The role of photorespiration in redox and energy balance of photosynthetic plant cells: a study with a barley mutant deficient in glycine decarboxylase, *Physiol. Plant*, 111(4), 427–438.

Jablonski, P.P. and Anderson, J.W. (1981). Light-dependent reduction of dehydroascorbate by ruptured pea chloroplasts, *Plant Physiol.*, 67(6), 1239–1244.

Jelesarov, I. and Bosshard, H.R. (1994). Thermodynamics of ferredoxin binding to ferredoxin: NADP+ reductase and the role of water at the complex interface, *Biochemistry*, 33(45), 13321–13328.

Josse, E.M., Simkin, A.J., Gaffe, J., Laboure, A.M., Kuntz, M. and Carol, P. (2000). A plastid terminal oxidase associated with carotenoid desaturation during chromoplast differentiation, *Plant Physiol.*, 123(4), 1427–1436.

Khristin, M.S. and Akulova, E.A. (1976). Two forms of pea leaf ferredoxin, *Biokhimiia*, 41(3), 500–505.

Kim, J.Y., Nakayama, M., Toyota, H., Kurisu, G. and Hase, T. (2016). Structural and mutational studies of an electron transfer complex of maize sulfite reductase and ferredoxin, *J. Biochem.*, 160(2), 101–109.

Kimata, Y. and Hase, T. (1989). Localization of ferredoxin isoproteins in mesophyll and bundle sheath cells in maize leaf, *Plant Physiol.*, 89(4), 1193–1197.

Kimata-Ariga, Y., Matsumura, T., Kada, S., Fujimoto, H., Fujita, Y., Endo, T., Mano, J., Sato, F. and Hase, T. (2000). Differential electron flow around photosystem I by two C4-photosynthetic-cell-specific ferredoxins, *EMBO J.*, 19(19), 5041–5050.

Kmiecik, P., Leonardelli, M. and Teige, M. (2016). Novel connections in plant organellar signalling link different stress responses and signalling pathways. *J. Exp. Bot.*, 67(13), 3793–3807.

Kocsy, G., Tari, I., Vankova, R., Zechmann, B., Gulyas, Z., Poor, P. and Galiba, G. (2013). Redox control of plant growth and development, *Plant Sci.*, 211, 77–91.

Kohchi, T., Mukougawa, K., Frankenberg, N., Masuda, M., Yokota, A. and Lagarias, J.C. (2001). The Arabidopsis HY2 gene encodes phytochromobilin synthase, a ferredoxin-dependent biliverdin reductase, *Plant Cell*, 13(2), 425–436.

Kramer, D.M. and Evans, J.R. (2011). The importance of energy balance in improving photosynthetic productivity, *Plant Physiol.*, 155(1), 70–78.

Krömer, S. and Scheibe, R. (1996). Function of the chloroplastic malate valve for respiration during photosynthesis, *Biochem. Soc. Trans.*, 24, 761–766.

Krömer, S., Stitt, M. and Heldt, H. (1988). Mitochondrial oxidative phosphorylation participating in photosynthetic metabolism of a leaf cell, *FEBS Lett.*, 226(2), 352–356.

Krueger, R.J. and Siegel, L.M. (1982). Spinach siroheme enzymes: isolation and characterization of ferredoxin-sulfite reductase and comparison of properties with ferredoxin-nitrite reductase, *Biochemistry*, 21(12), 2892–2904.

Kudoh, H. and Sonoike, K. (2002). Irreversible damage to photosystem I by chilling in the light: cause of the degradation of chlorophyll after returning to normal growth temperature, *Planta*, 215(4), 541–548.

Kurisu, G., Kusunoki, M., Katoh, E., Yamazaki, T., Teshima, K., Onda, Y., Kimata-Ariga, Y. and Hase, T. (2001). Structure of the electron transfer complex between ferredoxin and ferredoxin-NADP$^+$ reductase, *Nat. Struct. Biol.*, 8(2), 117–121.

Kurisu, G., Nishiyama, D., Kusunoki, M., Fujikawa, S., Katoh, M., Hanke, G.T., Hase, T. and Teshima, K. (2005). A structural basis of *Equisetum arvense* ferredoxin isoform II producing an alternative electron transfer with ferredoxin-NADP$^+$ reductase, *J. Biol. Chem.*, 280(3), 2275–2281.

Lancaster, J.R., Vega, J.M., Kamin, H., Orme-Johnson, N.R., Orme-Johnson, W.H., Krueger, R.J. and Siegel, L.M. (1979). Identification of the iron-sulfur center of spinach ferredoxin-nitrite reductase as a tetranuclear center, and preliminary EPR studies of mechanism, *J. Biol. Chem.*, 254(4), 1268–1272.

Lea, P.J. and Miflin, B.J. (1974). Alternative route for nitrogen assimilation in higher plants, *Nature*, 251(5476), 614–616.

Lee, Y.H., Ikegami, T., Standley, D.M., Sakurai, K., Hase, T. and Goto, Y. (2011). Binding energetics of ferredoxin-NADP$^+$ reductase with ferredoxin and its relation to function, *Chembiochem.*, 12(13), 2062–2070.

Lehtimaki, N., Lintala, M., Allahverdiyeva, Y., Aro, E.M. and Mulo, P. (2010). Drought stress-induced upregulation of components involved in ferredoxin-dependent cyclic electron transfer, *J. Plant Physiol.*, 167(12), 1018–1022.

Li, Z., Wakao, S., Fischer, B.B. and Niyogi, K. (2009). Sensing and responding to light, *Annu. Rev. Plant Biol.*, 60, 239–260.

Livingston, A.K., Cruz, J.A., Kohzuma, K., Dhingra, A. and Kramer, D.M. (2010). An Arabidopsis mutant with high cyclic electron flow around photosystem I (hcef) involving the NADPH dehydrogenase complex, *Plant Cell*, 22(1), 221–233.

Livingston, A.K., Kanazawa, A., Cruz, J.A. and Kramer, D.M. (2010). Regulation of cyclic electron flow in C(3) plants: differential effects of limiting photosynthesis at ribulose-1,5-bisphosphate carboxylase/oxygenase and glyceraldehyde-3-phosphate dehydrogenase, *Plant Cell Environ.*, 33(11), 1779–1788.

Majeran, W., Cai, Y., Sun, Q. and van Wijk, K.J. (2005). Functional differentiation of bundle sheath and mesophyll maize chloroplasts determined by comparative proteomics, *Plant Cell*, 17(11), 3111–3140.

Martinez-Julvez, M., Medina, M. and Velazquez-Campoy, A. (2009). Binding thermodynamics of ferredoxin:NADP$^+$ reductase: two different protein substrates and one energetics, *Biophys. J.*, 96(12), 4966–4975.

Matsumura, T., Kimata-Ariga, Y., Sakakibara, H., Sugiyama, T., Murata, H., Takao, T., Shimonishi, Y. and Hase, T. (1999). Complementary DNA cloning and characterization of ferredoxin localized in bundle-sheath cells of maize leaves, *Plant Physiol.*, 119(2), 481–488.

Matsumura, T., Sakakibara, H., Nakano, R., Kimata, Y., Sugiyama, T. and Hase, T. (1997). A nitrate-inducible ferredoxin in maize roots. Genomic organization and differential expression of two nonphotosynthetic ferredoxin isoproteins, *Plant Physiol.*, 114(2), 653–660.

Mehler, A.H. (1951). Studies on reactions of illuminated chloroplasts. II. Stimulation and inhibition of the reaction with molecular oxygen, *Arch. Biochem. Biophys.*, 34(2), 339–351.

Mellor, S.B., Nielsen, A.Z., Burow, M., Motawia, M.S., Jakubauskas, D., Moller, B.L. and Jensen, P.E. (2016). Fusion of ferredoxin and cytochrome P450 enables direct light-driven biosynthesis, *ACS Chem. Biol.*, 11(7), 1862–1869.

Mellor, S.B., Vavitsas, K., Nielsen, A.Z. and Jensen, P.E. (2017). Photosynthetic fuel for heterologous enzymes: the role of electron carrier proteins, *Photosynth. Res.*, doi: 10.1007/s11120-017-0364-0.

Mitchell, P. (1975). The protonmotive Q cycle: a general formulation, *FEBS Lett.*, 59(2), 137–139.

Miyake, C. and Asada, K. (1992). Thylakoid-bound ascorbate peroxidase in spinach-chloroplasts and photoreduction of its primary oxidation-product monodehydroascorbate radicals in thylakoids, *Plant Cell Physiol.*, 33, 541–553.

Montrichard, F., Alkhalfioui, F., Yano, H., Vensel, W.H., Hurkman, W.J. and Buchanan, B.B. (2009). Thioredoxin targets in plants: the first 30 years, *J. Proteom.*, 72(3), 452–474.

Müller, B. and Ziegler, H. (1969). Die lichtinduzierte Aktivitätssteigerung der NADP$^+$-abhängigen Glycerinaldehyd-3-phosphat-Dehydrogenase. IX. Die Reaktion in isolierten Chloroplasten, *Planta*, 85, 96–104.

Müller, B., Ziegler, H. and Ziegler, I. (1969). Lichtinduzierte, reversible Aktivitätssteigerung der NADP-abhängigen Glycerinaldehyd-3-phosphat-Dehydrogenase in Chloroplasten. Zum Mechanismus der Reaktion, *Eur. J. Biochem.*, 9, 101–106.

Munekage, Y., Hojo, M., Meurer, J., Endo, T., Tasaka, M. and Shikanai, T. (2002). PGR5 is involved in cyclic electron flow around photosystem I and is essential for photoprotection in Arabidopsis, *Cell*, 110(3), 361–371.

Muramoto, T., Tsurui, N., Terry, M.J., Yokota, A. and Kohchi, T. (2002). Expression and biochemical properties of a ferredoxin-dependent heme oxygenase required for phytochrome chromophore synthesis, *Plant Physiol.*, 130(4), 1958–1966.

Nakayama, M., Akashi, T. and Hase, T. (2000). Plant sulfite reductase: molecular structure, catalytic function and interaction with ferredoxin, *J. Inorg. Biochem.*, 82(1–4), 27–32.

Navrot, N., Finnie, C., Svensson, B. and Hagglund, P. (2011). Plant redox proteomics, *J. Proteom.*, 74(8), 1450–1462.

Nikkanen, L. and Rintamaki, E. (2014). Thioredoxin-dependent regulatory networks in chloroplasts under fluctuating light conditions, *Philos. Trans. R. Soc. Lond. B. Biol. Sci.*, 369(1640), p. 20130224.

Niyogi, K.K., Grossman, A.R. and Bjorkman, O. (1998). Arabidopsis mutants define a central role for the xanthophyll cycle in the regulation of photosynthetic energy conversion, *Plant Cell*, 10(7), 1121–1134.

Noctor, G. and Foyer, C.H. (2016). Intracellular redox compartmentation and ROS-related communication in regulation and signaling, *Plant Physiol.*, 171(3), 1581–1592.

Onda, Y., Matsumura, T., Kimata-Ariga, Y., Sakakibara, H., Sugiyama, T. and Hase, T. (2000). Differential interaction of maize root ferredoxin:NADP$^+$ oxidoreductase with photosynthetic and non-photosynthetic ferredoxin isoproteins, *Plant Physiol.*, 123(3), 1037–1045.

Park, Y.I., Chow, W.S., Osmond, C.B. and Anderson, J.M. (1996). Electron transport to oxygen mitigates against the photoinactivation of Photosystem II *in vivo*, *Photosynth. Res.*, 50(1), 23–32.

Peden, E.A., Boehm, M., Mulder, D.W., Davis, R., Old, W.M., King, P.W., Ghirardi, M.L. and Dubini, A. (2013). Identification of global ferredoxin interaction networks in *Chlamydomonas reinhardtii*, *J. Biol. Chem.*, 288(49), 35192–35209.

Pesaresi, P., Pribil, M., Wunder, T. and Leister, D. (2011). Dynamics of reversible protein phosphorylation in thylakoids of flowering plants: the roles of STN7, STN8 and TAP38, *Biochim. Biophys. Acta*, 1807(8), 887–896.

Petrov, V.D. and Van Breusegem, F. (2012). Hydrogen peroxide — a central hub for information flow in plant cells, *AoB Plants*, 2012, p. pls014.

Pierella Karlusich, J.J. and Carrillo, N. (2017). Evolution of the acceptor side of photosystem I: ferredoxin, flavodoxin, and ferredoxin-NADP$^+$ oxidoreductase, *Photosynth. Res.*, doi:10.1007/s11120-017-0338-2.

Pospisil, P., Arato, A., Krieger-Liszkay, A. and Rutherford, A.W. (2004). Hydroxyl radical generation by photosystem II, *Biochemistry*, 43(21), 6783–6792.

Pulido, P., Spịnola, M.C., Kirchsteiger, K., Guinea, M., Pascual, M.B., Sahrawy, M., Sandalio, L.M., Dietz, K.J., Gonzalez, M. and Cejudo, F.J. (2010). Functional analysis of the pathways for 2-Cys peroxiredoxin reduction in *Arabidopsis thaliana* chloroplasts, *J. Exp. Bot.*, 61, 4043–4054.

Reinbothe, C., Bartsch, S., Eggink, L.L., Hoober, J.K., Brusslan, J., Andrade-Paz, R., Monnet, J. and Reinbothe, S. (2006). A role for chlorophyllide a oxygenase in the regulated import and stabilization of light-harvesting chlorophyll *a/b* proteins, *Proc. Natl. Acad. Sci. USA*, 103(12), 4777–4782.

Ren, Y., Li, Y., Jiang, Y., Wu, B. and Miao, Y. (2017). Phosphorylation of WHIRLY1 by CIPK14 shifts its localization and dual functions in Arabidopsis, *Mol. Plant,* 10(5), 749–763.

Rodoni, S., Muhlecker, W., Anderl, M., Krautler, B., Moser, D., Thomas, H., Matile, P. and Hortensteiner, S. (1997). Chlorophyll breakdown in senescent chloroplasts (cleavage of pheophorbide a in two enzymic steps), *Plant Physiol.*, 115(2), 669–676.

Rumberg, B. and Siggel, U. (1969). pH changes in the inner phase of the thylakoids during photosynthesis, *Naturwissenschaften*, 56(3), 130–132.

Sacharz, J., Giovagnetti, V., Ungerer, P., Mastroianni, G. and Ruban, A.V. (2017). The xanthophyll cycle affects reversible interactions between PsbS and light-harvesting complex II to control non-photochemical quenching, *Nat. Plants*, 3, p. 16225.

Saitoh, T., Ikegami, T., Nakayama, M., Teshima, K., Akutsu, H. and Hase, T. (2006). NMR study of the electron transfer complex of plant ferredoxin and sulfite reductase: mapping the interaction sites of ferredoxin, *J. Biol. Chem.*, 281(15), 10482–10488.

Sakakibara, H. (2003). Differential response of genes for ferredoxin and ferredoxin:NADP$^+$ oxidoreductase to nitrate and light in maize leaves, *J. Plant Physiol.*, 160(1), 65–70.

Sakakibara, Y., Kimura, H., Iwamura, A., Saitoh, T., Ikegami, T., Kurisu, G. and Hase, T. (2012). A new structural insight into differential interaction of cyanobacterial and plant ferredoxins with nitrite reductase as revealed by NMR and X-ray crystallographic studies, *J. Biochem.*, 151(5), 483–492.

Satoh, K. (1970). Mechanism of photoinactivation in photosynthetic systems III. Site and mode of photoinactivation in photosystem I, *Plant Cell Physiol.*, 11(2), 187–197.

Scheibe, R. (1991). Redox-modulation of chloroplast enzymes: a common principle for individual control, *Plant Physiol.*, 96(1), 1–3.

Scheibe, R. (2004). Malate valves to balance cellular energy supply, *Physiol. Plant.*, 120(1), 21–26.

Scheibe, R., Backhausen, J.E., Emmerlich, V. and Holtgrefe, S. (2005). Strategies to maintain redox homeostasis during photosynthesis under changing conditions, *J. Exp. Bot.*, 56, 1481–1489.

Scheibe, R. and Beck, E. (1975). Formation of C-4 dicarboxylic acids by intact spinach chloroplasts, *Planta*, 125(1), 63–67.

Scheibe, R. and Dietz, K.-J. (2012). Reduction-oxidation network for flexible adjustment of cellular metabolism in photoautotrophic cells, *Plant Cell Environ.*, 35, 202–216.

Scheibe, R. and Jacquot, J.-P. (1983). NADP regulates the light activation of NADP-dependent malate dehydrogenase, *Planta*, 157, 548–553.

Scheibe, R. and Stitt, M. (1988). Comparison of NADP-malate dehydrogenase activation, Q_A reduction and O_2 evolution in spinach leaves, *Plant Physiol. Biochem.*, 26, 473–481.

Schmidt, H. and Heinz, E. (1990). Involvement of ferredoxin in desaturation of lipid-bound oleate in chloroplasts, *Plant Physiol.*, 94(1), 214–220.

Schuster, G., Timberg, R. and Ohad, I. (1988). Turnover of thylakoid photosystem II proteins during photoinhibition of *Chlamydomonas reinhardtii*, *Eur. J. Biochem.*, 177(2), 403–410.

Sétif, P. and Brettel, K. (1990). Photosystem I photochemistry under highly reducing conditions: study of the P700 triplet state formation from the secondary radical pair (P700$^+$–A$_1^-$), *Biochim. Biophys. Acta,* 1020(3), 232–238.

Setif, P., Fischer, N., Lagoutte, B., Bottin, H. and Rochaix, J.D. (2002). The ferredoxin docking site of photosystem I, *Biochim. Biophys. Acta*, 1555(1–3), 204–209.

Setif, P., Harris, N., Lagoutte, B., Dotson, S. and Weinberger, S.R. (2010). Detection of the photosystem I:ferredoxin complex by backscattering interferometry, *J. Am. Chem. Soc.*, 132(31), 10620–10622.

Setif, P., Hervo, G. and Mathis, P. (1981). Flash-induced absorption changes in photosystem I, radical pair or triplet state formation? *Biochim. Biophys. Acta*, 638(2), 257–267.

Setif, P., Mutoh, R. and Kurisu, G. (2017). Dynamics and energetics of cyanobacterial photosystem I:ferredoxin complexes in different redox states, *Biochim. Biophys. Acta*, 1858(7), 483–496.

Shen, Y.K., Chow, W.S., Park, Y.I. and Anderson, J.M. (1996). Photoinactivation of photosystem II by cumulative exposure to short light pulses during the induction period of photosynthesis, *Photosynth. Res.*, 47(1), 51–59.

Shin, M., Tagawa, K. and Arnon, D.I. (1963). Crystallization of ferredoxin-TPN reductase and its role in the photosynthetic apparatus of chloroplasts, *Biochem. Z.*, 338, 84–96.

Shinohara, F., Kurisu, G., Hanke, G., Bowsher, C., Hase, T. and Kimata-Ariga, Y. (2017). Structural basis for the isotype-specific interactions of ferredoxin and ferredoxin: NADP$^+$ oxidoreductase: an evolutionary switch between photosynthetic and heterotrophic assimilation, *Photosynth. Res.*, doi:10.1007/s11120-016-0331-1.

Shipton, C.A. and Barber, J. (1994). *In vivo* and *in vitro* photoinhibition reactions generate similar degradation fragments of D1 and D2 photosystem-II reaction-centre proteins, *Eur. J. Biochem.*, 220(3), 801–808.

Sonoike, K. (1995). Selective photoinhibition of photosystem I in isolated thylakoid membranes from cucumber and spinach, *Plant Cell Physiol.*, 36(5), 825–830.

Sonoike, K. (1996). Degradation of psaB gene product, the reaction center subunit of photosystem I, is caused during photoinhibition of photosystem I: possible involvement of active oxygen species, *Plant Sci.*, 115, 157–164.

Strand, D.D., Fisher, N., Davis, G.A. and Kramer, D.M. (2016). Redox regulation of the antimycin A sensitive pathway of cyclic electron flow around photosystem I in higher plant thylakoids, *Biochim. Biophys. Acta*, 1857(1), 1–6.

Strand, D.D., Livingston, A.K., Satoh-Cruz, M., Froehlich, J.E., Maurino, V.G. and Kramer, D.M. (2015). Activation of cyclic electron flow by hydrogen peroxide *in vivo*, *Proc. Natl. Acad. Sci. USA*, 112(17), 5539–5544.

Strodtkötter, I., Padmasree, K., Dinakar, C., Speth, B., Niazi, P.S., Wojtera, J., Voss, I., Do, P.T., Nunes-Nesi, A., Fernie, A.R., Linke, V., Raghavendra, A.S. and Scheibe, R. (2009). Induction of the AOX1D isoform of alternative oxidase in *A. thaliana* T-DNA-insertion lines lacking isoform AOX1A is insufficient to optimize photosynthesis when treated with antimycin A, *Mol. Plant*, 2, 284–297.

Stuhlfauth, T., Scheuermann, R. and Fock, H.P. (1990). Light energy dissipation under water stress conditions: contribution of reassimilation and evidence for additional processes, *Plant Physiol.*, 92(4), 1053–1061.

Suorsa, M., Jarvi, S., Grieco, M., Nurmi, M., Pietrzykowska, M., Rantala, M., Kangasjarvi, S., Paakkarinen, V., Tikkanen, M., Jansson, S. and Aro, E.M. (2012). PROTON GRADIENT REGULATION5 is essential for proper acclimation of Arabidopsis photosystem I to naturally and artificially fluctuating light conditions, *Plant Cell*, 24(7), 2934–2948.

Suzuki, A. and Knaff, D.B. (2005). Glutamate synthase: structural, mechanistic and regulatory properties, and role in the amino acid metabolism, *Photosynth. Res.*, 83(2), 191–217.

Takagi, D., Ishizaki, K., Hanawa, H., Mabuchi, T., Shimakawa, G., Yamamoto, H. and Miyake, C. (2017). Diversity of strategies for escaping reactive oxygen species production within photosystem I among land plants: P700 oxidation system is prerequisite for alleviating photoinhibition in photosystem I, *Physiol. Plant*, doi:10.1111/ppl.12562.

Tanaka, A., Ito, H., Tanaka, R., Tanaka, N.K., Yoshida, K. and Okada, K. (1998). Chlorophyll *a* oxygenase (CAO) is involved in chlorophyll *b* formation from chlorophyll *a*, *Proc. Natl. Acad. Sci. USA*, 95(21), 12719–12723.

Terauchi, A.M., Lu, S.F., Zaffagnini, M., Tappa, S., Hirasawa, M., Tripathy, J.N., Knaff, D.B., Farmer, P.J., Lemaire, S.D., Hase, T. and Merchant, S.S. (2009). Pattern of expression and substrate specificity of chloroplast ferredoxins from *Chlamydomonas reinhardtii*, *J. Biol. Chem.*, 284(38), 25867–25878.

Thomsen-Zieger, N., Pandini, V., Caprini, G., Aliverti, A., Cramer, J., Selzer, P.M., Zanetti, G. and Seeber, F. (2004). A single *in vivo*-selected point mutation in the active center of *Toxoplasma gondii* ferredoxin-NADP$^+$ reductase leads to an inactive enzyme with greatly enhanced affinity for ferredoxin, *FEBS Lett.*, 576(3), 375–380.

Van Aken, O. and Pogson, B.J. (2017). Convergence of mitochondrial and chloroplastic ANAC017/PAP-dependent retrograde signalling pathways and suppression of programmed cell death, *Cell Death Differ.*, 24(6), 955–960.

Vega, J.M. and Kamin, H. (1977). Spinach nitrite reductase. Purification and properties of a siroheme-containing iron-sulfur enzyme, *J. Biol. Chem.*, 252(3), 896–909.

Vicente, J.B., Gomes, C.M., Wasserfallen, A. and Teixeira, M. (2002). Module fusion in an A-type flavoprotein from the cyanobacterium Synechocystis condenses a multiple-component pathway in a single polypeptide chain, *Biochem. Biophys. Res. Commun.*, 294(1), 82–87.

Voss, I., Goss, T., Murozuka, E., Altmann, B., McLean, K.J., Rigby, S.E., Munro, A.W., Scheibe, R., Hase, T. and Hanke, G.T. (2011). FdC1, a novel ferredoxin protein capable of alternative electron partitioning, increases in conditions of acceptor limitation at photosystem I, *J. Biol. Chem.*, 286(1), 50–59.

Voss, I., Koelmann, M., Wojtera, J., Holtgrefe, S., Kitzmann, C., Backhausen, J.E. and Scheibe, R. (2008). Knock-out of major leaf ferredoxin reveals new redox-regulatory adaptations in *Arabidopsis thaliana*, *Physiol. Plant*, 133, 584–598.

Voss, I., Sunil, B., Scheibe, R. and Raghavendra, A.S. (2013). Emerging concept for the role of photorespiration as an important part of abiotic stress response, *Plant Biol.*, 56, 435–447.

Wlodarczyk, A., Gnanasekaran, T., Nielsen, A.Z., Zulu, N.N., Mellor, S.B., Luckner, M., Thofner, J.F., Olsen, C.E., Mottawie, M.S., Burow, M., Pribil, M., Feussner, I., Moller, B.L. and Jensen, P.E. (2016). Metabolic engineering of light-driven cytochrome P450 dependent pathways into Synechocystis sp. PCC 6803, *Metab. Eng.*, 33, 1–11.

Wojtera-Kwiczor, J., Groß, F., Leffers, H.-M., Kang, M., Schneider, M. and Scheibe, R. (2013). Transfer of a redox-signal through the cytosol by redox-dependent microcompartmentation of glycolytic enzymes at mitochondria and actin cytoskeleton, *Front. Plant Sci.* 3, p. 284.

Wolosiuk, R.A. and Buchanan, B.B. (1978). Activation of chloroplast NADP-linked glyceraldehyde-3-phosphate dehydrogenase by the ferredoxin/thioredoxin system, *Plant Physiol.*, 61(4), 669–671.

Xu, X., Chi, W., Sun, X., Feng, P., Guo, H., Li, J., Lin, R., Lu, C., Wang, H., Leister, D. and Zhang, L. (2016). Convergence of light and chloroplast signals for de-etiolation through ABI4-HY5 and COP1, *Nat. Plants*, 2(6), p. 16066.

Xu, X., Kim, S.K., Schurmann, P., Hirasawa, M., Tripathy, J.N., Smith, J., Knaff, D.B. and Ubbink, M. (2006). Ferredoxin/ferredoxin-thioredoxin reductase complex: complete NMR mapping of the interaction site on ferredoxin by Gallium substitution, *FEBS Lett.*, 580(28–29), 6714–6720.

Yamamoto, H., Kato, H., Shinzaki, Y., Horiguchi, S., Shikanai, T., Hase, T., Endo, T., Nishioka, M., Makino, A., Tomizawa, K. and Miyake, C. (2006). Ferredoxin limits cyclic electron flow around PSI (CEF-PSI) in higher plants-stimulation of CEF-PSI enhances non-photochemical quenching of Chl fluorescence in transplastomic tobacco, *Plant Cell Physiol.*, 47(10), 1355–1371.

Yamamoto, H., Peng, L., Fukao, Y. and Shikanai, T. (2011). An Src homology 3 domain-like fold protein forms a ferredoxin binding site for the chloroplast NADH dehydrogenase-like complex in Arabidopsis, *Plant Cell*, 23(4), 1480–1493.

Yang, W., Wittkopp, T.M., Li, X., Warakanont, J., Dubini, A., Catalanotti, C., Kim, R.G., Nowack, E.C., Mackinder, L.C., Aksoy, M., Page, M.D., D'Adamo, S., Saroussi, S., Heinnickel, M., Johnson, X., Richaud, P., Alric, J., Boehm, M., Jonikas, M.C., Benning, C., Merchant, S.S., Posewitz, M.C. and Grossman, A.R. (2015). Critical role of *Chlamydomonas reinhardtii* ferredoxin-5 in maintaining membrane structure and dark metabolism, *Proc. Natl. Acad. Sci. USA*, 112(48), 14978–14983.

Yonekura-Sakakibara, K., Onda, Y., Ashikari, T., Tanaka, Y., Kusumi, T. and Hase, T. (2000). Analysis of reductant supply systems for ferredoxin-dependent sulfite reductase in photosynthetic and nonphotosynthetic organs of maize, *Plant Physiol.*, 122(3), 887–894.

Yoshida, K., Hara, S. and Hisabori, T. (2015). Thioredoxin selectivity for thiol-based redox regulation of target proteins in chloroplasts, *J. Biol. Chem.*, 290, 14278–14288.

Yoshida, K. and Hisabori, T. (2017). Distinct electron transfer from ferredoxin-thioredoxin reductase to multiple thioredoxin isoforms in chloroplasts, *Biochem. J.*, 474, 1347–1360.

Zachgo, S., Hanke, G.T. and Scheibe, R. (2013). Plant cell microcompartments: a redox-signaling perspective, *Biol. Chem.*, 394, 203–216.

Zhang, L.T., Zhang, Z.S., Gao, H.Y., Meng, X.L., Yang, C., Liu, J.G. and Meng, Q.W. (2012). The mitochondrial alternative oxidase pathway protects the photosynthetic apparatus against photodamage in Rumex K-1 leaves, *BMC Plant Biol.*, 12, p. 40.

Zivcak, M., Brestic, M., Kunderlikova, K., Sytar, O. and Allakhverdiev, S.I. (2015). Repetitive light pulse-induced photoinhibition of photosystem I severely affects CO_2 assimilation and photoprotection in wheat leaves, *Photosynth Res.*, 126(2–3), 449–463.

Chapter 14

Cyclic Electron Flow in Cyanobacteria and Eukaryotic Algae

A. W. D. Larkum*,§, M. Szabó*,¶, D. Fitzpatrick† and J. A. Raven*,‡

*Climate Change Cluster,
University of Technology Sydney, Broadway, NSW 2007 Australia
†Molecular Plant Biology, University of Turku, Finland
‡Division of Plant Sciences,
University of Dundee at the James Hutton Institute,
Invergowrie, Dundee DD2 5DA, UK
†duncan.fitzpatrick@utu.fi
‡j.a.raven@dundee.ac.uk
§a.larkum@sydney.edu.au
¶Milan.Szabo@uts.edu.au

In oxygenic photosynthesis light energy is largely captured in linear electron flow (LEF) between the photosystems and drives ATP formation *via* a thylakoid proton-driven ATP synthase. In addition, for over 50 years there has been good evidence that an additional cyclic electron flow (CEF) around photosystem I (PSI) is harnessed to provide extra ATP in addition to that produced by LEF. The evidence comes from all oxygenic organisms, cyanobacteria, eukaryotic algae and embryophytic plants. However, the CEF mechanism has been difficult to investigate because of the cyclic nature of the EF and confusion with other pathways not using oxygen as a terminal electron acceptor, and the MAPS, flavodiiron and chlororespiration pathways to oxygen. This article discusses the current evidence for CEF in all oxygenic organisms and suggests future experiments by which the situation can be clarified.

1. Introduction

The first description of photophosphorylation, *i.e.* light-energized ADP phosphorylation by isolated *Spinacia oleracea* thylakoids, was by Arnon *et al.* [1954], and "cyclic photophosphorylation" was coined by Arnon *et al.* [1958] for ATP production by thylakoids with no net synthesis of an oxidant and a reductant. Tagawa *et al.* [1963] demonstrated the occurrence of cyclic photophosphorylation by isolated thylakoids with ferredoxin as the only added redox co-factor; ferredoxin is now recognised as an *in vivo* co-factor in cyclic electron flow (CEF) coupled to H^+ flux into the thylakoid lumen and hence, *via* H^+ flux out of the lumen through the CF_0CF_1 ATP synthase, ADP phosphorylation [Peltier *et al.*, 2010; Labs *et al.*, 2016; Peltier *et al.*, 2016; Yamori and Shikanai, 2016; Govindjee *et al.*, 2017].

Most of the subsequent mechanistic work on CEF has focused on a few species of flowering plants (*e.g. Arabidopsis thaliana*) and eukaryotic algae (*e.g. Chlamydomonas reinhardtii*) as well as some cyanobacteria [Peltier *et al.*, 2010, 2016; Yamori and Shikanai 2016]. Despite this, there are a number of areas of uncertainty, *e.g.* the contribution of CEF during steady state photosynthesis, the mechanism of proton gradient regulation-like photosynthetic phenotype 1 (PGRL1) and proton gradient regulation 5 (PGR5) (see below), and the NDH pathway where it occurs as well as PGRL1-PGR5, the factors determining the relative importance of the two pathways. As well as considering these questions in the context of cyanobacteria and eukaryotic algae, this paper considers the algae in which CEF has been sought, the energy-requiring processes that CEF powers when linear electron flow (LEF) photosynthesis is blocked, and also the cases where other energy sources can substitute for CEF, or CEF is not involved. A further area that is considered is environmental influences on CEF by phenotypic acclimation and genetic adaptation.

2. *In Vivo* Evidence of Cyclic Photophosphorylation in Algae

2.1. *Overview*

The early and, more generally, non-molecular genetic work on cyclic electron flow and cyclic photophosphorylation in eukaryotic algae and cyanobacteria involved three main categories of *in vivo* studies. Most of the publications involve studies of ATP-requiring processes that were light stimulated and were characterized by one or more of (1) independence of the presence of CO_2 and O_2, (2) a higher ratio of the rate in radiation >700 nm to the rate in radiation <700 nm than is the case for photosynthesis and (3) less inhibition by DCMU than is the case for LEF photosynthesis [reviewed by Teichler-Zallen and Hoch, 1967; Simonis and Urbach, 1973; Raven 1976a, 1976b, 1984; Falkowski and Raven, 2007]. These studies involved a wide range of processes, and classes of algae (Table 1). A smaller number of papers deal with turnover of c-type cytochromes and, especially, P700 in radiation >700 nm and/or in a wide range of wavelengths in the presence of DCMU at a concentration that completely inhibits LEF photosynthesis (*e.g.* Biggins, 1973; Maxwell and Biggins, 1976; see Table 1). Three

Table 1. Non-molecular evidence of cyclic electron flow and associated phosphorylation in cyanobacteria and eukaryotic algae.

Division/Class	Electron flow	Photoacoustics	Photophosphorylation
Cyanobacteria	Teichler-Zahlen and Hoch [1967]; Berla *et al.* [2015]		Urbach and Simonis [1973]; Der-Vertanian *et al.* [1981]; Ogawa *et al.* [1985a, 1985b]
Rhodophyta: Bangiophyceae	Biggins [1973]; Maxwell and Biggins [1976]	Herbert *et al.* [1990]	Raven *et al.* [1990]
Rhodophyta: Floridiophyceae			
Chlorophyta: Prasinophyceae			
Chlorophyta: Chlorophyceae	Teichler-Zahlen and Hoch [1967]		Urbach and Simonis [1973]
Chlorophyta: Trebouxiophyceae		Cha and Mauzerall [1992]	Urbach and Simonis [1973]
Chlorophyta: Ulvophyceae		Herbert *et al.* [1990]	Urbach and Simonis [1973]
Streptophyta: Charophyceae			Urbach and Simonis [1973]
Dinophyta			
Ochrophyta: Bacillariophyceae	Bailleul *et al.* [2015] Coldman *et al.* [2015]		
Ochrophyta: Eustigmatophyceae			
Ochrophyta: Phaeophyceae		Herbert *et al.* [1990] Fork *et al.* [1990]	
Cryptophyta			
Haptophyta	Zhang *et al.* [2014]		Paasche [1964]; Paasche [1966].

publications [Herbert *et al.*, 1990; Fork *et al.*, 1991; Cha and Mauzerall 1992] deal with photoacoustic estimates of energy storage under a range of wavelengths (see Table 1).

Considering *in vivo* ATP-requiring processes as an assay for cyclic photophosphorylation, the reviews cited above show that acetate and glucose uptake and metabolism, active (against an electrochemical potential gradient) ion transport, and diazotrophy in heterocystous cyanobacteria, photoreduction using H_2 in the absence of photosystem II (PSII), and inhibition of fermentation and respiration. Table 1 shows the diversity of cyanobacteria and algae in one or more of these processes. Raven [1976a, 1976b] showed that the capacity for cyclic photophosphorylation is similar to that of oxidative phosphorylation in the green algae examined.

However, none of the ATP-requiring processes examined in the charophycean (Characeae) *Chara corallina* (formerly *Chara australis*) used cyclic photophosphorylation as the ATP source [Smith and Raven, 1974]. In this alga light-stimulated ATP-requiring processes apparently involve LEF with oxygen as the terminal electron acceptor *via* the Mehler ascorbate peroxidase (MAP) or the flavodiiron reactions [Smith and Raven, 1974; Chaux *et al.*, 2017]. The processes examined include some that can be powered by cyclic photophosphorylation in another member of the Characeae, *Nitella translucens.* The reason for the inability (so far) to find processes energized by cyclic photophosphorylation in *Chara corallina* is not yet known. One major energy-requiring process in *Chara corallina* is the ATP-powered H^+ extrusion at the plasmalemma that is light-stimulated and inhibited by DCMU at the concentration that just inhibits LEF photosynthesis [Keifer and Spanswick 1978], and thence HCO_3^- use as part of the CO_2 concentrating mechanism [CCM] [Walker *et al.*, 1980]. In some other eukaryotic algae CCMs are powered by oxidative phosphorylation [*Nannochloropsis*: Huertas *et al.*, 2002] or pseudocyclic photophosphorylation, *i.e.* ADP phosphorylation coupled to H^+ gradients across the thylakoid lumen generated by LEF from H_2O to O_2 in the Mehler reaction involving both PSI and PSII [*e.g. Chlamydomonas*: Sultemeyer *et al.*, 1993]. In the cyanobacteria, *Anabaena variabilis* [Ogawa and Ogren, 1985] and *Anacystis nidulans* [Ogawa *et al.*, 1985a, 1985b] action spectra show a contribution of photosystem I (PSI) to the CCM, although interpretation is complicated by the involvement of energized, unidirectional, CO_2 to HCO_3^- conversion by the $NDH-I_3$ and $NDH-I_4$ as well as ATP-dependent active HCO_3^- influx in the CCM [Price, 2011].

Another major ATP-requiring process is N_2 fixation [Bottomley and Stewart, 1977]. Diazotrophy is restricted to archaea and bacteria, and cyanobacteria are the only N_2-fixing oxygenic organisms. The O_2 sensitivity of nitrogenase is accommodated in different cyanobacteria by limiting diazotrophy to the dark phase of the diel cycle, or having it occur in the light phase in colonial or multicellular cyanobacteria in cells that have no, or very limited, oxygenic photosynthesis. This second possibility has been most investigated in the filamentous cyanobacteria that have diazotrophy confined to heterocysts, which are cells that lack PSII and autotrophic CO_2 assimilation and obtain organic carbon from the neighbouring, symplasmically linked, vegetative cells [Bottomley and Stewart, 1977]. As would be expected, light-dependent N_2 fixation in heterocystous cyanobacteria is ATP-dependent [Bottomley and Stewart, 1977] and has the action spectrum of PSI with contributions from phycobilin light-harvesting complexes [Tyagi *et al.*, 1981; Staal *et al.*, 2003]. The symbiotic (with marine planktonic prymnesiophycean algae) diazotrophic unicellular cyanobacterium UCYN-A is effectively a hetero-cyst, lacking PSII and autotrophic CO_2 assimilation [Zehr *et al.*, 2008; Thompson *et al.*, 2012; Hagino *et al.*, 2013].

The "obvious" means of directly estimating CEF-related photophosphoryla-tion and other forms of photophosphorylation, is to measure the net increase of ATP in a dark-light transition under appropriate conditions [Bottomley and Stewart, 1976, 1977]. However, the rapid turnover of ATP relative to the time needed to quench microalgae is a major problem; similar problems occur with the use of externally supplied ^{32}P-HPO$_4^{2-}$ as a tracer [Raven, 1973].

Returning to the evidence of electron flow through P_{700}, cytochrome c_6 and cytochrome *f* in the presence of DCMU and/or under long wavelength irradiation mainly absorbed by PSI, an alternative explanation is net electron flow through PSI from organic carbon to an acceptor such as O_2. One indicator of this process is the Kok effect, *i.e.* a larger increment of net photosynthesis per unit increase in photon flux density at lower than at higher incident photon flux densities at less than half of the saturating photon flux density [Kok, 1949]. The Kok effect is widespread in cyanobacteria and eukaryotic algae [Healey and Myers, 1971; Peltier and Sarrey, 1988; Peltier *et al.*, 2010]. The Kok effect is a result of decreased respiration rather than an increase in gross photosynthesis [Hoch *et al.*, 1963; Bunt, 1965; Healey and Myers, 1971; Peltier and Sarrey, 1988]. Hoch *et al.* [1963] suggested that inhibition of (mitochondrial) respiration resulted from a higher free energy of hydrolysis of ATP generated by photophosphorylation than by oxidative phosphorylation, and oxidative phosphorylation close to equilibrium with the mitochondrial proton-motive force. However, the fact that the addition of an uncoupler (CCCP) at concentrations partially inhibiting photosynthetic and oxidative phosphorylation does not alter the Kok effect shows that the interaction is in terms of electrons from respiration being fed into PSI [Healey and Myers, 1971]. Subsequently, with the discovery of chlororespiration, this latter process rather than mitochondrial respiration is considered as the source of electrons [Peltier *et al.*, 2010]. O_2 as a sink for electrons from PSI could not explain the effect on net photosynthesis measured as O_2 production.

Interception by PSI of electrons from chlororespiration is, in the absence of additional evidence, a possible alternative to cyclic electron flow as an explanation of electron flow through PSI and the immediate electron donors to PSI [*e.g.* Huan *et al.*, 2014; Gao *et al.*, 2016]. However, the other lines of evidence, *i.e.* light dependent ATP-requiring processes and the absence of PSII and in the absence of electron acceptors, and photo-acoustic estimates of energy storage in the absence of PSII, do not suffer from the possibility of electron input to PSI from organic carbon.

A further interaction of cyclic photophosphorylation with dark energy metabolism is with glycolysis in the absence of O_2. Kandler and Haberer-Liesenkötter [1963] used *Chlorella*, Hirt *et al.* [1971] used *Scenedesmus* wild type and mutants 11 and 8 lacking, respectively, PSII and photophosphorylation, and Gfeller and Gibbs [1984], and Gfeller and Gibbs [1985] used *Chlamydomonas* and the inhibitors DCMU and FCCP. Light inhibits glycolytic fermentation by a process that is independent of PSII but is prevented by the protonophore uncoupler FCCP, showing that the inhibition involves cyclic photophosphorylation. This is analogous to the inhibition of fermentation by oxidative phosphorylation (Pasteur effect), but contrasts with the diversion of chlororespiratory electrons to PSI operating in a linear mode.

2.2. *Energetics of the assimilation of exogenous glucose and acetate powered by CEF and their mechanistic implications*

The relevant investigations here involved two genera of the Chlorophyta, *i.e. Chlorella pyrenoidosa* and *Chlorella vulgaris* (Trebouxiophyceae) and *Chlamydobotrys stellata* (Chlorophyceae), and two purple non-sulfur photosynthetic proteobacteria, *Rhododospirillum rubrum* and *Rhodopseudomonas capsulata*.

Tanner *et al.* [1965] measured the rate of glucose assimilation by *Chlorella pyrenoidosa* under aerobic conditions in the dark (energized by oxidative phosphorylation) and under N_2 in the light (energized by cyclic photophosphorylation). The dark-aerobic rate of glucose assimilation is 10–20% greater than the light-N_2 rate [Tanner *et al.*, 1965]. Tanner *et al.* [1968] measured the cost in absorbed photons of assimilation of glucose by *Chlorella vulgaris* under anaerobic conditions, and of photosynthetic $^{14}CO_2$ assimilation with illumination at 658 nm (absorbed by PSI and PSII) and 712 nm (almost all absorbed by PSI). Glucose uptake and assimilation requires 6.0 mol and 4.1 mol absorbed photons per mol glucose at 658 nm and 712 nm respectively; the corresponding values for photosynthesis are

12.5 mol and 40 mol absorbed photons per mol carbon dioxide at 658 nm and 712 nm respectively [Tanner *et al.*, 1968]. Wiessner [1966a] examined the relative absorbed photon costs of anoxic glucose photoassimilation in *Chlorella pyrenoidosa*: the values are relative because, although the photon absorption was measured, the glucose assimilation was given as counts per minute of ^{14}C-labelled glucose rather than as mol glucose. The relative photon cost of glucose assimilation at wavelengths with minimal PSII absorption is 70% of that at wavelengths where both PSI and PSII absorb [Wiessner, 1966a]; these findings are in qualitative agreement with the findings of Tanner *et al.* [1968] on *Chlorella vulgaris.*

Assimilation of one mol glucose into the main products, oligo- and poly-saccharides, requires 2 mol ATP. Influx of glucose in *Chlorella vulgaris* is by 1 proton: 1 glucose symport [Komor and Tanner, 1974]. The proton electrochemical gradient across the plasmalemma is maintained by an energized, ATP-powered proton efflux with a mol proton: mol ATP ratio of 1 [Raven, 1984]. This adds 1 mol ATP per mol glucose to the cost of converting external glucose into oligo- and poly-saccharide, making a total of 3 mol ATP per mol exogenous glucose assimilated; with 4.1 mol photon per mol glucose, each mol ATP corresponds to 1.37 mol absorbed photon. This can be accommodated by the following mechanism. One mol electron moved is moved from the oxidizing to the reducing end of PSI per mol photon absorbed by PSI. One mol electron moves from the reducing end of PSI through NDH1/2 and then PQ-cyt b_6/f (Figure 1A) and the oxidizing end of PSI with movement of 4 mol protons from stroma to thylakoid lumen [Raven *et al.*, 2014; Strange *et al.*, 2016]. Finally, 1 mol ADP phosphorylated per 4 mol protons moving from the thylakoid lumen to the stroma was measured [Peterson *et al.*, 2012] (Table 2). The outcome is 1 mol ATP synthesized per mol photon absorbed by PSI. This can readily accommodate the observed 1.37 mol photon absorbed per mol ATP. With the H$^+$:ATP of 14/3, or 4.67, predicted from structural biology 1 mol ATP from cyclic photophosphorylation requires 1.17 mol photon, again compatible with the observed 1.37 mol photon absorbed per mol ATP (Table 2).

However, NDH2 does not occur in the plastid genome of *Chlorella variabilis* [Cardol *et al.*, 2011]; this also seems to be the case for the other algae examined other than the Charophyceae and some members of the Prasinophyceae [Turmel *et al.*, 2009; Martín and Sabatier, 2010; Civán *et al.*, 2014]. The work of Tanner *et al.*, [1965] on *Chlorella pyrenoidosa* shows that CEF is inhibited by antimycin A, consistent with PGR5/PGRL1 (refer to Section 3.1. below) transferring electrons from ferredoxin to PQ [Labs *et al.*, 2016]. PGR5/PGRL1 occurs in *Chlorella variabilis* [Cardol *et al.*, 2011; Hertle *et al.*, 2012]. The PGR5/PGRL1 is not known to pump protons, and the proton:electron ratio of CEF lacking NDH2 or its H$^+$-pumping equivalent the mol photon per mol ATP is 2 [H$^+$:ATP of 4] or 2.33 (H$^+$:ATP of 4.67), neither of which is compatible with the *Chlorella* glucose

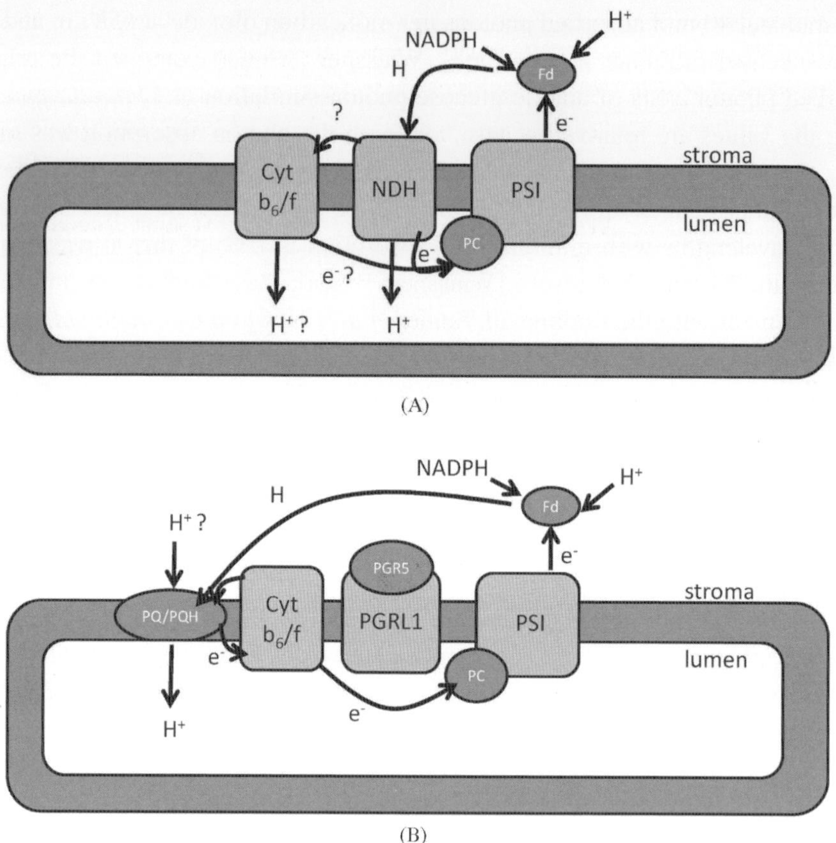

Figure 1. Diagrams of the proposed cyclic electron flow pathways in the thylakoid membranes of cyanobacteria and eukaryotic algae. (A) The NDH reductase pathway which is reduced by reduced ferredoxin, and which delivers protons to the thylakoid lumen. Note, if a Q cycle operates then extra protons are delivered to the thylakoid lumen. However, this would have to involve the operation of the cytochrome b_6/f complex. (B) The PGR5/PGRL-1 controlled pathway. Note that it is assumed that ferredoxin donates hydride to the plastoquinone pool, which then reduces the cytochrome b_6/f complex and delivers a proton to the thylakoid lumen. It is assumed that a Q cycle operates in which case 2 protons are delivered to the lumen for each hydride reducing PQ. In this formulation the PGR5/PGRL-1 complex is assumed to take on a sensing/controlling role.

assimilation value of 1.60 mol photon absorbed per mol ATP (4.67 mol H+ per mol ATP in the CF_0CF_1 ATP synthase) or 1.37 mol photon absorbed per mol ATP (4 mol H+ per mol ATP in the CF_0CF_1 ATP synthase) (Table 2).

If the mol proton:mol ATP ratio for the plasmalemma proton ATPase is 2, as is often thermodynamically possible [Raven, 1984], only 2.5 mol ATP per mol are required per mol exogenous glucose assimilated. With 4.1 mol absorbed photon per mol glucose, each mol ATP used in assimilation of exogenous glucose

Table 2. Comparison of measured *in vivo* photon yield of ATP production from CEF with predictions from CEF mechanisms with NDH2 (eukaryotes) or with PGR5-PGRL1 (eukaryotes) and UQH$_2$ as the first stable reduced product of photochemistry. See Section 2.2 of text for references.

Organism	*In vivo* PSI absorbed photons per ATP produced by CEF	PSI absorbed photons per ATP in CEF using NDH2	PSI absorbed photons per ATP in CEF using PGR5-PGRL1 (eukaryotes), UQH$_2$ first stable reduced product of photochemistry
Chlorella	1.37[1] [1.64][2]	1.0[4][1.17][3]	2.0[4][2.33][5]
Chlamydobotrys	16.7[1] [1.95][2]	1.0[4][1.17][3]	2.0[4][2.33][5]
Rhodospirillum Rubrum	4.4–5.13[3]	Not applicable	1.65[6]
Rhodopseudomonas capsulata	5.3–5.63[3]	Not applicable	1.65[6]

[1]H$^+$: ATP = 1 for plasmalemma H$^+$ ATPase
[2]H$^+$: ATP = 2 for plasmalemma H$^+$ ATPase
[3]No allowance for energized entry of acetate
[4]H$^+$: ATP = 4 for CF$_0$CF$_1$ATP synthase
[5]H$^+$: ATP = 4 67 for CF$_0$CF$_1$ATP synthase
[6]H$^+$: ATP = 3 33 4 for F$_0$F$_1$ATP synthase

References
Cardol *et al.* [2011]; Feniouk and Junge [2009]; Hertle *et al.* [2012]; Klamt *et al.* [2008]; Komor *et al.* [1968]; Labs *et al.* [2016]; Peterson *et al.* [2012]; Raven [1984]; Raven *et al.* [2014]; Soga *et al.* [2017]; Strange *et al.* [2016]; Tanner *et al.* [1965]; Tanner *et al.* [1968]; Wiessner [1966a]; Wiessner [1966b]; Wiessner and Gaffron [1964]; Yang *et al.* [2015].

corresponds to 1.91 mol photon absorbed per mol ATP (4.67 mol H$^+$ per mol ATP in the CF$_0$CF$_1$ ATP synthase) or 1.64 mol photon absorbed per mol ATP (4 mol H$^+$ per mol ATP in the CF$_0$CF$_1$ ATP synthase). These are still inconsistent with a non H$^+$-PGRL1-PGR5 catalysing all of the electron flow from ferredoxin to PQ, although the photon cost with the 4.67 mol H$^+$ per mol ATP in the CF$_0$CF$_1$ ATP synthase comes close (Table 2).

Wiessner [1965, 1966b] investigated the absorbed photon cost of the DCMU-insensitive aerobic acetate photoassimilation in *Chlamydobotrys*. With 723 nm and 716 nm incident irradiance the photon requirements are respectively 8.3 and 8.8. mol photon per mol acetate assimilated; at 450–680 nm incident irradiance the cost is 11.8–12.3 mol photon per mol acetate assimilated [Wiessner, 1965]. The ATP cost of acetate assimilation is 3.25 mol ATP per mol intracellular acetate, so the photon cost of the required ADP phosphorylation is 2.55–2.71 mol photon per mol ATP produced. This is consistent, assuming an H$^+$ per mol ATP ratio of 4 for the CF$_0$CF$_1$ ATP synthase, with the 2 mol absorbed photon per mol ATP if PGR5/PGRL1 catalyses all of the electron flow from ferredoxin to PQ, with no necessary involvement of NDH2. If the H^+/ATP ratio is 4.67 for the CF_oCF_1 ATP synthase, the 2.33 absorbed photons per mol ATP for the PGR5/PGRL1 is still compatible with data of Wiessner [1965, 1966b] (Table 2). However, Wiessner [1965, 1996b] did not include any energy costs associated with moving acetate across the plasmalemma. Yang *et al.*, [2015] suggest that acetate entry in chlamydomonad chlorophyceans is energized, adding another 0.5 or 1 mol ATP to the energy cost of assimilation of one mol of external acetate. This correction yields *in vivo* values of 1.95–2.34 mol photons per mol ATP for an H^+/ATP ratio of 4 for the CF_oCF_1 ATP synthase; the lower value is not, granted the variability in the data, significantly different from 2.0, so there is again no requirement for the involvement of NDH2 in CEF. For an H^+/ATP ratio of 4.67 for the CF_oCF_1 ATP synthase the lower of the *in vivo* values of 1.67–2.0 is more challenging for the absence of involvement of NDH2 in CEF (Table 2).

The situation for acetate assimilation in *Chlamydobotrys* seems more complicated than that for glucose assimilation in *Chlorella*; the rate of acetate photoassimilation in the absence of oxygen is only 24% of that with normal atmospheric oxygen, and the anoxic rate is restored to a rate approximating the rate with oxygen by the presence of carbon dioxide [Wiessner and Gaffron, 1964]. These findings are possibly related to redox balance in the glyoxylate cycle used in converting acetate into hexose [Wiessner and Gaffron, 1964].

Wiessner [1966b] also measured the absorbed photon cost of acetate photoassimilation in the purple photosynthetic bacteria *Rhodospirillum rubrum* and *Rhodopseudomonas capsulata*. The bacteria use a PSII-like (but non-oxygenic) reaction centre to energize CEF involving a UQ cytochrome bc_1 proton pump but, mechanistically and energetically, no involvement of the NDH is found in these organisms [Nicholls and Ferguson, 2013]. The absorbed photon costs of acetate assimilation at wavelengths corresponding to the first excited state of the bacteriochlorophylls (795–893 nm) involved in photon harvesting and photochemistry in these organisms are as follows: for *Rhodospirillum rubrum* the values are 8.8–10.1 mol absorbed photons per mol acetate, and for *Rhodopseudomonas capsulata* they are 10.6–11.1 [Wiessner, 1966b]. In the purple bacteria assimilation of internal acetate requires 2 mol ATP per mol acetate [Wiessner, 1966b], so each mol ATP produced is associated with absorption of 4.4–5.6 mol photons (Table 2).

Purple photosynthetic bacteria have the equivalent of NDH2 that is involved in oxidative phosphorylation under dark aerobic growth conditions, and the reduction of NAD^+ by UQH_2 driven by the proton gradient generated by CEF during autotrophic CO_2 assimilation [Klamt *et al.*, 2008; Nicholls and Ferguson, 2013]. However, the NDH2 equivalent is not involved in CEF in purple photosynthetic bacteria [Klamt *et al.*, 2008; Nicholls and Ferguson, 2013]. With a CEF photon:electron ratio of 1, a proton:electron ratio of the purple bacterial UQ cytochrome bc_1 proton pump of 2 [Klamt *et al.*, 2008; Nicholls and Ferguson,

2013], and a proton:ATP ratio of 3.3 for the bacterial ATP synthase [Klamt *et al.,* 2008; Feniouk and Junge, 2009; Nicholls and Ferguson, 2013; Soga *et al.,* 2017] the predicted photon cost of ATP synthesis is 1.65 mol absorbed photon per mol ATP. This predicted photon cost is much less than the observed *in vivo* 4.4–5.6 mol absorbed photons per mol ATP for assimilation of intracellular acetate, so any energy cost of acetate flux into the cells could be very readily accommodated (Table 2). This contrasts with the much closer similarity of predicted and measured values in *Chlamydobotrys.*

3. Detailed Mechanism of Cyclic Electron Transport and ATP Formation in Oxygenic Photosynthetic Organisms

3.1. *Overview*

Having established the strong historical evidence for cyclic photophosphorylation and cyclic electron transport around PSI in cyanobacteria and eukaryotic algae, proposals for the detailed mechanism are now discussed.

Two well-established mechanisms for CEF exist across all the oxygenic photosynthetic organisms, both of which probably involve the cytochrome b_6/f complex and plastoquinone, but the exact mechanism is not well-established (see Figure 1):

(A) An NDH dehydrogenase (NADH/NADPH dehydrogenase) system, driven by electrons from reduced ferredoxin, which ferries electrons and protons inwards to the inner compartment of the thylakoid, the lumen (Figure 1A). Most commonly this is an NDH-1 system but in eukaryotic algae apparently there is an NDH-2 system. The NDH-1/2 system, which is the dominant one in cyanobacteria, is found in some green eukaryotic algae, especially streptophytes (see Section 4.3) and is a minor protein component in most embryophytic plants but lacking in several clades [Ruhlman *et al.,* 2015].

(B) An electron/proton transport system involving at least two protein complexes with many subcomplexes (Figure 1B). The two complexes are PGR5 and PGRL1 complexes (see definitions on the first page of the Introduction). Reduction of PGR5 on the stroma/plasma side of the thylakoid membrane is by ferredoxin. There is some evidence for these complexes in cyanobacteria, but there they play a minor role [Allahverdiyeva *et al.,* 2013]. They are found in some eukaryotic algae particularly in the green algal line e.g. *Chlamydomonas reinhardtii* and are more abundant than in cyanobacteria. In land plants the NDH system apparently plays only a minor role and the PGR5/PGRL1 system predominates. Thus there seems to be an evolutionary drift from the NDH dehydrogenase system (utilizing Fd[H]) to the PGR5/PGRL1 system as one goes from cyanobacteria to land plants but the reason(s) for this are at present not clear.

3.2. *Non-cyclic (= linear) electron flow, cyclic electron flow, the Q cycle and ATP generation*

In non-cyclic flow or LEF, electrons from the oxidation of water are passed from the reaction centre of PSII to Q_A/Q_B in a semiquinone/quinone reaction. The electrons are then passed on to plastoquinone, which is reduced to plastoquinol by a plastoquinone reductase on the stroma side of the thylakoid membrane or the cytoplasmic side of the cyanobacterial plasma membrane. This is then coupled to electron donation to the Q_1 site of the cytochrome b_6/f complex and simultaneous deposition of a H^+ in the lumen of the thylakoids (of algal plastids and many cyanobacteria) or the periplasmic space of the cyanobacterium *Gloeobacter*. From there the cytochrome b_6/f complex takes part in a modified (activated) Q cycle [Mulkidjanian, 2010] whereby an electron is returned to the outer side of the membrane picks up a proton and hydrogenates a second molecule of plastoquinone. The second cycle moves another H^+ from the cytoplasm/stroma to the lumen for each e^-, so that a total of 2 H^+ are moved per e^- participating in the cycle or 8 H^+ per 4 e^-. To summarize, for each pair of water molecules split in the oxygen evolving complex 4 e^- are moved by PSII and 4 H^+ are deposited in the lumen, which with the operation of the Q cycle deposits 12 protons in the thylakoid lumen per 4 e^-. The number of ATP that are generated depends on the number of H^+/ATP involved in the CF_oCF_1 ATP synthase in the thylakoid/cyanobacteria plasma membrane; this ratio is usually taken to be ~4, although the structural biology prediction is 14/3 or 4.67 [Steigmiller *et al.*, 2008].

In cyclic electron flow much less is clear. We know that there are membrane dehydrogenases. Some of these are clearly allied to Type 1 dehydrogenases and drive 3 – 4 protons across the active membrane per $2e^-$. These almost certainly interact with the cytochrome b_6/f complex but it is not clear whether a Q cycle is involved [Joliot and Johnson, 2011; Johnson *et al.*, 2014], and if it is whether it is an activated Q cycle [Mulkidjanian, 2000]. The PGR system of green algae and embryophyte plants, seems more reliably to interact with the cytochrome b_6/f complex (Figure 1B) and there is better evidence that a Q cycle is involved there (and e^- flow is generally but not always inhibited by antimycin A (see below)).

Recently, the role of PGR5/PGRl-1 in controlling CEF has been questioned. As the name "proton gradient regulator" implies these protein complexes were discovered, first in *Arabidopsis thaliana* and then in *Chlamydomonas reinhardtii*, (see Section 4), as essential for CEF to occur. However, doubts have been raised as to whether they actually control or modify the ΔpH gradient. Or whether they play a much more subtle role in influencing the thylakoid ATPase and in modulating the tie-up between ΔpH and NPQ in photoprotection [Armbruster *et al.*, 2017; Kanazawa *et al.*, 2017]. As a result of these new ideas it is also necessary to place on the table the question of the exact role of the NDH proteins in CEF.

In summary, CEF can generate ATP. How much ATP is generated from CEF has not been fully established and would depend on which dehydrogenase system is used, the interconnections of the membrane complexes and the ATP synthase involved (and this is considered in Section 2.2). To further complicate the situation, reductive processes close to PSI are also the result of a balance of metabolic processes and, if CEF is deleted by genetic manipulation, mitochondrial respiration, the MAP pathway, chlororespiration and other processes can potentially come into play [Dang *et al.*, 2014; Johnson *et al.*, 2014; Chaux *et al.*, 2017].

3.3. *Early evolution of NAD dehydrogenases — Bacteria, cyanobacteria, plastids and mitochondria*

The complex protein assemblage, which constitutes the NDH complex of cyanobacteria and plastids clearly shares an evolutionary history with complex 1 of bacteria and mitochondria. Much more is known of the structure and functioning of complex 1 than of the NDH complex, although the detailed structure of the NDH complex, which has been elucidated, links it intimately with complex 1. It is known that complex 1 has a two-part structure: (1) a peripheral hydrophilic arm abutting the aqueous phase outside the membrane, which mediates electron transfer through an FMN molecule and 10 Fe/S clusters; (2) a hydrophobic membrane arm that mediates proton translocation across the membrane by an unknown mechanism [Friedrich, 2014]. In complex 1, the binding site for the NADH is a ubiquinone, which lies at the junction of the hydrophilic and hydrophobic parts. It was previously thought that one NADH oxidation couples the transfer of two electrons to the translocation of $4H^+$; however, Wikstrom and Hummer [2012] argued cogently for a coupling of $2e^-$ per $3H^+$. In any case the respiratory complex 1 is an energy converting NADH-ubiquinone oxidoreductase.

Phylogenetic trees based on the NuoH polypepetides [*e.g.* Moparthi and Hagerhall, 2011] indicate that complex 1 has a long history in ancient cells back at least to LUCA (the last common universal ancestor), in which the complex 1 genes share affinities with archaea and bacteria and where the genes of anoxygenic photosynthetic bacteria are well separated from cyanobacteria.

3.4. *Mechanisms for CEF in cyanobacteria*

As cyanobacteria are the oldest known oxygenic photosynthesizers, it is logical to assume that the major dehydrogenase, NDH dehydrogenase complex (now utilizing Fd[H]), is the major mechanism for cyclic electron transport, and came by lateral gene transfer from similar dehydrogenases in other non-photosynthetic organisms [Vermaas *et al.*, 2010; Shih *et al.*, 2016; Soo *et al.*, 2017]. Whether or not NDH was the earliest CEF system to evolve is not known, but certainly it is the most abundant system and occurs in all extant cyanobacteria that have been assayed [Peltier *et al.*, 2016]. However it should be remembered that the thylakoid membrane of cyanobacteria is, evolutionarily, an invaginated extension of the plasma membrane (and indeed cyanobacteria in the genus *Gloeobacter* lack thylakoids altogether): and respiratory enzymes, for example, are present in both membranes [Mullineaux, 2014] and were all probably evolved by lateral gene transfer borrowing (see Vermaas *et al.*, 2010).

The current nomenclature for the chloroplast NDH-dehydrogenase-like complex (utilizing Fd[H]) was first given by Ifuku *et al.*, [2011] and is centred on the complex in eukaryotic organisms, but this nomenclature can be used for cyanobacteria with just a few exceptions.

The development of proteomic procedures and single particle electron microscopy led to the first indications of the molecular structures of complexes involved in CEF. NDH complexes are present in a wide variety of photosynthetic organisms from cyanobacteria to embryophytic plants [Peltier *et al.*, 2016]. However, in eukaryotic algae and in embryophytic plants the role and abundance of NDH complexes is currently obscure. NDH complexes seem to have evolved to regulate and control the electron transport reactions of photosynthetic organisms [Howitt *et al.*, 1999; Howitt *et al.*, 2001; Peltier *et al.*, 2016; Strand *et al.*, 2016] and it seems that these reactions drew on reactions, which had been established in bacteria for respiratory purposes (see previous section).

In cyanobacteria the NDH-1 complex would appear to be the major protein complex involved in CEF [Arteni *et al.*, 2006; Battchikova *et al.*, 2011a; Deák *et al.*, 2014: Ohkawa *et al.*, 2001: Xu *et al.*, 2008; Zhang *et al.*, 2004, 2005]; a protein complex that even at the most basic level (NAD-1M) is made up of 14 subunits. For the physiological function, which may be control of electron flow or redox regulation (see Section 3.1) [Peltier *et al.*, 2016] it is necessary to have several other subunits, which form a hydrophilic domain and several subunits, which form a hydrophobic domain. One NDH subunit of the hydrophilic domain, Ndhs, is responsible for interacting with ferredoxin, which has recently been shown to be responsible for reducing the active complex and initiating CEF [Battchikova *et al.*, 2011b; He *et al.*, 2015], rather than NADH or NADPH.

In cyanobacteria there are two forms of NDH-1: (1) NDH-1_3 contains the so-called NdhD3 and NdhF3, with, in addition, two additional subunits, which together form a region in the distal arm of the complex. (2) NDH-1_4 has the same NDH-1m component but, however, incorporates different subunits in the distal arm of the complete complex.

Cyanobacteria also harbor another set of protein complexes (PGR5/PGRL1 or homologues), which are discussed below [Allahverdiyeva *et al.*, 2013]. There seems to be no evolutionary relationship between these two sets of proteins [Peltier *et al.*, 2010]. Both NDH and PGR proteins seem to have found their evolutionary way through the green algae, which colonized the land, to embryophytes [Hori *et al.*, 2014].

3.5. *Detailed mechanism and inhibitors of the NDH system*

The detailed mechanism of CEF in cyanobacteria involves an NAD complex with a hydrophilic arm incorporating several FeS ligands interacting through a quinone (ubiquinone) with NADH (see Section 3.3). This interaction results in a concomitant translocation of the two protons for every electron across the membrane [Peltier *et al.*, 2016], implying that the NDH-1 complex has a function in electron flow as well as in regulation and control.

There is a lack of good inhibitors for NAD-1 and NAD-2. Based on the similarity of complex 1 of other bacteria, and mitochondria, with those complexes of cyanobacteria, then piericidin A, an inhibitor of quinone binding sites, should be appropriate. However, there is no report of an effect of this inhibitor, which interacts at the site of coenzyme Q reduction, *i.e.* the quinone-binding sites in mitochondria (along with amytal and rotenone); in chloroplasts, piericidin interacts with the reduction of Q_B in PSII [Ikegawa *et al.*, 2002] and this may have deterred any attempts to look for effects on CEF; rotenone does inhibit the cyanobacterial complex 1 [Berger *et al.*, 1999], which should encourage the use of rotenone and piericidin in studies of cyanobacterial CEF.

The lack of functional inhibitors for NDH-1 or NDH-2 has limited research into the detailed mechanism of CEF in cyanobacteria, in comparison with the strong inhibitory action of antimycin A on CEF in green algae and higher plants. A single point mutation in PGR5 in *Chlamydomonas* can confer resistance to this inhibitor [Sugimoto *et al.*, 2013]; and there is evidence that a similar base shift in *Synechocystis* sp. PCC6808 (ssr2016) confers resistance/sensitivity to antimycin A [Yeremenko *et al.*, 2005].

An alternative line of investigation is to use strains in which these complexes have been impaired by the use of deletion mutants; this was first achieved in the M55 strain of *Synechocystis* 36803 [Ogawa, 1991] and since then in a range of other mutants [Gao *et al.*, 2016a, 2016b]. Use of these mutants has shown that deletion of the linker protein CpcG2 is strongly inhibitory of CEF under strong light, indicating that the NDH-1L-CpcG2- PSI supercomplex facilitates PSI-CEF *via* NDH-1L. So far, however, these mutants have been little used to probe the mechanism of CEF. Initial results [Fitzpatrick, 2016] agree with previous inferences [Howitt *et al.*, 2001; Battchikova *et al.*, 2011a; Jia *et al.*, 2015; Chen *et al.*, 2016] that CEF is used in Cyanobacteria to generate extra ATP through a proton pumping activity across the thylakoid membrane. One caveat here is that the supercomplex may also be involved in other reductive processes, when the electron transport [ET] from PSII is inhibited, *e.g.* by high temperatures [Fitzpatrick, 2016].

Another line of investigation has been to delete PSII and respiratory oxidases in a cyanobacterium. This was done by Howitt *et al.* [2001] where respiratory and PSII mutants of *Synechocystis* sp. PCC6803, grown on glucose showed growth proportional to photosynthetically active radiation up to 25 μmol m^{-2} s^{-1} and which suggested that CEF around PSI was the major provider of energy under these circumstances. Unfortunately, there has been little follow up of these interesting experiments.

To summarize, in cyanobacteria there is a major NDH-1 complex which is strongly implicated in CEF. There is also an NDH-2 complex for which there is much less evidence of its function. In addition there seem to be PGR (Proton Gradient Regulation complexes; discussed below) but again the evidence for their general presence and functional significance is generally lacking; and there is even evidence that inhibition by antimycin A, a major feature of PGR complexes in land plants, is only prevented by a single point mutation. In eukaryotic algae PGR5/PGRL1 proteins are widespread [Peltier *et al.*, 2010; Thamatrakoln *et al.*, 2013]. NDH-1 and NDH-2 are restricted to many of the charophycean members of the Streptophyta, and some basal Chlorophyta [Peltier *et al.*, 2010; DePriest *et al.*, 2013; Leliaert *et al.*, 2016].

3.6. *Towards a better understanding of cyanobacterial CEF*

In spite of decades of research, accurate measurements to determine rates of CEF under steady state, physiologically relevant conditions, in cyanobacteria and eukaryotic algae, have proven elusive [Fan *et al.*, 2016].

Work on CEF in cyanobacteria has been hindered by the lack of a specific CEF inhibitor and in eukaryotic algae, apart from *Chlamydomonas reinhardtii*, where a specific inhibitor is also lacking, work has been very sparse. Recently, Fitzpatrick [2016] and Fitzpatrick *et al.* (in preparation) have worked to integrate membrane inlet mass spectrometry (MIMS) [Beckmann *et al.*, 2009] with a P700-based method to estimate CEF, published by Kou *et al.* [2013], and also described in Chapter 12 of this book, to measure CEF in cyanobacteria. The CEF estimate, termed ΔFlux, is based on the difference in rates of electrons that pass through P680, the reaction center chlorophylls of PSII (determined using rates of gross O$_2$

evolution and requiring the addition of the stable $^{18}O_2$ isotope to discriminate O_2 consumption from O_2 evolving reactions) and P700, the reaction center chlorophylls of PSI. The integrated system was used to investigate the response of *Synechocystis* PCC6803 cells to acute high temperature stress. Although the dataset is preliminary, it has been included as an example of a cyanobacterial response to high temperature stress. Presented in Figure 2, clear differences in trends were observed when comparing the WT (wild type) cells, with those of the M55 NDH-1 mutant [Ogawa, 1991], which lacks the predominant cyanobacterial CEF pathway. This mutant was used to determine *f*1, the fraction of absorbed light partitioned to PSI (required for calculating electron transport through PSI, ETR1), in the absence of a specific CEF inhibitor in cyanobacteria, as *f*1 must be determined under conditions where CEF is absent.

In Figure 2 the rates of ETR2(O_2) between the WT and the mutant exhibited very similar trends in response to higher temperature incubations: dropping at 50°C and showing almost complete inhibition of ETR2(O_2) at 55°C. However, a large difference could be observed between the two samples in terms of the calculated rates of ΔFlux. Whereas the M55 samples showed no ΔFlux, the WT samples exhibited increasing rates of ΔFlux with increasing temperature, with a marked

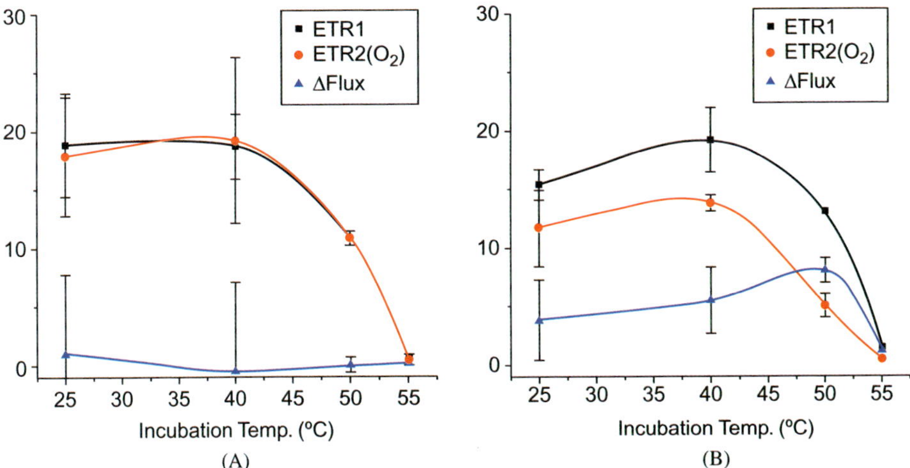

Figure 2. ΔFlux, an estimated rate of cyclic electron transport around PSI in samples of (A) the NDH-1 inhibited mutant of *Synechocystis* PCC6803, [M55], and (B) wild type samples of the same species, after samples were incubated for 10 minutes in the dark at each of the specified temperatures. DFlux was determined as the difference between two simultaneously collected values: ETR2[O_2] based on gross O_2 evolution rates measured with MIMS utilizing $^{18}O_2$ and ETR1, based on the Y[I] values from P700⁺ measurements. Values were collected during steady state photosynthesis at 100 μmol photons.m^{-2}.s^{-1} with cyanobacterial cells immobilized on glass fiber discs with saturating total carbon. Cells were cultivated at 30°C in BG-11 medium pH 8.0 in 2% CO_2 enriched air Error bars = standard deviation of three repeats.

increase as ETR2(O_2) became inhibited after exposure to elevated temperature (50°C). Although some aspects of the method are still being optimized, the data suggest that this integrated approach holds some promise for estimating rates of CEF under steady state conditions in samples where it is possible to estimate the partitioning of absorbed energy between the two photosystems (*i.e.* using inhibitors or a mutant to inhibit CEF without impacting upon linear electron transfer between PSII and PSI — see Section 3.5).

Significantly, the use of MIMS is an improvement on original published methods, especially for experiments with cyanobacteria where rates of gross O_2 evolution can be underestimated by as much as 50% due to the activity of flavodiiron 1,3 proteins [Allahverdiyeva *et al.*, 2013]. The power of MIMS to utilize stable isotopes to measure true gross gas fluxes may also be useful to deconvolute potential alternative donors to the thylakoid membrane, which could complicate the interpretation of $P700^+$ reoxidation kinetics, a commonly used estimate of CEF rates.

It is to be hoped that use of the MIMS technique combined with P700 and other measurements will allow precise quantification of CEF activity in cyanobacteria and eukaryotic algae in the future; and furthermore allow for the establishment of the interaction of this pathway with other pathways, such as LEF, respiration, glycolytic reactions, reactions to oxygen such as the MAP and flavodiiron pathways and chlororespiration.

4. Cyclic Electron Transport Complexes of Chlamydomonas and Other Green Algae

4.1. *Introduction*

While it is clear that NDH complexes do occur throughout the eukaryotic algae and in higher plants (in some eukaryotic algae it has been argued that NDH-2 replaces NDH-1 [Peltier and Cournac, 2002]) and were obviously inherited through the endosymbiosis of a cyanobacterial line that led to the evolution of plastids [Larkum *et al.*, 2006], another set of membrane protein complexes shares a major role in the plastids of many eukaryotic algae and higher plants. These are the PGR5 and PGRL1 complexes [Yamori and Shikanai, 2016].

As mentioned in the early sections of this review, CEF was discovered in higher plant thylakoids in a study of ferredoxin-dependent ATP synthesis that could be inhibited by antimycin A [Tagawa *et al.*, 1963]. Antimycin A, the key to this early work on embryophyte chloroplasts, was discovered as an inhibitor of mitochondrial respiration, specifically electron transport and proton transport, powering oxidative phosphorylation. In eubacteria and mitochondria, antimycin A binds to the Q_i site of the cytochrome bc_1 cytochrome c reductase; this inhibits the oxidation of ubiquinone at the Q_i site of ubiquinol and prevents turnover of the Q cycle [Labs *et al.*, 2016]. In photosynthetic LEF a similar cytochrome oxido-reductase operates between the two photosystems and in embryophytes antimycin A plays a similar role inhibiting the cytochrome b_6/f complex, which also takes part in a Q cycle [Cramer *et al.*, 2011]. The cytochrome b_6/f complex is also the probable route for the PGR dependent CEF (see below and Figure 1). Thus, antimycin A is generally a potent inhibitor of CEF in those organisms that have a strong component of PGR-dependent CEF and, where it is active, can be used to measure the involvement of CEF in photosynthetic reactions. However, a single point mutation in cyanobacteria and *Chlamydomonas* confers resistance to antimycin A [Yermenko *et al.*, 2001; Sugimoto *et al.*, 2013]; and in other eukaryotic algae the inhibition of AA is sporadic.

4.2. *PGR5/PGRL1 complexes in thylakoids*

As mentioned above, the naming of these complexes relates to their role in sensing and/or controlling the proton gradient (ΔpH) set up by the reduction of the plasto-quinone pool and electron flow through the cytochrome b_6/f complex: hence PGR (proton gradient regulation) across the thylakoid membrane of chloroplasts (and which in turn drives the formation of ATP) [Fisher and Kramer, 2011]. Their relationship to NDH-generated proton gradients still needs to be determined.

The machinery of the PGR5/PGRL1 complexes has been elucidated using molecular genetics, mainly in *Arabidopsis* and *Chlamydomonas reinhardtii* [Yamori and Shikanai, 2016]. PGR5 was first identified from a series of *Arabidopsis* mutants found to suffer a major disruption in their ability in generating ΔpH across the thylakoid membrane. Later PGR5 was identified in *Chlamydomonas* and other algae and also in cyanobacteria (where it is present in the form of a homologue, Ssl0352) [Yeremenko *et al.*, 2005]. The other functionally related complex PGRL1 was also been identified first in *Arabidopsis* [Dal Corso *et al.*, 2008], and then in *Chlamydomonas* [Merchant *et al.*, 2007; Johnson *et al.*, 2014]; but PGRL1 is only found in eukaryotes and is absent from cyanobacteria [Dal Corso *et al.*, 2008; Peltier *et al.*, 2010].

The structure of the PGR5 and PGRL1 complexes is still not fully resolved. It is clear that PGR5 is a small protein component on the outer (stromal) side of the thylakoid membrane, abutting ferredoxin and the calcium sensing (CAS) phosphoprotein. The major transmembrane component is made up of the PGRL1 protein, for which there are usually two gene copies. The PGR5/PGRL1 complex forms a supercomplex with PSI and the cytochrome b_6/f complex in *Chlamydomonas*. This differentiates it from *Arabidopsis* where the cytochrome b_6/f complex is separate. Nevertheless proton pumping occurs in both organisms through the cytochrome b_6/f complex. The exact mechanism of AA inhibition in PGR5/PGRL1 is still a matter of speculation [Leister and Shikanai, 2013; Labs *et al.*, 2016]: in *Arabidopsis* evidence suggests that AA acts in conjunction with PGRL1 in preventing plastoquinone from interacting with the cytochrome b_6/f complex and inhibiting CEF (see below). In *Chlamydomonas* the situation, with a very large supercomplex, is less clear, although AA is not inhibitory of CEF here. However, this can be due to a single point mutation and does not have wide evolutionary significance [Sugimoto *et al.*, 2013].

The evolutionary phylogeny of PGR genes is unclear. They occur in green algae and higher plants [Peltier *et al.*, 2010], and homologues occur across the algal divisions and even in cyanobacteria: PGR5 genes or homologues (Ssl0352) occur in cyanobacteria. So it appears that the two systems, NDH and PGR, both evolved at the earliest stages of oxygenic photosynthesis and have been inherited through the algal and embryophyte systems. Even the action of antimycin A can be traced back to cyanobacteria [Yeremenko *et al.*, 2005]. PGR5 is the smaller of these complexes and there are no clear indications of how this protein evolved. It has some short sequence homology with PGRL1, the larger protein [Larkum, unpublished], which may indicate an evolutionarily shared origin, but the broader origins and even the complete function of these two protein complexes are frustratingly unresolved.

4.3. *Reductases of thylakoid membranes and their role in CEF*

It was earlier assumed that NADPH was the coenzyme of choice for reductase reactions of cyanobacteria and plastids, especially in interactions with the reducing side of PSI. In terms of CEF it was always assumed that an NADPH dehydrogenase would interact with a thylakoid membrane component to donate electrons through a membrane ET carrier towards the intrathylakoid space (or the cell wall or periplasmic space of some cyanobacteria). This enzyme has recently been named NADH dehydrogenase-like complex because it has been discovered that it accepts electrons from ferredoxin [Yamamoto and Shikanai, 2013].

There has been much research effort involved in identifying just what enzyme system is involved. Two hypotheses emerged over the years. On the one hand there is a proposal for a plastoquinone reductase (PQR), as proposed by Bendall and coworkers [Moss and Bendall, 1984; Cleland and Bendall, 1992; Bendall and Manasse, 1995]. According to their hypotheses antimycin A did not inhibit a cytochrome, *i.e.* cytochrome b_6, as in mitochondria, and therefore they suggested that AA inhibited the PQR. Unfortunately, testing of this hypothesis has been difficult, especially the route for reducing equivalents across the thylakoid membrane towards the intrathylakoid space [Labs *et al.*, 2016]. Alternatively, Hertle *et al.* [2013] suggested that the route is through PGRL1 and that this is the elusive PQR. However, this proposal has also not been universally accepted to date. In *Chlamydomonas*, there is a supercomplex made up of PGRL1/PGR5, cytochrome b_6/f, plastoquinone, PSI and Lhca5 and 6 reduced by ferredoxin [Peltier *et al.*, 2016]; and somewhere in there the reducing equivalents cross the membrane to the inside. However, the complexity of this supercomplex makes the establishment of the exact route problematical. It has even been suggested that the PGRL1/PGR5 proteins act as regulatory agents rather than as oxidoreductase components [Peltier *et al.*, 2016].

The action of an NADPH-like dehydrogenase, which in fact has been shown to utilize Fd(H), as the primary reductant has found much greater support; however, even here much more evidence is needed. And it is well-established that this route even when it occurs accounts for only a small fraction of the CEF activity [Yamori and Shikanai, 2016].

Both hypotheses might be accommodated based on the proposal from studies on tobacco and *Arabidopsis* that there are two CEF pathways. One involving the NAD-1 complex and the other involving an antimycin A sensitive Fd-PQR [Joet *et al.*, 2001; Hashimoto *et al.*, 2003]. It has been proposed that the PGRL1 acts as the antimycin A sensitive pathway; and the discovery that NDH-1 complex receives electrons from Fd suggests that it is the antimycin-insensitive Fd-PQR pathway [Peng *et al.*, 2011; Yamamoto *et al.*, 2011]. Recently studies of knockout-mutants have concluded that lack of NDH-1 significantly lowers the generation of ΔpH in *Arabidopsis* at low light [Yamori at al., 2011], while the PGR5/PGRL1 pathway operates at high light [Wang *et al.*, 2015]. Two CEF pathways also seem to operate in *Chlamydomonas* [Ravenal *et al.*, 1994]. Since the mutant *Chlamydomonas* in these studies uses NADH, it is speculated that a transhydrogenase, catalyzing NADH/NADPH interconversion and driven by ΔpH, exists in the thylakoid membrane.

Cyanobacteria, lack the PGR5/PGRL1 complex, although, as mentioned already a homologous protein to PGR5 but not to PGRL1, exists in some cyanobacteria [Allahverdiyeva *et al.*, 2013]. Here CEF seems to rely on the NDH-1 complexes: $NDH-1_1$ and $NDH-1_2$, whilst $NDH-1_4$ complex seems to be involved in a CO_2 concentrating mechanism [Peltier *et al.*, 2016].

The situation is complicated further in cyanobacteria because the plasma and thylakoid membranes are coextensive. However, it seems that while some proteins may be initiated on the plasma membrane, such as respiratory proteins, and can migrate to the thylakoid membrane, the reverse is not true and chlorophyll proteins for example are restricted to the thylakoid membrane, except in *Gloeobacter* which has no thylakoid membranes. Nevertheless this means that thylakoid membranes of cyanobacteria can simultaneously be carrying: (i) respiratory proteins, (ii) photosynthetic proteins and (iii) chlororespiratory proteins of the PTOX-dependent system as well as other important enzymes [Mullineaux, 2014]. The situation is somewhat simpler in plastids but in many algal divisions there is still much "cross talk" between cytosol, mitochondria and plastids and there is a need for further investigation and clarification. The presence of mitochondria, which can generate reduced compounds such a succinate, malate (and other acids of the tricarboxylic acid cycle), NADH, NADPH, especially in the dark under aerobic conditions, and which can be imported into plastids, needs special consideration. These issues are discussed in the following sections.

5. Evidence for Cyclic Electron Flow in Eukaryotic Algae

5.1. *Pyrrophyta, Dinophyta (Dinoflagellata)*

5.1.1. *Symbiodinium sp., the coral endosymbiont algae*

The first observations indicated that *Symbiodinium* exhibits remarkable light-dependent O_2 uptake activity [Jones *et al.*, 1998; Leggat *et al.*, 1999] which has been assigned later on as a Mehler reaction as the most important alternative electron transport and photoprotective process in several *Symbiodinium* clades [Tchernov *et al.*, 2004; Suggett *et al.*, 2008; Roberty *et al.*, 2014] and recently it was found that light-dependent O_2 uptake along with other photophysiological parameters reflects the acclimatory capacity to vertical light gradients in corals [Einbinder *et al.*, 2016]. However, other studies indicate that the MAP pathway is not the only significant alternative electron transfer pathway in *Symbiodinium*, as chlororespiration [Reynolds *et al.*, 2008], and CEF around PSI [Aihara *et al.*, 2016] likely play photoprotective roles, *e.g.* under heat stress, recently reviewed by Warner and Suggett [2016]. Measured on intact corals, it was found that PSI is quite robust under various stress conditions as compared to PSII, and the elevated PSI activity under these conditions may play a role in CEF [Hoogenboom *et al.*, 2012]. The existence of strong chlororespiratory activity in corals [Jones and Hoegh-Guldberg, 2001; Hill and Ralph, 2008] and the inhibition of Calvin–Benson cycle [Hill *et al.*, 2014] results in reduction of the PQ pool, which might also be the initiator of CEF, although direct evidence for CEF has not been shown in these studies.

Although several hypotheses have been suggested on the role of CEF as a stress avoidance mechanism in *Symbiodinium*, currently there is no evidence about the existence and operation of the "typical" CEF components (such as PTOX, PGR5 and NDH) in *Symbiodinium*. [Aihara *et al.*, 2016] reported that amino acid sequence homologues of PGR5/PGRL1, PTOX and NDH2 in *C. reinhardtii* exist in the genome of *Symbiodinium*. However, to date there is no physiological confirmation which of these components/pathways may be operational in *Symbiodinium*. Moreover, considering the high diversity of the *Symbiodinium* genetic types, it is likely that the significance and components of alternative electron transfer pathways and other photophysiological processes vary in different *Symbiodinium* types [Suggett *et al.*, 2015; Aihara *et al.*, 2016]. Therefore clarification of the role and function of CEF warrants further proteomics, molecular biology and bioinformatics investigations, possibly in various genetic types. A promising new direction is functional genomics, which might facilitate our knowledge on several physiological/metabolic functions in dinoflagellates [Murray *et al.*, 2016]. Creating knock-out mutants of specific CEF components would facilitate our understanding of the role of alternative electron flow in *Symbiodinium*, as a response to environmental stress.

5.2. *Haptophyta*

5.2.1. *Prymnesiophyceae*

Zhang *et al.* [2014] found differences in the CEF:(CEF + LEF) ratio at high light between two closely related strains of *Isochrysis galbana*. Paasche [1964, 1966] showed that light-dependent production of the $CaCO_3$-containing coccoliths in the coccolithophorid *Coccolithus* (now *Emiliana*) *huxleyi* was much less inhibited by DCMU than was photosynthesis, consistent with cyclic photophosphorylation involvement in transport of Ca^{2+} and HCO_3^- as substrates for calcification in an intracellular compartment. The interpretation is complicated by the involvement of polysaccharides, whose production needs DCMU-inhibited LEF, in the formation of coccoliths [Lavoie *et al.*, 2016; Monteiro *et al.*, 2016].

5.3. *Ochrophyta*

5.3.1. *Bacillariophyceae*

Bailleul *et al.* [2015] and Goldman *et al.* [2015] showed that CEF (+DCMU) was about 5% of CEF + LEF (–DCMU) in the non-psychrophilic marine *Fragilaria pinnata*, *Phaeodactylum tricornutum*, *Thalassiosira pseudonana* and *Thassiosira weissflogii* at light saturation, but up to 30% at low light. Goldman *et al.* [2015] showed that CEF (+DCMU) was about 35% of CEF + LEF (–DCMU) in the psychrophilic marine *Fragilariopsis pinnata*. Thamatrakoln *et al.* [2013] report predicted genes with sequence similarity to PGR5 in the genomes of *Fragilariopsis cylindrus*, *Phaeodactylum tricornutum* and *Thalassiosira pseudonana*, but also have "death-specific proteins" (DSPs) that may also act as proton gradient

     A. W. D. Larkum et al.

regulators in the "red" lineage of chloroplasts. Thamatrakoln *et al.* [2013] show that over-expression of the *Thalassiosira pseudonana* DSP (*trDSP1*) has no effect of CEF in Fe-replete cells, but increases CEF by 61% when cells are deficient in Fe.

5.3.2. *Eustigmatophyceae*

Simionato *et al.* [2013] used P700 turnover measurements to show that control (N-replete) *Nannochloropsis gaditana* CEF [+DCMU] was 0.04 of CEF plus LEF (–DCMU). Also using P700 turnover in the presence and absence of DCMU, CEF in *Nannochloropsis gaditana* occurs at up to 0.08 of the rate of CEF plus LEF [Meneghesso *et al.*, 2016]. The function of CEF in *Nannochloropsis* is not known; HCO_3^- entry in the CO_2 concentrating mechanism is energized by mitochondrial respiration rather than ATP produced by CEF [Huertas *et al.*, 2002].

5.3.3. *Phaeophyceae*

There is evidence of the occurrence of a CEF in brown algal photoacoustic measurements on *Macrocystis pyrifera* [Herbert *et al.*, 1990; Fork *et al.*, 1991], and from measurements of PSI electron transport rates in the presence and absence of two DCMU concentrations indicating that CEF is 10% or less than CEF + LEF in *Sargassum fusiforme* [Gao *et al.*, 2016]. Coughlan [1977] found that 10 mmol m^{-3} DCMU inhibited light-stimulated active influx of SO_4^{2-} in *Fucus serratus* to almost the dark rate, though the absence of other concentrations of DCMU and of comparisons with DCMU inhibition of photosynthesis means that the absence of energization of SO_4^{2-} by CEF needs further testing.

5.4. *Rhodophyta*

5.4.1. *Bangiophyceae*

The available data concern estimates of electron transport through PSI in the absence of PSII, *i.e.* CEF, and of CEF + LEF, so that CEF:[CEF + LEF] can be computed for growth under conditions yielding high growth rates [Biggins, 1973; Maxwell and Biggins, 1976; Gao and Wang, 2012], and when growth is limited by desiccation [Gao and Wang 2012]. The Bangiophyceae, and the Cyanidiophyceae, lack chloroplast NDH [references in DePriest *et al.*, 2013].

5.4.2. *Floridiophyceae*

The Floridiophyceae (references in DePriest *et al.* [2013]) lack chloroplast NDH.

6. The Role of CEF in Eukaryotic Algae in a Range of Habitats

6.1. *Desiccation*

Gao *et al.* [2011] investigated the effect of water loss from the thallus of the ulvophycean marine green alga *Ulva* on the electron transport rate through PSII and PSI. The PSI electron transport rate in the presence and absence of 10 mmol DCMU m^{-3} shows that the cyclic electron flow through PSI as a fraction of total electron flow is 0.08 in the fully hydrated controls (100% hydration), 0.21 at 65% hydration, and 0.93 at 22% hydration. Gao and Wang [2012] found similar outcomes in the bangiophycean red alga *Porphyra* [now *Pyropia*] *yezoensis*, with PSI as a fraction of total electron flow of 0.09 in the fully hydrated controls [100% hydration], 0.15 at 69% hydration, and 0.98 at 2% hydration.

6.2. *Variations in salinity*

Gao *et al.* [2016] found a higher PSI activity in the presence of DCMU to block PSII under higher and lower salinity than in controls of the low intertidal brown alga *Sargassum fusiforme*. The measured PSI turnover cannot necessarily be related to CEF in view of the possibility of net electron flow through PSI when DCMU blocks PSII, with electrons supplied by chrysolaminarin breakdown *via* NADPH [Gao *et al.*, 2016]. Huan *et al.* [2014] examined the effect of increased salinity in the intertidal ulvophycean green alga *Ulva prolifera* on PSI turnover in the presence of DCMU. The increased PSI turnover at high salinity seems to involve net electron flow rather than CEF in view of the decreased starch and sugar content of the thalli and the activity of the oxidative pentose phosphate pathway enzyme glucose-6-phosphate dehydrogenase.

6.3. *Low temperature*

The low temperature-adapted *Chlamydomonas raudensis* UWO241 from Lake Bonney in Taylor Valle, Antarctica is a psychrophile that cannot grow above 16°C, and is halotolerant [Dolhi *et al.*, 2013]. *Chlamydomonas raudensis* SAG49.72 isolated from a pond in the Czech Republic has an identical ITS sequence UWO241 but is a mesophile with an optimal growth temperature of 29°C [Dolhi *et al.*, 2013]. This phylogenetic analysis supercedes the taxonomy used in earlier work on the CEF [*e.g.* Morgan-Kiss *et al.*, 2002]. The psychrophilic *Chlamydomonas raudensis* UWO241 has a CEF rate three times that of the mesophilic *Chlamydomonas raudensis* SAG49.72 [Morgan-Kiss *et al.*, 2002; Dolhi *et al.*, 2013; Szyska-Mroz *et al.,* 2015]. A similar difference is found in marine diatoms; the psychrophilic *Fragilariopsis cylindrus* has a CEF:(CEF + LEF) ratio more than four times that in the mesophilic *Thalassiosira weissflogii* [Goldman *et al.,* 2015]. Borla *et al.* [2015] show that CEF is 7% and 12% of CEF + LEF in wild-type cells of *Synechocystis* PCC6803 at 30°C and 20°C respectively; for an alkane-producing strain the values are, respectively, 10% and 21%.

6.4. *High temperature*

Aihara *et al.* [2016] showed that temperature above the normal tolerance range induced what is probably antimycin A-insensitive PSI CEF in the symbiotic dino-flagellate *Symbiodinium*. This apparent increase in CEF seemed to be an inevitable consequence of the higher temperatures rather than a protective or repair response. The main alternative photosynthetic electron transport pathway in *Symbiodinium* is the PSI-dependent MAP pathway [Roberty *et al.,* 2014].

6.5. *High light*

Niu *et al.* [2016] showed the role of CEF in decreasing non-photochemical quenching during high light treatments of the bangiophycean red alga *Pyropia* (formerly *Porphyra*) *yezoensis*.

6.6. *Nitrogen deficiency*

N-deficient *Nannochloropsis gaditana* had CEF [+DCMU] that was 0.25 of CEF + LEF (–DCMU) as compared to 0.08 in N-replete cells [Simionato *et al.,* 2013].

6.7. *Iron deficiency*

PSI has a higher Fe content than other thylakoid complexes and the PSI content is usually decreased relative to the other Fe complexes under Fe deficiency [Raven *et al.,* 1999; Ivanov *et al.,* 2000, 2004; Marchionetti and Maldonado, 2016; Raven and Beardall, 2017] and cells adapted genetically to low Fe habitats also have a lower PSI content [Strzepek and Harrison, 2004]. Ivanov *et al.* [2000] showed that the PSII:PSI ratio increased from 0.53 to 0.9 under Fe deficiency and that CEF increased relative to $NADP^+$ reduction in *Synecchococcus* sp. PCC 7942. The decreased electron flux to $NADP^+$ under Fe deficiency is (partly at least) offset by electron flow from PSII to O_2 *via* PTOX rather than the Mehler reaction involving

PSI [Bailey *et al.*, 2008]. Thamatrakoln *et al.* [2013] found that CEF per PSI reaction centre was unchanged by Fe limitation in *Thalassiosira pseudonana*. A further CEF-related aspect of the response to Fe deficiency is that PGRL1, involved in one mechanism of CEF, is also related to remodeling the photosynthetic apparatus in *Chlamydomonas reinhardtii* [Petroutsos *et al.*, 2009].

7. Summary and Conclusions

This article outlines the strong evidence, in cyanobacteria and eukaryotic algae for CEF in the generation of extra proton motive force, and therefore extra ATP from PSI activity, *i.e.* an increase in the amount of ATP produced per e⁻ transported from PSII to PSI. The early *in vitro* evidence for this was established over 50 years ago, but despite strong evidence *in vivo* early on this has remained inferential. The use of mutants over recent years in *Arabidopsis thaliana* and *Chlamydomonas reinhardtii* has strengthened the evidence for CEF, but because of overlapping pathways of electron transport close to PSI the exact proof has remained frustratingly elusive. The situation is not helped by the fact that there are at least two dehydrogenase pathways involved (NDH and PGR) in all oxygenic PS organisms (cyanobacteria, eukaryotic algae and embryophytes). Also, in oxygenic phototrophs, there is no good inhibitor for the NDH pathway, and the only known inhibitor for the PGR pathway, antimycin A, is patchy in its effectiveness (active in PS bacteria, inactive in cyanobacteria, inactive in many algae including *Chlamydomonas,* and active in embryophytes).

Cyanobacteria have a strong NDH-linked CEF which clearly contributes significantly to ATP production for the Calvin–Benson Cycle and other energy dependent processes such a nitrogen fixation, nitrogen metabolism and chlorophyll synthesis. The eukaryotic algae have a great variety of structural and metabolic variation; and there is evidence for CEF in all of the algal divisions where it has been put to the test.

The case for CEF is strongest where stress effects lower the input of energy from LEF: in such situations as temperature stress, salt stress, and dehydration stress (such as in intertidal algae), *etc.* A few examples of such stresses are given but these examples will undoubtedly grow in the coming years.

Finally, we need better techniques to elucidate exactly what mechanisms contribute to CEF and what other electron transport pathways, such as the MAP pathway, oxidative respiration, glycolysis, and chlororespiration, intersect with CEF and how these interactions are controlled. Use of MIMS (membrane inlet mass spectrometry) and stable isotopes of oxygen and carbon are a promising means of improving precise knowledge of the CEF pathway.

Acknowledgements

The University of Dundee is a registered Scottish charity, Number SC 015096. This work was supported by funds from the Climate Change Cluster, University of Technology Sydney, Australia. DF acknowledges funding for a PhD scholarship provided by the Research School of Biology at the Australian National University, Canberra, Australia.

References

Aihara, Y., Takahashi, S. and Minagawa, J. (2016). Heat induction of cyclic electron flow around photosystem I in the symbiotic dinoflagellate *Symbiodinium*, *Plant Physiol.*, 171, 522–529.

Allahverdiyeva, Y., Mustilla, H., Emmakova, M., Bersanini, L., Richar, P., Aljani, G., Battchikova, N., Cournac, L. and Aro, E.-M. (2013). Flavodiiron proteins FLv1 and FLv3 enable growth and photosynthesis under fluctuating light, *Proc. Natl. Acad. Sci. USA*, 110, 4111–4116.

Armbruster, U., Galvis, V.C., Kunz, H.-H. and Strand, D.D. (2017). The regulation of the chloroplast proton motive force plays a key role for photosynthesis in fluctuating light, *Curr. Opin. Plant Biol.*, 37, 56–62.

Arnon, D.I., Allen, M.B. and Whatley, F.R. (1954). Photosynthesis by isolated chloroplasts, *Nature*, 174, 394–396.

Arnon, D.I., Whatley, F.R. and Allen, M.B. (1958). Assimilatory power in photosynthesis: photosynthetic phosphorylation by isolated chloroplasts is coupled with TPN reduction, *Science*, 127, 1026–1034.

Arteni, A.A., Zhang, P., Battchikova, N., Ogawa, T., Aro, E.-M. and Boekema, E.J. (2006). Structural characterization of NDH-1 complexes of *Thermosynechococcus elongatus* by single particle electron microscopy, *Biochim. Biophys. Acta*, 1757, 1469–1475.

Bailey, S., Melis, A., MacKey, K.R.M., Cardol, P., Finazzi, G., van Dijkjen, G., Berg, G.M., Arrigo, K., Schrager, G. and Grosman, A. (2007). Alternative photosynthetic electron flow to oxygen in marine *Synechococcus*, *Biochim. Biophys. Acta*, 1777, 269–276.

Bailleul, B., Berne, N., Murik, O., Petroutsos, D., Prihoda, J., Tanaka, A., Villanova, V., Bligny, R., Flori, S., Falconet, D., Krieger-Liszkay, A., Santabarbara, S., Rappaport, F., Joliot, P., Tirichine, L., Falkowski, P.G., Cardol, P., Bowler, C and Finazzi, G. (2015). Energetic coupling between plastids and mitochondria drive CO_2 assimilation in diatoms, *Nature*, 524, 366–369.

Battchikova, N., Eisenhut, M. and Aro, E.-M. (2011a). Cyanobacterial NDH-1 complexes: novel insights and remaining puzzles, *Biochim. Biophys. Acta*, 1807, 935–944.

Battchikova, N., Wei, L., Du, L., Bersanini, L., Aro, E.-M. and Ma, W. (2011b). Identification of novel Ssl0352 protein (NdhS), essential for efficient operation of cyclic electron transport around photosystem I, in NADPH: plastoquinone oxidoreductase (NDH-1) complexes of *Synechocystis* sp. PCC6803, *J. Biol. Chem.*, 286, 36992–37001.

Beckmann, K., Messinger, J., Badger, M.R., Wydrzynski, T. and Hillier, W. (2009). On-line mass spectrometry: membrane inlet sampling, *Photosynth. Res.*, 102, 511–522.

Bendall, D.S. and Manasse, R.S. (1995). Cyclic photophosphorylation and electron transport, *Biochim. Biophys. Acta*, 1229, 23–38.

Berger, S., Ellersiek, U. and Steinmueller, K. (1991). Cyanobacteria contain a mitochondrial complex 1-homologous NADH-dehydrogenase, *FEBS Lett.*, 286, 129–132.

Berla, B.M., Saha, R., Maranas, C.D. and Pakrasi, H.B. (2015). Cyanobacterial alkanes modulate photosynthetic cyclic electron flow to assist growth under cold stress, *Sci. Rep.*, 5, 14894, doi:10.1038/srep14894.

Biggins, J. (1973). Kinetic behaviour of cytochrome *f* in cyclic and non-cyclic electron transport in *Porphyridium cruentum*, *Biochemistry*, 12, 1165–1169.

Bottomley, P.J. and Stewart, W.D.P. (1976). ATP pools and transients in the blue-green alga, *Anabaena cylindrical*, *Arch. Microbiol.*, 108, 249–258.

Bottomley, P.J. and Stewart, W.D.P. (1977). ATP and nitrogenase activity in nitrogen-fixing heterocystous blue-green algae, *New Phytol.*, 79, 625–638.

Cardol, P., Forti, G. and Finazzi, G. (2011). Regulation of electron transport in microalgae, *Biochim. Biophys. Acta*, 1807, 912–918.

Cha, Y. and Mauzerall, D.C. (1992). Energy storage of linear and cyclic electron flows in photosynthesis, *Plant Physiol.*, 100, 1869–1877.

Chaux, F., Burlacot, A., Mekhalfi, M., Auroy, P., Blangy, S., Richaud, P. and Peltier, G. (2017). Flavidiiron proteins promote fast and transient O_2 photoreduction in *Chlamydomonas*, *Plant Physiol.* 174(3), 1825–1836, doi:10.1104/pp.17.00421.

Chen, X., He, Z., Xu, M., Peng, L. and Mi, H. (2016). NdhV subunit regulates the activity of type-1 NAD(P)H dehydrogenase under high light conditions in cyanobacterium *Synechocystis* sp. PCC 6803, *Sci. Rep.*, 6, 28361.

Civán, P., Foster, P.G., Embley, M.T., Séneca, A. and Cox, C.M. (2014) Analyses of charophyte chloroplast genomes help characterise the ancestral chloroplast genome of land plants, *Genome Biol. Evol.*, 6, 897–911.

Cleland, R.E. and Bendall, D.S. (1992). Photosystem I cyclic electron transport: measurement of ferredoxin-plastoquinone reductase activity, *Photosynth. Res.*, 34, 409–418.

Coughlan, S. (1977). Sulphate uptake in *Fucus serratus*. *J. Exp. Bot.*, 28, 1207–1215.

Cramer, W.A., Hasan, S.S. and Yamashita, E. (2011). The Q cycle of cytochrome *bc* complexes: a structure perspective, *Biochim. Biophys. Acta*, 1807, 788–802.

Dal Corso, G., Pesaresi, P., Masiero, S., Aseeva, E., Schuenemann, D., *et al.* (2008). A complex containing PGRL1 and PGR5 is involved in the switch between linear and cyclic electron flow in *Arabidopsis*, *Cell*, 132, 273–285.

Dang, K.-V., Plet, J., Tolleter, D., Jokel, M., Cuiné, S., Carrier, P., Auroy, P., Richaud, P., Johnson, X., Alric, J. and Allahverdiyeva, Y. (2014). Combined increases in mitochondrial cooperation and oxygen photoreduction compensate for deficiency in cyclic electron flow in *Chlamydomonas reinhardtii*, *Plant Cell*, 26, 3036–3050.

De Priest, M.S., Bhattacharya, D. and López-Bautista, J.M. (2013). The plastid genome of the red macroalga *Grateloupia taiwanensis* (Halymeniaceae), *PLOS ONE*, 8, p. e68246.

Deák, Z., Sass, L., Kiss, É. and Vass, I. (2014). Characterization of wave phenomena in the relaxation of flash-induced chlorophyll fluorescence yield in cyanobacteria, *Biochim. Biophys. Acta*, 1837, 1522–1532.

Der-Vertanian, M., Joset-Espardellier, F. and Astier, C. (1981). Contributions of respiratory and photosynthetic pathways during growth of a facultative photoautotrophic cyanobacterium, *Aphanocapsa* 6714, *Plant Physiol.*, 68, 974–998.

Dolhi, J.M., Maxwell, D.P. and Morgan-Kiss, R.M. (2013). Review: the Antarctic *Chlamydomonas raudensis*: an emerging model for cold adaptation of photosynthesis, *Extremophiles*, 17, 711–722.

Einbinder, S., Gruber, D.F., Salomon, E., Liran, O., Keren, N. and Tchernov, D. (2016). Novel adaptive photosynthetic characteristics of mesophotic symbiotic microalgae within the reef-building coral, *Stylophora pistillata*, *Front. Mar. Sci.*, 3, 195.

Falkowski, P.G. and Raven, J.A. (2007). *Aquatic Photosynthesis.* (Princeton University Press, Princeton).

Fan, D., Fitzpatrick, D., Oguchi, R., Ma, W., Kou, J. and Chow, W. (2016) Obstacles in the quantification of the cyclic electron flux around photosystem I in leaves of C_3 plants, *Photosynth. Res.*, 129, 239–251.

Fisher, N. and Kramer, D.M. (2014). Non-photochemical reduction of thylakoid photosynthetic redox carriers *in vitro*: relevance to cyclic electron flow around photosystem I? *Biochim. Biophys. Acta*, 1837, 1944–1954.

Fitzpatrick, D. (2016). Energetic responses to transient high temperature stress in cyanobacteria. PhD Thesis, Australian National University.

Fork, D.C. and Herbert, S. (1993). Electron transport and photophosphorylation by photosystem I *in vivo* in plants and cyanobacteria, *Photosynth. Res.*, 36, 149–168.

Fork, D.C., Herbert, D.K. and Malkin, S. (1991). Light-energy distribution in the brown alga *Macrocystis pyrifera* (giant kelp), *Plant Physiol.*, 95, 731–739.

Friedrich, T. (2014). On the mechanism of respiratory complex I, *J. Bioenerg. Biomembr.*, 46, 255–268.

Gao, S., Huan, L., Lu, X.-P., Jin, W.-H., Wang, X.-L., Wu, M.-J. and Wang, G.-C. (2016). Photosynthetic responses of the low intertidal macroalga *Sargassum fusiforme* (Sargassaceae) to saline stress, *Photosynthetica*, 54, 430–437.

Gao, S., Shen, S., Wang, G., Niu, J., Lin, A. and Pan, G. (2011). PSI-driven cyclic electron flow allows intertidal macroalgae *Ulva* sp. (Chlorophyta) to survive in desiccated conditions, *Plant. Cell Physiol.*, 52, 885–893.

Gao, F., Zhao, J., Chen, L., Battchikova, N., Ran, Z., *et al.* (2016). The NDH-1L-PSI supercomplex is important for efficient cyclic electron transport in cyanobacteria, *Plant Physiol.*, 172, 1451–1464.

Gao, F., Zhao, J., Wang, X., Qin, S., Wei, L. and Ma, W. (2016). NdhV is a subunit of NADPH dehydrogenase essential for cyclic electron transport in *Synechocystis* sp. strain PCC 6803, *Plant Physiol.*, 170, 752–760.

Gfeller, R.P. and Gibbs, M. (1984). Fermentative metabolism of *Chlamydomonas reinhardtii*. I. Analysis of fermentative products from starch in dark and light, *Plant Physiol.*, 75, 212–218.

Gfeller, R.P. and Gibbs, M. (1985). Fermentative metabolism of *Chlamydomonas reinhardtii*. II. Role of plastoquinone, *Plant Physiol.*, 77, 509–511.

Goldman, J.A.L., Kranz, S.A., Young, J.N., Tortell, P.D., Stanley, R.H.R., Bender, M.L. and Morel, F.M.M. (2015). Gross and net production during the spring bloom along the Western Antarctic Penninsula, *New Phytol.*, 205, 182–191.

Govindjee, Sherela, D. and Björn, L.O. (2017). Evolution of the Z-scheme of photosynthesis: a perspective, *Photosynth. Res.*, 133, 5–15, doi:10.1007/s11120-06-0533-z.

Hagino, K., Onuma, R., Kawachi, M. and Horiguchi, T. (2013). Discovery of an endosymbiotic nitrogen-fixing cyanobacterium UCYN-A in *Braarudoaphaera bigelowii* (Pryrmnesiophyceae), *PLOS ONE*, 8, p. e81749.

He, Z., Zheng, F., Wu, Y., Li, Q., Lv, J., *et al.* (2015). NDH-1L interacts with ferredoxin *via* the subunit NdhS in *Thermosynechococcus elongatus*, *Photosynth. Res.*, 126: 341–349.

Healey, F.P. and Myers, J. (1971). The Kok effect in *Chlamydomonas reinhardti*, *Plant Physiol.*, 47, 373–379.

Herbert, S.K., Fork, D.C. and Malkin, S. (1990). Photoacoustic measurements *in vivo* of energy storage by cyclic electron flow in algae and higher plants, *Plant Physiol.*, 94, 926–934.

Hertle, A.P., Blunden, T., Wurden, T., Pesaresi, P., Pribil, M., Armbruster, U. and Leister, D. (2012). PGRL1 is the elusive ferredoxin-plastoquinone reductase in photosynthetic cyclic electron flow, *Mol. Cell*, 49, 511–523.

Hill, R. and Ralph, P.J. (2008). Dark-induced reduction of the plastoquinone pool in zooxanthellae of scleractinian corals and implications for measurements of chlorophyll *a* fluorescence, *Symbiosis*, 46, 45–56.

Hill, R., Szabó, M., ur Rehman, A., Vass, I., Ralph, P.J. and Larkum, A.W. (2014). Inhibition of photosynthetic CO_2 fixation in the coral *Pocillopora damicornis* and its relationship to thermal bleaching, *J. Exp. Biol.*, 217, 2150–2162.

Hirt, G., Tanner, W. and Kandler, O. (1971). Effects of light on the rate of glycolysis in *Scenedesmus obliquus*, *Plant Physiol.*, 47, 841–843.

Hoogenboom, M.O., Campbell, D.A., Beraud, E., Dezeeuw, K. and Ferrier-Pages, C. (2012). Effects of light, food availability and temperature stress on the function of photosystem II and photosystem I of coral symbionts, *PLOS ONE*, 7(1), p. e30167. doi:10.1371/journal.pone.0030167 PONE-D-11-13070 (p ii).

Hori, K., Maruyama, F., Fujisawa, T., Togashi, T., Yamamoto, N., *et al.* (2014). *Klebsormidium flaccidum* genome reveals primary factors for plant terrestrial adaptation, *Nat. Commun.*, 5, p. 3978.

Howitt, C.A., Cooley, J.W., Wiskich, J.T. and Vermaas, W.F.J. (2001). A strain of *Synechocystis* sp. PCC 6803 without photosynthetic oxygen evolution and respiratory oxygen consumption: implications for the study of cyclic photosynthetic electron transport, *Planta*, 214, 46–56.

Howitt, C., Udall P. and Vermaas W. (1999). Type 2 NADH dehydrogenases in the cyanobacterium *Synechocystis* sp. strain PCC 6803 are involved in regulation rather than respiration, *J. Bacteriol.*, 181, 3994–4003.

Huan, L., Xie, X., Zheng, Z., Sun, F., Wu, S., Li, M., Gao, S., Gu, W. and Wang, G. (2014). Positive correlation between PSI response and oxidative pentose phosphate pathway activity during salt stress in an intertidal macroalga, *Plant. Cell Physiol.*, 55, 1395–1403.

Huertas, I.E., Colman, B. and Espie, G.S. (2002). Mitochondria-driven bicarbonate transport supports photosynthesis in a marine microalga, *Plant Physiol.*, 130, 284–291.

Ifuku, K., Endo, T., Shikanai, T. and Aro, E.M. (2011) Structure of the chloroplast NADH dehydrogenase-like complex: nomenclature for nuclear encoded subunits, *Plant. Cell Physiol.*, 52, 1560–1568.

Ikezawa, N., Ifuku, K., Endo, T. and Sato, F. (2002). Inhibition of photosystem II of spinach by the respiration inhibitors piericidin A and thenoyltrifluoroacetone, *Biosci. Biotechnol. Biochem.*, 66, 1925–1929.

Ivanov, A.G., Park, Y.I., Miskiewicz, E., Raven, J.A., Huner, N.P.A. and Öquist, G. (2000). Iron stress restricts photosynthetic intersystem electron transport in *Synechococcus* sp. PCC 7942, *FEBS Lett.*, 485, 173–177.

Ivanov, A.G., Sane, P.V., Simidjiev, I., Park, Y.I., Huner, N.P.A. and Öquist, G. (2000). Restricted capacity for PSI cyclic electron flow in Δ*petE* mutant compromises the ability for acclimation to iron stress in *Sunechococcus* sp. PCC 7942, *Biochim. Biophys. Acta*, 1817, 1277–1284.

Iwai, M., Takiawa, K., Tokotsu, R., Okamuro, A., Takahashi, Y. and Minagawa, J. (2010). Isolation of the elusive supercomplex that drives cyclic electron flow in photosynthesis, *Nature*, 464, 1210–1213.

Jia, X.-H., Zhang, P.-P., Shi, D.-J., Mi, H.-L., Zhu, J.-C., *et al.* (2015). Regulation of *pepc* gene expression in *Anabaena* sp. PCC 7120 and its effects on cyclic electron flow around photosystem I and tolerances to environmental stresses, *J. Integr. Plant Biol.*, 57, 468–476.

Joet, T., Cournac, L., Horvath, E.M., Medgyesy, P. and Peltier, G. (2001). Increased sensitivity of photosynthesis to antimycin A induced inactivation of the chloroplast *ndhB* gene. Evidence for the participation of the NADH-dehydrogenase complex to cyclic electron transport around Photosystem I, *Plant Physiol.*, 125, 1919–1929.

Jones, R. and Hoegh-Guldberg, O. (2001). Diurnal changes in the photochemical efficiency of the symbiotic dinoflagellates (Dinophyceae) of corals: photoprotection, photoinactivation and the relationship to coral bleaching, *Plant Cell. Environ.*, 24, 89–99.

Jones, R.J., Hoegh-Guldberg, O., Larkum, A.W.D. and Schreiber, U. (1998). Temperature-induced bleaching of corals begins with impairment of the CO_2 fixation mechanism in zooxanthellae, *Plant Cell. Environ.*, 21, 1219–1230, doi:10.1046/j.1365-3040.1998.00345.x.

Johnson, X., Steinbeck, J., Dent, R.M., Takahashi, H., Richaud P., *et al.* (2014). Proton gradient regulation 5-mediated cyclic electron flow under ATP- or redox-limited conditions: a study of delta ATPase pgr5 and delta rbcL pgr5 mutants in the green alga *Chlamydomonas reinhardtii, Plant Physiol.*, 165, 438–452.

Joliot, P. and Johnson, G.N. (2011). Regulation of cyclic and linear electron flow in higher plants, *Proc. Natl. Acad. Sci. USA*, 108, 13317–13322.

Kanazawa, A., Ostendorf, E., Kohzuma, K., Hoh, D., Strand, D.D., Sato-Cruz, M., Savage, L., Cruz, J.A., Fisher, N. and Froehlich, J.E. (2017). Chloroplast ATP synthase modulation of the thylakoid proton motive force: implications for photosystem I and photosystem II photoprotection, *Front. Plant Sci.*, 8, 719.

Kandler, O. and Haberer-Liesenkötter, I. (1963). Über den Zussamenhang zwischen Phosphathaushalt und Photosyntheses, V. Regulation der Glycolyse durch die Lichtphosphorylierung bei *Chlorella. Z, Naturforsch.*, 18B, 718–730.

Keifer, D.W. and Spanswick, R.M. (1978). Activity of the electrogenic pump in *Chara coralline* as inferred from measurements of the membrane potential, conductance and potassium permeability, *Plant Physiol.*, 62, 653–666.

Kok, B. (1949). On the interrelation of respiration and photosynthesis in green plants, *Biochim. Biophys. Acta*, 3, 623–631.

Komor, E. and Tanner, W. (1974). The hexose-proton symport system of *Chlorella vulgaris*. Specificity, stoichiometry and energetics of sugar-induced proton uptake, *Eur. J. Biochem.*, 44, 219–223.

Kou, J., Takahashi, S., Oguchi, R., Fan, D., Badger, M. and Chow, W. (2013). Estimation of the steady-state cyclic electron flux around PSI in spinach leaf discs in white light, CO_2-enriched air and other varied conditions, *Funct. Plant Biol.*, 40, 1018–1028.

Labs, M., Rűhle, T. and Leister, D. (2016). The antimycin A-sensitive pathway of cyclic electron flow: from 1963 to 2015, *Photosynth. Res.*, 129, 231–238.

Lavoie, M., Raven, J.A. and Levasseur, M. (2016). Energy cost and putative benefits of cellular mechanisms modulating buoyancy in a flagellate marine phytoplankton, *J. Phycol.*, 52, 239–251.

Leggat, W., Badger, M.R. and Yellowlees, D. (1999). Evidence for an inorganic carbon-concentrating mechanism in the symbiotic dinoflagellate *Symbiodinium* sp., *Plant Physiol.*, 121, 1247–1255.

Leister, D., Shikanai T. (2013). Complexities and protein complexes in the antimycin A-sensitive pathway of cyclic electron flow in plants, *Front. Plant Sci.*, 4, 161.

Leliaert, F., Tronholm, A., Lemieux, C., Turmel, M., DePriest, M.S., Bhattacharya, D., Karol, K.G., Fredericq, S., Zechman, F.W. and Lopez-Bautista, M. (2016). Chloroplast phylogenomic analyses reveal the deepest-branching lineage of the Palmophylophyceae class. nov, *Sci. Rep.*, 6, 25367.

Marchionetti, A. and Maldonado, M.I. (2016). Iron. In *The Physiology of Microalgae*, Borowitzka, M.A., Beardall, J. and Raven, J.A., eds. (Heidelberg: Springer), pp. 233–279.

Martín, M. and Sabatier, B. (2010) Plastid *ndh* genes in plant evolution, *Plant Physiol. Biochem.*, 48, 636–645.

Maxwell, P.C. and Biggins, J. (1976). Role of cyclic electron transport in photosynthesis as measured by the photoinduced turnover of P_{700} *in vivo*, *Biochemistry*, 15, 3975–3981.

Meneghesso, A., Simionato, D., Gertto, C., La Rocca, N., Finazzi, G. and Morosinotto, T. (2016). Photoacclimation of photosynthesis in the Eustigmatophycean *Nannochloropsis gaditana*, *Photosynth. Res.*, 129, 291–305.

Merchant, S.S., Prochnik, S.E., Vallon, O., Harris, E.H., Karpowicz, S.J., Witman, G.B., *et al.* (2007). The *Chlamydomonas* genome reveals the evolution of key animal and plant functions, *Science*, 318, 245–250.

Merchant, S.S., Prochnik, S.E., Vallon, O., Harris, E.H., Karpowicz, S.J., Witman, G.B., Terry, A., Salamov, A., Fritz-Laylin, L.K., Maréchal-Drouard, L., Marshall, W.F., *et al.* (2007). The *Chlamydomonas* genome reveals the evolution of key animal and plant functions, *Science*, 318, 245–251.

Monteiro, F.M., Bach, L.T., Brownlee, C., Bowen, P., Rickaby, R.E.M., Poulton, A.J., Tyrell, T., Beaufort, L., Dutkiewicz, S., Gibbs, S., Gutowska, M.A., Lee, R., Riebesell, U., Young, J. and Ridgwell, A. (2016). Why marine phytoplankton calcify, *Sci. Adv.*, 2, p. e1501822.

Moparthi, V.K. and Hagerhall, C. (2011). The evolution of respiratory chain complex I from a smaller last common ancestor consisting of 11 protein subunits, *J. Mol. Evol.*, 72, 484–497.

Morgan-Kiss, R.H., Ivanov, A.G. and Huner, N.F.A. (2002). The Antarctic psychrophile, *Chlamydomonas sucordata* is deficient in state I–state II transitions, *Planta*, 241, 435–445.

Morosinotto, T. (2016). Photoacclimation of photosynthesis in the Eustigmatophycean *Nannochloropsis gaditana*, *Photosynth. Res.*, 129, 291–305.

Moss, D.A. and Bendall, D.S. (1984). Cyclic electron transport in chloroplasts, the Q-cycle and the site of action of antimycin A, *Biochim. Biophys. Acta*, 767, 389–395.

Mulkidjanian, A.Y. (2010). Activated Q-cycle as a common mechanism for cytochrome *bc*(1) and cytochrome *b*(6)*f* complexes, *Biochim. Biophys. Acta*, 1797, 1858–1868.

Mullineaux, C.W. (2014). Co-existence of photosynthetic and respiratory activities in cyanobacterial thylakoid membranes, *Biochim. Biophys. Acta*, 1837, 503–511.

Munekage, Y., Hojo, M., Meurer, J., Endo, T., Tasaka, M. and Shikanai, T. (2002). PGR5 is involved in cyclic electron flow around photosystem I and is essential for photoprotection in *Arabidopsis*, *Cell*, 110, 361–371.

Munekage, Y., Hashimoto, M., Miyake, C., Tomizawa, K.-I., Endo, T., *et al.* (2004). Cyclic electron flow around photosystem I is essential for photosynthesis, *Nature*, 429, 579–582.

Murray, S.A., Suggett, D.J., Doblin, M.A., Kohli, G.S., Seymour, J.R., Fabris, M., *et al.* (2016). Unravelling the functional genetics of dinoflagellates: a review of approaches and opportunities, *Perspect. Phycol.*, 3, 37–52.

Nicholls, D.G. and Ferguson, S.J. (2013). *Bioenergetics*, 4th edn. (Academic Press, Amsterdam).

Ogawa, T., Miyano, A. and Inoye, Y. (1985a). Photosystem I-driven inorganic carbon transport in the cyanobacterium, *Anacystis nidulans*, *Biochim. Biophys. Acta*, 88, 77–84.

Ogawa, T. and Ogren, W.L. (1985). Action spectra for accumulation of inorganic carbon in the cyanobacterium, *Anabaena variabilis*, *Photochem. Photobiol.*, 41, 583–587.

Ogawa, T., Omata, T., Miyano, A. and Inoye, Y. (1985b). Photosynthetic reactions involved in the CO_2-concentrating mechanism in the cyanobacterium, *Anacystis nidulans*. In *Inorganic Carbon Uptake by Aquatic Photosynthetic Organisms*, Lucas, W.J. and Berry, J.A., eds. (Rockville: American Society of Plant Physiologists), pp. 287–304.

Ohkawa, H., Sonoda, M., Shibata, M. and Ogawa, T. (2001). Localization of NAD(P)H dehydrogenase in the cyanobacterium *Synechocystis* sp. strain PCC 6803, *J. Bacteriol.*, 183, 4938–4939.

Paasche, E. (1964). A tracer study of the inorganic carbon uptake during coccolith formation and photosynthesis in the coccolithophorid *Coccolithus huxleyi*, *Physiol. Plant.*, (Suppl. 3), 1–81.

Paasche, E. (1966). Action spectrum of coccolith formation, *Physiol. Plant.*, 19, 770–777.

Peltier, G., Aro, E.-M. and Shikanai, T. (2016). NDH-1 and NDH-2 plastoquinone reductase in oxygenic photosynthesis, *Ann. Rev. Plant Biol.*, 67, 55–80.

Peltier, G. and Cournac, L. (2002). Chlororespiration, *Annu. Rev. Plant Biol.*, 53, 523–550.

Peltier, G. and Sarrey, F. (1988). The Kok effect and the inhibition of chlororespiration in *Chlamydomonas reinhardtii*, *FEBS Lett.*, 228, 259–262.

Peltier, G., Tolleter, D., Billon, E. and Cournac, L. (2010). Auxilliary electron transport pathways in chloroplasts of microalgae, *Photosynth. Res.*, 106, 19–31.

Peng, I., Yamamoto, H. and Shikanai, T. (2011). Structure and biogenesis of the chloroplast NAD(P)H dehydrogenase complex, *Biochim. Biophys. Acta*, 1807, 945–953.

Petroutsos, D., Terauchi, A.M., Busch, A., Hirschmann, I., Merchant, S.S., Finazzi, G. and Hippler, M. (2009). PGRL1 participates in iron-induced remodelling of the photosynthetic apparatus and in energy metabolism in *Chlamydomonas reinhardtii*, *J. Biol. Chem.*, 284, 32770–32781.

Price, G.D. (2011). Inorganic carbon transporters of the cyanobacterial CO_2 concentrating mechanism, *Photosynth. Res.*, 109, 47–57.

Raven, J.A. (1973). Letter: caloric recalculation, *Biophys. J.*, 13, 1002–1003.

Raven, J.A. (1976a). Division of labour between chloroplast and cytoplasm. In *The Intact Chloroplast*, Barber, J., ed. (Amsterdam: Elsevier-North Holland), pp. 403–443.

Raven, J.A. (1976b). The rate of cyclic and non-cyclic photophosphorylation and oxidative phosphorylation, and regulation of the rate of ATP consumption in *Hydrodictyon africanum*, *New Phytol.*, 76, 205–212.

Raven, J.A. (1984). *Energetics and Transport in Aquatic Plants*, (A.R. Liss, New York).

Raven, J.A. and Beardall, J. (2017). Genotypic loss and phenotypic regulation of Complex I in mitochondria, *J. Exp. Bot.*, 68, 2683–2692.

Raven, J.A., Beardall, J. and Giordano, M. (2014). Energy costs of carbon dioxide concentrating mechanisms in aquatic organisms, *Photosynth. Res.*, 121, 111–124.

Raven, J.A., Evans, M.C.W. and Korb, E. (1999). The role of trace metals in photosynthetic electron transport in O_2-evolving organisms, *Photosynth. Res.*, 60, 111–119.

Raven, J.A., Johnston, A.M. and MacFarlane, J.J. (1990). Carbon metabolism. In *The Biology of Red Algae*, Cole, K.M. and Sheath, R.G., eds. (Cambridge: Cambridge University Press), pp. 172–202.

Ravenel, J., Peltier, G. and Havaux, M. (1994). The cyclic electron pathways around photosystem I in *Chlamydomonas reinhardtii* as determined *in vivo* by photoacoustic measurements of energy storage, *Planta*, 193, 351–359.

Reynolds, J.M., Bruns, B.U., Fitt, W.K. and Schmidt, G.W. (2008). Enhanced photoprotection pathways in symbiotic dinoflagellates of shallow-water corals and other cnidarians, *Proc. Natl. Acad Sci. USA*, 105, 13674–13678, doi:10.1073/pnas.0805187105 0805187105 (pii).

Roberty, S., Bailleul, B., Berne, N., Franck, F. and Cardol, P. (2014). PSI Mehler reaction is the main alternative photosynthetic electron pathway in *Symbiodinium* sp., symbiotic dinoflagellates of cnidarians, *New Phytol.*, 204(1), 81–91. doi:10.1111/nph.12903.

Ruhlman, T.A., Chang, W.J., Chen, J.J.W., Huang, Y.T., Chan, M.T., Zhang, J., Liao, D.-C., Blazier, J.C., Shih, M.-C., Janson, R.K. and Lin, C.-S. (2015). NDH expression marks major transitions in plant evolution and reveals coordinate intracellular gene loss, *BMC Plant Biol.*, 15, article 100.

Simionato, D., Block, M.A., La Rocca, N., Jouhet, J., Maréchal, E., Finazzi, G. and Morisinotto, T. (2013). The response of *Nannochloropsis gaditana* to nitrogen starvation includes *de novo* biosynthesis of triacylglycerols, a decrease of chloroplast galactolipids, and reorganization of the photosynthetic apparatus, *Eukaryot. Cell*, 12, 665–676.

Simonis, W. and Urbach, W. (1973). Photophosphoryation *in vivo*, *Annu. Rev. Plant Physiol.*, 24, 89–114.

Smith, F.A. and Raven, J.A. (1974). Energy-dependent processes in *Chara corallina*: absence of ligh stimulation when only photo-system one is operative, *New Phytol.*, 73, 1–12.

Soja, N., Kimura, K., Kinistota, K., Jr., Yoshida, M. and Suzuki, T. (2017). Perfect chemo-mechanical coupling of F_OF_1-ATP synthase, *Proc. Nat. Acad. Sci. USA*, 114(19), 4960–4965, doi:10.1073/pnas.1700801114.

Soo, R.M., Hemp, J., Parks, D.H., Fischer, W.W. and Hugenholtz, P. (2017). On the origins of oxygenic photosynthesis and aerobic respiration in Cyanobacteria, *Science*, 355, 1436–1439.

Staal, M., Stal, I.J., Te Lintel Hekkert, S. and Harven, F.J.M. (2003). Light action spectra of N_2 fixation by heterocystous cyanobacteria from the Baltic Sea, *J. Phycol.*, 39, 668–677.

Steigmiller, S., Turina, P. and Graeber, P. (2008). The thermodynamic H^+/ATP ratios of the H^+-ATPsynthases from chloroplasts and *Escherichia coli*, *Proc. Nat. Acad. Sci. USA*, 105, 3745–3750.

Strand, D.D., Fisher, N. and Kramer, D. (2016a). Distinct energetics and regulatory functions of the two major cyclic electron flow pathways in chloroplasts. In *Chloroplasts:*

Current Research and Future Trends, Kirchhoff, H., ed. Chap. 4 (Norfolk: Caister Academic Press), pp. 89–100.

Strand, D.D., Fisher, N. and Kramer, D.M. (2016b). The higher plant plastid complex I (NDH) is a reversible proton pump that increases ATP production by cyclic electron flow around photosystem I, *bioRxiv*, doi:10.1101/049759. Preprint first posted online April 22, 2016.

Strzepek, R.F. and Harrison, P.J. (2004). Photosynthetic architecture in coastal and oceanic diatoms, *Nature*, 431, 689–692.

Suggett, D.J., Goyen, S., Evenhuis, C., Szabó, M., Pettay, D.T., Warner, M.E., *et al.* (2015). Functional diversity of photobiological traits within the genus *Symbiodinium* appears to be governed by the interaction of cell size with cladal designation, *New Phytol.*, 208, 370–381.

Suggett, D.J., Warner, M.E., Smith, D.J., Davey, P., Hennige, S. and Baker, N.R. (2008). Photosynthesis and production of hydrogen peroxide by *Symbiodinium* (Pyrrhophyta) phylotypes with different thermal tolerances, *J. Phycol.*, 44(4), 948–956, doi:10.1111/j.1529-8817.2008.00537.x.

Sugimoto, K., Okegawa, Y., Tohri, A., Long, T.A., Covert, S.F., *et al.* (2013). A single amino acid alteration in PGR5 confers resistance to antimycin A in cyclic electron transport around PSI, *Plant Cell Physiol.*, 54, 1525–1534.

Sultemeyer, D., Biehler, K. and Fock, H.P. (1993). Evidence for the contribution of pseudocyclic photophosphorylation for the mechanism for concentrating inorganic carbon in *Chlamydomonas*, *Planta*, 189, 235–242.

Suorsa, M.J.S., Grieco, M., Nurmi, M., Pietrzykowska, M., Rantala, M., Kangasjarvi, S., Paakkarinen, V., Tikkanen, M., Jansson, S. and Aro, E.M. (2012). Proton gradient regulation5 is essential for proper acclimation of *Arabidopsis* photosystem I to naturally and artificially fluctuating light conditions, *Plant Cell*, 24, 2934–2948.

Szyska-Mroz, B., Pittock, P., Ivanov, A.G., Lajoie, G. and Huner, N.P.A. (2015). The Antarctic psychrophile *Chalmydomonas* sp. UWO 241 preferentially phosphorylates photosystem I — cyctochrome b_6/f supercomplex, *Plant Physiol.*, 169, 717–736.

Tagawa, K., Tsujimoto, H.Y. and Arnon, D.I. (1963a). Role of chloroplast ferredoxin in the energy conversion process of photosynthesis, *Proc. Natl. Acad. Sci. USA*, 49, 567–572.

Tagawa, K., Tsujimoto, H.Y. and Arnon, D.I. (1963b). Role of chloroplast ferredoxin in the energy conversion process of photosynthesis, *Proc. Natl. Acad. Sci. USA*, 49, 567–572.

Tanner, W., Dächel, L. and Kandler, O. (1965). The effects of DCMU and antimycin A on photoassimilation of glucose in *Chlorella*, *Plant Physiol.*, 40, 1151–1156.

Tanner, W., Loos, E., Klob, W. and Kandler, O. (1968) The quantum requirement for light dependent anaerobic glucose assimilation by *Chlorella vulgaris*. Z, *Pflanzenphysiol*, 59, 301–303.

Tchernov, D., Gorbunov, M.Y., de Vargas, C., Yadav, S.N., Milligan, A.J., Haggblom, M., *et al.* (2004). Membrane lipids of symbiotic algae are diagnostic of sensitivity to thermal bleaching in corals, *Proc. Natl. Acad. Sci. USA*, 101, 13531–13535, doi:10.1073/pnas.0402907101.

Teichler-Zallen, D. and Hoch, G.E. (1967). Cyclic electron transport in algae, *Arch. Biochem. Biophys.*, 120, 227–230.

Thamatrakoln, K., Bailleul, B., Brown, C.M., Gorbunov, M.Y., Kustka, A.B., Frada, M., Joliot, P.A., Falkowski, P.G. and Bidle, K.D. (2013). Death-specific protein in a marine diatom regulates photosynthetic responses to iron and light availability, *Proc. Natl. Acad. Sci. USA*, 110, 20123–20128.

Thomson, A.W., Foster, R.A., Kruke, A., Carter, B.J., Musat, N., Vaulot, D., Kuypers, M.M. and Zehr, J.P. (2012). Unicellular cyanobacterium symbiotic with single-celled eukaryotic alga, *Science*, 237, 1546–1550.

Turmel, M., Gagnon, M.C., O'Kelly, C.J., Otis, C. and Lemieux, C. (2009). The chloroplast genomes of the green algae *Pyramimonas*. *Monomastix*, and *Pycncoccus* shed light on the evolutionary history of prasinophytes and the origin of the secondary plastids of euglenids, *Mol. Biol. Evol.*, 26, 631–648.

Tyagi, V.V.S., Ray, T.B., Mayne, B.C. and Peters, G.A. (1981). The *Azolla-Anabaena azollae* relationship. XI. Phycobiliproteins in the action spectrum for nitrogenase-catalysed acetylene reduction, *Plant Physiol.*, 68, 1479–1484.

Wagner, G. (1974). Fluxes and compartmentation of K and Cl in the green alga *Mougeottia*, *Planta*, 118, 145–157.

Walker, N.A., Smith, F.A. and Cathers, I.R. (1980). Bicarbonate assimilation by freshwater charophytes and higher plants. I. Membrane transport of bicarbonate is not proven, *J. Membrane Biol.*, 58, 51–58.

Wang, C., Yamamoto, H. and Shikanai, T. (2015). Role of cyclic electron transport around photosystem I in regulating proton motive force, *Biochim. Biophys. Acta*, 1847, 931–938.

Warner, M.E. and Suggett, D.J. (2016). The photobiology of *Symbiodinium* spp.: linking physiological diversity to the implications of stress and resilience. In *The Cnidaria, Past, Present and Future*, Goffredo, S. and Dubinsky, Z., eds. (Switzerland: Springer), pp. 489–509.

Wiessner, W. (1966a). Relative quantum yields for anaerobic photoassimilation of glucose, *Nature*, 212, 403.

Wiessner, W. (1966b). Vergleichende Studien zum Quantenbedarf der Photoassimilation von Essigsaure durch photoheterotrophe Purpurbacterien und Grunalgen, *Ber. Deut, Bot. Ges.*, 79, 58–62.

Wiessner, W. and Gaffron, H. (1964). Role of photosynthesis in the light-induced assimilation with acetate by *Chlamydobotrys*, *Nature*, 201, 725.

Xu, M., Ogawa, T., Pakrasi, H.B. and Mi, H. (2008). Identification and localization of the CupB protein involved in constitutive CO_2 uptake in the cyanobacterium, *Synechocystis* sp. strain PCC 6803, *Plant. Cell Physiol.*, 49, 994–997.

Yamamoto, H., Peng, I., Fukao, Y. and Shikanai, T. (2011). An Src homology 3 domain-like fold protein forms a ferredoxin binding site for the chloroplast NADH dehydrogenase-like complex in *Arabidopsis*, *Plant Cell*, 23, 1480–1493.

Yamori, W. and Shikanai, T. (2016). Physiological functions of cyclic electron around photosystem I in sustaining photosynthesis and plant growth, *Annu. Rev. Plant Biol.*, 67, 81–106.

Yang, W., Catalonolti, C., Wittkopf, T.M., Posewitz, M.C. and Grossman, A.R. (2015). Algae after dark: mechanism to cope with anoxic/hypoxic conditions, *Plant J.*, 82, 481–503.

Yeremenko, N., Jeanjean, R., Prommeenate, P., Krasikov, V., Nixon, P., Vermaas, W., Havaux, M. and Matthijs, H. (2005). Open reading frame ssr2016 is required for antimycin A-sensitive photosystem I-driven cyclic electron flow in the cyanobacterium *Synechocystis* sp. PCC 6803, *Plant. Cell Physiol.*, 46, 1433–1436.

Zehr, J.P., Bench, S.R., Carterm B.J., Heirsen, I., Niazi, F., Shi, T., Tripp, H.J. and Affourtit, J.P. (2008). Globally distributed uncultivated oceanic N_2-fixing cyanobacteria lack oxygenic photosystem II, *Science*, 322, 1110–1112.

Zhang, P.P., Battchikova, N., Jansen, T., Appel, J., Ogawa, T. and Aro, E.-M. (2004). Expression and functional roles of the two distinct NDH-1 complexes and the carbon acquisition complex NdhD3/NdhF3/CupA/Sll1735 in *Synechocystis* sp. PCC 6803, *Plant Cell*, 16, 3326–3340.

Zhang, P.P., Battchikova, N., Paakkarinen, V., Katoh, H., Iwai, M., *et al.* (2005). Isolation, subunit composition and interaction of theNDH-1complexes from *Thermosynechococcus elongatus* BP-1, *Biochem. J.*, 390, 513–520.

Zhang, L., Li, L. and Liu, J. (2014). Comparison of the photosynthetic characteristics of two *Isochrysis galbana* strains under high light, *Bot. Mar.*, 57, 477–482.

Index

AAA+ proteins, 168, 169, 171, 173, 176

acclimation, 249, 250, 252, 258

acetate assimilation, 314

aggregation, 203–205, 207–211

allosteric regulation, 290

alternative electron transfer pathways, 327

antenna, 189, 191, 192, 194, 195, 197–203, 205–207, 210, 211

antimycin A, 265, 267, 269–273, 311, 316, 319, 320, 322–325, 330, 331

Arabidopsis, 196–202, 306, 316, 323–325, 331

artificial photosynthesis, 117, 119, 136

aspartate, 205, 207

assimilation, 277, 281, 282, 285, 287

ATP binding, 36, 37, 40, 41, 43, 44, 46, 49, 50

ATP formation, 305, 315

Baccillariophyceae, 307, 327

biosynthesis, 277, 281, 282, 289

Bronsted slope, 43, 46, 49

calcium sensing phosphoprotein, 323

Calvin–Benson cycle, 326, 331

CFoCF1 ATP synthase, 191, 306, 312–314, 316

charge separation, 81, 82, 86, 102, 110, 113

Chlamydomonas, 306, 308, 310, 315, 316, 319, 320, 322–325, 329, 331

chlorophyll, 191, 194, 196–200, 204, 207–209, 211, 212

chlorophyll fluorescence, 196, 200, 204

chloroplasts, 319, 322, 323, 328

chlororespiration, 305, 309, 310, 317, 322, 326, 331

clustering, 197–199, 201–203, 205, 207, 210

CO_2 assimilation, 309, 314

CO_2 fixation, 161, 163, 164, 166, 174

complex 1, 317, 319

controlled rotation, 35–37, 39, 40, 43, 44–51

corals, 326

CP24, 192, 194, 197, 199, 201, 202, 205

CP26, 192, 193, 202, 205

CP29, 192, 193, 201, 205, 206

cross-linker, 196

cryo-electron microscopy, 195

crystallinity, 197–199

crystal structure, 149–151, 155

cyanobacteria, 305–312, 315–326, 331

cyclic electron flow (CEF), 266, 270, 271, 305, 306, 310, 311, 313–331
cyclic photophosphorylation, 306, 308, 310, 311, 315, 327
cytochrome b_6/f complex, 312, 315, 316, 323–325
cytochrome c oxidase, 55–57, 59, 60

D1 protein, 190, 196
D-channel, 60
dicyclohexycarbodiimide, DCCD, 206, 207
DCMU, 306–310, 313, 327–330
desiccation, 328, 329
diaminodurene, DAD, 205, 208
diatoms, 330
dielectric channels, 74
Dinoflagellata, 326
Dinophyta, 307, 326
dodecylmaltoside, 204, 205

efficiency, 65–68, 70
electron microscopy, 194, 195, 197, 201, 202, 205, 206
electron partitioning, 280, 281, 285– 287
electron transfer, 7, 10, 12, 13, 19, 22, 24, 26, 66, 67, 71
electron transport, 189, 191, 195, 277, 279–282, 285–287, 289–291
electron transport rate (ETR), 321, 322, 331
equation, 210
eukaryotic algae, 305–309, 312, 315, 318, 320, 322, 323, 326, 329, 331
Eustigmatophyceae, 307, 328
evolution, 243, 245, 247, 251
exciton dynamics, 81–83, 85, 86, 99

F1-ATPase, 35, 36, 40–42, 50
femtosecond infrared crystallography, 81, 83, 99
ferredoxin, 277, 280, 281, 283, 286, 288, 306, 311–313, 315, 318, 322–325

flavodiiron pathway, 322
fluorescence recovery after photobleaching, FRAP, 196
free rotation, 36–38, 41, 42, 46–50
freeze-fracture, 194, 197, 198, 202

glucose assimilation, 310, 311, 314
glutamate, 205, 207
grana, 191, 193–196, 198, 199, 201, 202, 207
grana thylakoids, 221–226, 228, 230, 231, 234
group transfer theory, 35, 40, 50

Haptophyta, 307, 327
H-channel, 55–60
heme-copper oxidases, 55–57, 65, 67–69
high light, 325, 327, 330
high temperature, 320, 321, 330
homeostasis, 291
H-parameter, 208, 209

iron deficiency, 330

77 K fluorescence, 203–205
Kok effect, 309

lateral diffusion, 221, 227, 228
Lhca, 192, 194
Lhcb, 192, 197, 199, 201, 206, 211
LHCII, 189–211
light harvesting, 189–192, 194, 195, 199, 201, 210, 243–246, 257
light harvesting complex, 149, 151, 154, 155
light memory, 210, 211
light reactions, 243, 244, 246, 247, 256, 258
linear electron flow (LEF), 305–308, 316, 322, 323, 327, 328, 330, 331
lipid diffusion, 196
low temperature, 329
lutein, 191, 192, 194, 201, 207–209

magnesium, 201, 207
manganese, 127, 129, 132, 136
MAP pathway, 317, 326, 330, 331
maquette strategy, 1, 3, 5, 14, 22
margins, 194, 199
mechanism, 65, 67, 69, 70, 128, 130, 135
membrane dynamics, 191, 196–199
membrane inlet mass spectrometry
 (MIMS), 320–322, 331
mesoscopic level, 222, 230
metabolism, 281, 290, 291
mitochondria, 55–57, 59
mobile fraction, 196–198, 202, 203
molecular chaperones, 159

NDH dehydrogenase, 315, 317
NDH reductase pathway, 312
negative stain, 195
neoxanthin, 191, 192, 194, 208
nicotinamide adenine dinucleotide
 dehydrogenase-like complex, 265, 270,
 271, 273
nitrogen deficiency, 330
nitrogen fixation, 331
NMR, 208
non-photochemical quenching, NPQ,
 200–212

$^{18}O_2$ isotope, 321
Ochrophyta, 307, 327
oriented crystals, 81, 82–84, 86, 87, 91, 113
oxygen evolution, 125, 131, 137
oxygenic photosynthesis, 305, 309, 324

P700, 307, 309, 320–322, 328
PGR5/PGRL1 complexes, 323
Phaeophyceae, 307, 328
photoinhibiton, 190
photosynthesis, 7, 10, 11, 117–120, 123,
 136, 150, 159, 160, 166, 174–176, 243,
 247, 259
photosynthetic membrane, 189–191, 195,
 196, 198–200, 210

photosystem I (PSI), 149–155, 191–195,
 199, 200, 265–268, 272, 277–287, 291
photosystem II (PSII), 81–86, 91, 100,
 103, 105–107, 110–114, 117, 119–121,
 123–139, 189, 190–202, 205, 210, 212,
 222–239, 231, 233, 308–311, 316, 319,
 320, 322, 326, 328–331
photosystems, 150
piericidin A, 319
pigment protein complexes, 190
pK, 205, 208–211
plastoquinol terminal oxidase (PTOX),
 326, 330
plastoquinone, 312, 315, 316, 323–325
plastoquinone reductase (PQR), 324, 325
pleochroism, 84
protein design, 2
protein diffusion, 197
protein order, 223, 226, 229
proteomics, 327
proton gradient regulation-like
 photosynthetic phenotype 1 (PGRL1),
 306, 313, 315, 322–325, 331
proton gradient regulation 5 (PGR5), 267,
 269–273, 306, 315, 319, 322–327
PGR5/PGRL1 complexes, 323
PGR5/PGRL1 pathway, 325
proton pump, 57, 59, 60, 320, 323
proton transfer, 65, 70, 74
proton translocation, 55, 56, 59, 189, 199
PsbS, 197, 199–207, 209–211
PSII supercomplex, 194, 195, 199, 210
pull-down assay, 206
Pyrrophyta, 326

Q cycle, 312, 316, 323
quencher, 201, 207, 209, 211, 212

reaction center, 189, 194, 200
reactive oxygen species, 278
redox regulation, 279, 318
regulation of photosynthesis, 247
Resonance Raman, 208

Rhodophyta, 307, 328
rotenone, 319
Rubisco, 159–169, 171–176

salinity, 329
single molecule imaging, 39
single particle electron microscopy, 318
stalling experiments, 38–41, 43, 44, 47, 51
state transitions, 191, 194, 196, 198, 199,
    201, 210
stroma, 191, 193, 194, 196, 198, 199, 210
structure-based theory, 81, 83, 101, 113
supercomplexes, 72, 73, 75
supernumerary subunits, 67, 71, 72, 75
Symbiodinium, 326, 327, 330

template, 67
thylakoid, 191, 193, 195, 197–199, 202,
    203, 205, 206, 210
thylakoid membrane, 221–226, 228–235,
    243–248, 250–258, 312, 315, 316, 318,
    320, 322–325
time-resolved fluorescence, 201, 203

violaxanthin, 192, 194, 202, 204–206, 208

xanthophylls, 191, 192, 201, 208, 211

yeast, 65, 66, 68–75

zeaxanthin, 192, 200–211